Multidisciplinary Applications of AI Robotics and Autonomous Systems

Tanupriya Choudhury
Graphic Era University, India

Anitha Mary X.
Karunya Institute of Technology and Sciences, India

Subrata Chowdhury
Sreenivasa Institute of Technology and Management Studies, India

C. Karthik
Jyothi Engineering College, India

C. Suganthi Evangeline
Sri Eshwar College of Engineering, India

A volume in the Advances in Computational Intelligence and Robotics (ACIR) Book Series

Published in the United States of America by
IGI Global
Engineering Science Reference (an imprint of IGI Global)
701 E. Chocolate Avenue
Hershey PA, USA 17033
Tel: 717-533-8845
Fax: 717-533-8661
E-mail: cust@igi-global.com
Web site: http://www.igi-global.com

Library of Congress Cataloging-in-Publication Data

CIP Pending
ISBN: 979-8-3693-5767-5
EISBN: 979-8-3693-5769-9

This book is published in the IGI Global book series Advances in Computational Intelligence and Robotics (ACIR) (ISSN: 2327-0411; eISSN: 2327-042X)

British Cataloguing in Publication Data
A Cataloguing in Publication record for this book is available from the British Library.

For electronic access to this publication, please contact: eresources@igi-global.com.

Advances in Computational Intelligence and Robotics (ACIR) Book Series

Ivan Giannoccaro
University of Salento, Italy

ISSN:2327-0411
EISSN:2327-042X

MISSION

While intelligence is traditionally a term applied to humans and human cognition, technology has progressed in such a way to allow for the development of intelligent systems able to simulate many human traits. With this new era of simulated and artificial intelligence, much research is needed in order to continue to advance the field and also to evaluate the ethical and societal concerns of the existence of artificial life and machine learning.

The **Advances in Computational Intelligence and Robotics (ACIR) Book Series** encourages scholarly discourse on all topics pertaining to evolutionary computing, artificial life, computational intelligence, machine learning, and robotics. ACIR presents the latest research being conducted on diverse topics in intelligence technologies with the goal of advancing knowledge and applications in this rapidly evolving field.

COVERAGE

- Algorithmic Learning
- Synthetic Emotions
- Neural Networks
- Cognitive Informatics
- Computational Logic
- Evolutionary Computing
- Artificial Life
- Computational Intelligence
- Automated Reasoning
- Intelligent Control

IGI Global is currently accepting manuscripts for publication within this series. To submit a proposal for a volume in this series, please contact our Acquisition Editors at Acquisitions@igi-global.com or visit: http://www.igi-global.com/publish/.

Titles in this Series

For a list of additional titles in this series, please visit:
www.igi-global.com/book-series/advances-computational-intelligence-robotics/73674

AI Algorithms and ChatGPT for Student Engagement in Online Learning
Rohit Bansal (Vaish College of Engineering, India) Aziza Chakir (Faculty of Law, Economics, and Social Sciences, Hassan II University, Casablanca, Morocco) Abdul Hafaz Ngah (Faculty of Business Economics and Social Development, Universiti Malaysia, Terengganu, Malaysia) Fazla Rabby (Stanford Institute of Management and Technology, Australia) and Ajay Jain (Shri Cloth Market Kanya Vanijya Mahavidyalaya, Indore, ndia)
Information Science Reference • © 2024 • 292pp • H/C (ISBN: 9798369342688) • US $265.00

Applications, Challenges, and the Future of ChatGPT
Priyanka Sharma (Swami Keshvanand Institute of Technology, Management, and Gramothan, Jaipur, India) Monika Jyotiyana (Manipal University Jaipur, India) and A.V. Senthil Kumar (Hindusthan College of Arts and Sciences, ndia)
Engineering Science Reference • © 2024 • 309pp • H/C (ISBN: 9798369368244) • US $365.00

Modeling, Simulation, and Control of AI Robotics and Autonomous Systems
Tanupriya Choudhury (Graphic Era University, India) Anitha Mary X. (Karunya Institute of Technology and Sciences, India) Subrata Chowdhury (Sreenivasa Institute of Technology and Management Studies, India) C. Karthik (Jyothi Engineering College, India) and C. Suganthi Evangeline (Sri Eshwar College of Engineering, India)
Engineering Science Reference • © 2024 • 295pp • H/C (ISBN: 9798369319628) • US $300.00

Explainable AI Applications for Human Behavior Analysis
P. Paramasivan (Dhaanish Ahmed College of Engineering, India) S. Suman Rajest (Dhaanish Ahmed College of Engineering, India) Karthikeyan Chinnusamy (Veritas, USA) R. Regin (SRM Institute of Science and Technology, India) and Ferdin Joe John Joseph (Thai-Nichi Institute of Technology, Thailand)
Engineering Science Reference • © 2024 • 369pp • H/C (ISBN: 9798369313558) • US $300.00

Bio-Inspired Intelligence for Smart Decision-Making
Ramkumar Jaganathan (Sri Krishna Arts and Science College, India) Shilpa Mehta (Auckland University of Technology, New Zealand) and Ram Krishan (Mata Sundri University Girls College, Mansa, India)
Information Science Reference • © 2024 • 334pp • H/C (ISBN: 9798369352762) • US $385.00

AI and IoT for Proactive Disaster Management
Mariyam Ouaissa (Chouaib Doukkali University, Morocco) Mariya Ouaissa (Cadi Ayyad University, Morocco) Zakaria Boulouard (Hassan II University, Casablanca, Morocco) Celestine Iwendi (University of Bolton, UK) and Moez Krichen (Al-Baha University, Saudi Arabia)
Engineering Science Reference • © 2024 • 299pp • H/C (ISBN: 9798369338964) • US $355.00

701 East Chocolate Avenue, Hershey, PA 17033, USA
Tel: 717-533-8845 x100 • Fax: 717-533-8661
E-Mail: cust@igi-global.com • www.igi-global.com

Table of Contents

Detailed Table of Contents

Chapter 1

N. Dheerthi, Sri Ramakrishna Engineering College, India
A. Kishore Kumar, Sri Ramakrishna Engineering College, India
S. Sarveswaran, Sri Ramakrishna Engineering College, India
A. Murugarajan, Sri Ramakrishna Engineering College, India

In recent years, there has been a growing interest in the development of soft robotic technologies inspired by biological systems for various applications, particularly in healthcare. The study encompasses a multidisciplinary approach, integrating principles from biomechanics, robotics, and materials science to design and characterize the soft robotic arm. By examining the mechanical structure, actuation mechanisms, and control strategies, this research aims to elucidate the advantages and challenges associated with deploying bio-inspired soft robotic arms in healthcare settings. Furthermore, the investigation delves into the materials selection process, considering factors such as biocompatibility, durability, and flexibility to ensure safe and effective interaction with biological tissues. Overall, this comprehensive analysis contributes to advancing the understanding of bio-inspired soft robotics and highlights its potential transformative impact on healthcare by offering innovative solutions for improving patient care, surgical outcomes, and quality of life.

Chapter 2

S. Sarveswaran, Sri Ramakrishna Engineering College, India
Kishore Kumar Arjunsingh, Sri Ramakrishna Engineering College, India
N. Dheerthi, Sri Ramakrishna Engineering College, India
A. Murugarajan, Sri Ramakrishna Engineering College, India

Delta robots, known for their unique design featuring parallel linkages and a stationary base, have emerged as transformative tools in various industries, including healthcare. In surgery, delta robots enable minimally invasive procedures with enhanced precision and shorter recovery times. They facilitate targeted therapies in rehabilitation, promoting better outcomes for patients with neurological and musculoskeletal conditions. Delta robots also improve medication dispensing accuracy in pharmacies and automate repetitive tasks in laboratories, increasing efficiency and reducing errors. Additionally, the chapter explores the potential of delta robots in specialized fields such as orthopedic, neuro, and cardiac surgery, as well as their role in enhancing medical imaging accuracy and guiding interventional procedures in real-time. It also discusses the future of AI-powered diagnostics and personalized medicine, envisioning a healthcare landscape where delta robots play a central role in improving patient outcomes and shaping the future of healthcare delivery.

Chapter 3

A. Madhesh, Karpagam Academy of Higher Education, India
Clara Barathi Priyadharshini, Karpagam Academy of Higher Education, India

Earlier methods focused on reducing the forecast uncertainty for individual agents and avoiding this unduly cautious behavior by either employing more experienced models or heuristically restricting the predictive covariance. Findings indicate neither the individual prediction nor the forecast uncertainty have a major impact on the frozen robot problem. The result is that dynamic agents can solve the frozen robot problem by employing joint collision avoidance and clear the way for each other to build feasible pathways. Potential paths for safety evaluation are ranked according to the likelihood of collisions with known objects and those that happen outside the planning horizon. The whole collision probability is examined. Monte Carlo sampling is utilized to approximate the collision probabilities. Designing and selecting routes to reach the intended location, this approach aims to provide a navigation framework that reduces the likelihood of collisions.

Chapter 4

K. Yogesh, Karpagam Academy of Higher Education, India
R. Gunasudari, Karpagam Academy of Higher Education, India

Machine vision systems have emerged as a viable non-invasive approach for investigating the connection between fruit visual traits and physicochemical qualities at varying ripening degrees, and have been used in recent research efforts to identify the stages. The current study aims to develop an intelligent algorithm that can estimate various physical properties, such as firmness and soluble solid content, as well as three chemical properties, namely starch, acidity, and titratable acidity. A hybrid approach was used to further optimise the physicochemical estimation method of PSO with CNN. This method was applied to the evaluation parameters in order to describe their classification behaviour. The sample accuracy was 95.84% when using the different parameters to characterise them. A second set of apples was utilised for validation after the first set was used as trial samples in PSO+CNN.

Chapter 5

Johnwesily Chappidi, VIT-AP University, India
Divya Meena Sundaram, VIT-AP University, India

In the field of conserving wildlife, the utilization of autonomous systems equipped with computer vision holds tremendous promise. This research explores the potential of integrating YOLO v7, a cutting-edge object recognition model, with stochastic gradient descent (SGD) optimization techniques to bolster wild animal conservation efforts. The primary objective is to enhance the precision, accuracy, and scalability of autonomous systems in detecting and monitoring wild animals across diverse habitats. The experimental results showcase substantial advancements, demonstrating the efficacy of the YOLOv7-SGD amalgamation in autonomous systems. The model exhibits superior detection accuracy and robustness in identifying a multitude of wild animal species across diverse landscapes.

Chapter 6

Aruna Kasinathan, Karunya Institute of Technology and Science, India
Shrilatha Sampath, Christian Medical College, India
Hemalatha Sampath, University of Maryland, USA

The study aims to assess customer satisfaction and trust in autonomous artificial intelligence (AAI) systems within the banking sector. Its primary objectives include exploring factors contributing to customer trust in AAI, investigating preferences for AI-driven features in banking, and determining the impact of AAI on perceived service quality. The research, adopting a descriptive design, employs both qualitative and quantitative methods. A survey, distributed to customers of leading banks in India, particularly in Tamil Nadu, with a sample size of 213, utilizes simple random and convenient sampling. Results highlight customer preferences for customized services, financial advice, and automation in banking. The implementation of AAI is perceived positively, especially in terms of transparency in processes like loans, account management, and more. Practical implications include helping banks understand customer expectations, identify weaknesses in AAI features, and enhance service quality in Tamil Nadu.

Chapter 7

Xavier Arockiaraj Santhappan, Adhiyamaan College of Engineering, India
Ronica Bis, Sathyabama Institute of Science and Technology, India

Ultrasound is a conventional diagnostic instrument employed in prenatal care to track the progression and advancement of the fetus. In routine clinical obstetric assessments, the standard planes of fetal ultrasound hold considerable importance in evaluating fetal growth metrics and identifying abnormalities. In this work, a method to detect FFSP using deep convolutional neural network (DCNN) architecture to improve detection efficiency is presented. Squeeze net, 16 convolutional layers with small 3x3 large kernel, and all three layers form the proposed DCNN. The final pooling layer uses global average pooling (GAP) to reduce inconsistency in the network. This helps reduce the problem of overfitting and improves the performance from different training data. To improve cognitive performance, data augmentation methods developed specifically for FFSP are used in conjunction with adaptive learning strategies. Extensive testing shows that the proposed method gives accuracy of 96% which outperforms traditional methods, and DCNN is an important tool to identify FFSP in clinical diagnosis.

Chapter 8

Abhishek Choubey, Sreenidhi Institute of Science and Technology, Hyderabad, India
Shruti Bhargava Choubey, Sreenidhi Institute of Science and Technology, Hyderabad, India

Biorobotics and nanobots represent a cutting-edge area of biotechnology with tremendous promise to transform scientific research, environmental monitoring, and healthcare delivery. In this chapter, the authors explore their cutting-edge ideas, applications, and advances, showing their ability to radically change future industries like industry medicine. This chapter's primary objective is to explore both existing and emerging applications of bio-robots and nanobots in healthcare, environmental monitoring, environmental inspection, and materials science research. These technological advancements offer real-time diagnostics, minimally invasive surgery, and targeted drug delivery as well as environmental quality evaluation through bio-robot water quality evaluation and nanobot pollution detection at unprecedented scales. Furthermore, bio-robots and nanobots have proven invaluable for scientific study fields like neuroscience, synthetic biology, and materials science.

Chapter 9

V. Saran, Karpagam Academy of Higher Education, India
R. Chennappan, Karpagam Academy of Higher Education, India

Machine learning (ML), deep learning, fuzzy logic, and traditional neural networks are just a few of the subsets that make up artificial intelligence (AI). These subgroups possesses unique qualities and skills that could improve the effectiveness of modern medical sciences. Human intervention in clinical diagnostics, medical imaging, and decision-making is facilitated by these clever solutions. The development of information technology, the concept of intelligent healthcare has become more and more popular. Intelligent healthcare is a revolutionary approach to healthcare that leverages state-of-the-art technology such as AI and the internet of things (IoT) to improve overall efficacy, convenience, and personalisation of the medical system.

Chapter 10

R. Gokulakrishnan, Karpagam Academy of Higher Education, India
C. Balakumar, Karpagam Academy of Higher Education, India

This research provides a method for gesture recognition that integrates two separate recognizers. These two recognizers use the CAR equation to ascertain the hands sign. The robot's two main parts are its sending and receiving ends. Within the process of developing the same, three domains were specifically combined: biomedicine, which involved registering biosignals using analog channels composed of instrumental amplifiers; software development, involving microcontrollers, core processing (DSP), and the resulting control of the robot hand; PC software for tracking the registered biosignals; and mechatronics, involving the design and mechanical construction of the robot hand. The hand can control how much pressure is given to things because of the force sensor (FSR) in each finger. While developing a hand and wrist prototype that can rotate in response to EMG signal pulses, this was discovered.

Chapter 11

N. Nissi Angel, Department of ECE, Velagapudi Ramakrishna Siddhartha Engineering College, Vijayawada, India
Gunnam Suryanarayana, Department of ECE, Velagapudi Ramakrishna Siddhartha Engineering College, Vijayawada, India
Siva Ramakrishna Pillutla, School of Electronics Engineering, VIT-AP University, Amaravati, India
Kathik Chandran, Jyothi Engineering College, India

Image steganography methods use manual features for hiding payload data in cover images. These manual features allow less payload capacity and also cause image distortion. In this chapter, the authors detail a CNN-based network image steganography. The major contributions are twofold. First, they presented a CNN-based encoder-decoder architecture for hiding image. Secondly, they introduce a loss function, which checks joint end-to-end encoder-decoder networks. They evaluate this architecture on publicly available datasets CIFAR10. The results indicate an increase in payload capacity with high peak signal-to-noise ratio and structural similarity index values.

Nalla Bhanu Teja, *Department of Mechanical Engineering, Aditya College of Engineering, Surampalem, India*
V. Kannagi, *Department of Electronics and Communication Engineering, R.M.K. College of Engineering and Technology, Puduvoyal, India*
A. Chandrashekhar, *Department of Mechanical Engineering, Faculty of Science and Technology, ICFAI Foundation for Higher Education, Hyderabad, India*
T. Senthilnathan, *Department of Applied Physics, Sri Venkateswara College of Engineering, Sriperumbudur, India*
Tarun Kanti Pal, *Department of Mechanical Engineering, College of Engineering and Management, Kolaghat, India*
Sampath Boopathi, *Department of Mechanical Engineering, Muthayammal Engineering College, Namakkal, India*

The integration of nanotechnology into robotics has revolutionized the design, manufacturing, and performance of robotic systems. Nano-materials, with their unique properties at the nanoscale, enhance strength, flexibility, and functionality, revolutionizing the construction and operation of robots. Nano fluids, with their superior heat transfer properties, address overheating issues, improving performance, extended operational lifespans, and increased adaptability in diverse environmental conditions. The chapter also explores the environmental impact of robotics, highlighting the integration of nano-materials and nano fluids in eco-friendly solutions. The chapter delves into the challenges and future directions of the synergy between nanotechnology and robotics, discussing potential breakthroughs, ethical considerations, and the need for ongoing research. It provides a comprehensive analysis of the impacts of nano-materials and nano fluids on the robot industry and their environments.

Ahamed Thaiyub, *KPR Institute of Engineering and Technology, India*
Akshay Bhuvaneswari Ramakrishnan, *SASTRA University, India*
Shriram Kris Vasudevan, *Intel Corporation, India*
T. S. Murugesh, *Government College of Engineering, Srirangam, India*
Sini Raj Pulari, *Bahrain Polytechnic, Bahrain*

Organizations face enormous issues when it comes to employee turnover, which is why they need to develop accurate predictive models for retention. The purpose of this chapter is to present a three-tiered machine learning approach for predicting employee turnover that makes use of resume parsing, performance analysis, and advanced algorithms. In addition, the authors make use of Intel oneAPI, which is a unified programming model that is increasingly becoming the industry standard, in order to improve the scalability and performance of the solution. The system that is offered delivers full HR (human resource) analytics, which enables firms to make educated decisions regarding recruiting and retention tactics. The results of the experimental evaluation show that the solution is effective in providing an accurate forecast of attrition, which paves the way for proactive retention measures. The approach enhances system performance by utilizing oneAPI, which in turn ensures that it is scalable over a variety of different hardware architectures.

Chapter 14

D. Faridha Banu, Sri Eshwar College of Engineering, Coimbatore, India
P. T. Kousalya, Sri Eshwar College of Engineering, Coimbatore, India
Kavin Varsha, Sri Eshwar College of Engineering, Coimbatore, India
C. Keerthi Prashanth, Sri Eshwar College of Engineering, Coimbatore, India
P. Madhumohan, Sri Eshwar College of Engineering, Coimbatore, India
S. Meivel, M. Kumarasamy College of Engineering, Karur, India

Any attempt to explain the relevance of data by putting it in a visual context is referred to as data visualization. With the aid of data visualization software, patterns, trends, and correlations that could go unnoticed in text-based data can be exposed and identified more easily. The graphical presentation of quantitative information is known as data visualization. In other words, data visualizations convert big and small data sets into pictures that the human brain can comprehend and digest more readily. In our daily lives, data visualizations are surprisingly prevalent, yet they frequently take the shape of recognizable charts and graphs. It can be applied to find unknown trends and facts. When communication, data science, and design come together, good data visualizations are produced. When done well, data visualizations provide important insights into complex data sets in clear, understandable ways. The authors talk about data visualization, its significance, tools for data visualization, etc. in this chapter.

Chapter 15

A. Kishore Kumar, Sri Ramakrishna Engineering College, India
S. Sarveswaran, Sri Ramakrishna Engineering College, India
N. Dheerthi, Sri Ramakrishna Engineering College, India
A. Murugarajan, Sri Ramakrishna Engineering College, India

In the rapidly evolving landscape of Healthcare 4.0/5.0, the integration of artificial intelligence (AI) and robotics has shown immense potential in transforming patient care. However, the deployment of these technologies in human-robot interactions (HRI) demands a delicate balance between efficiency and transparency. This chapter explores the research directions and challenges associated with the implementation of Explainable AI (XAI) in the context of HRI for the advancement of healthcare services. The authors delve into the critical aspects of ensuring transparency and interpretability in AI-driven robotic systems, emphasizing the need for explainability to foster trust and collaboration between healthcare professionals, patients, and intelligent robotic entities. The chapter highlights key challenges, proposes potential research directions, and suggests methodologies to address the complexities in deploying XAI within the healthcare ecosystem.

The chapter intends to create a system that alerts the victim of theft in real time. The current system does not distinguish between people and objects; instead, it uses a methodology to identify the burglar after the theft has taken place. The internet of things and advancements in wireless sensor networks make it possible to create a smart, safe home that can detect burglars in real time and notify the homeowner while the theft is occurring. The suggested approach is better than the current ones that use CCTV cameras for surveillance.

Nano robotics is a rapidly developing technology that operates at microscopic scales, revolutionizing fields like medicine and manufacturing. However, it faces numerous challenges, including technical, ethical, and practical issues. These include precision engineering, control mechanisms, and power sources, as well as ethical concerns about autonomy, safety, and societal impact. The chapter explores the future of nano robotics, highlighting its potential in various fields such as medicine and manufacturing. It highlights the potential of nano robots in enhancing durability and functionality, offering targeted drug delivery, minimally invasive surgeries, and precise diagnostics. The chapter also addresses technical challenges, ethical considerations, and potential developments, aiming to make the seemingly impossible achievable at the tiniest scales, emphasizing the need for further advancements.

Preface

Welcome to the realm of Intelligent Robotics and Autonomous Systems (IRAS), a captivating intersection of robotics, artificial intelligence (AI), and control systems. As editors, it is our pleasure to present to you this edited reference book, *Multidisciplinary Applications of AI Robotics and Autonomous Systems*, curated by Tanupriya Choudhury, Anitha Mary X, Subrata Chowdhury, C. Karthik, and C. Suganthi Evangeline.

Intelligent Robotics and Autonomous Systems (IRAS) is a dynamic field where innovation knows no bounds. Within its realm, system modeling, simulation, and control with Artificial Intelligence (AI) are indispensable pillars. These facets empower engineers and researchers to craft intelligent machines capable of navigating complex tasks autonomously, thus reshaping industries and redefining possibilities.

System modeling entails the creation of mathematical representations that capture the essence of robotic systems, encompassing their dynamics, kinematics, sensors, actuators, and interrelations. Simulation, on the other hand, provides a virtual playground for testing these models rigorously, fostering insights and foresight without the constraints of the physical realm. Complementing these, control algorithms infused with AI breathe life into robotic systems, endowing them with the autonomy to make informed decisions based on sensory input.

The applications of these methodologies are boundless. From autonomous driving systems revolutionizing transportation to industrial automation streamlining production processes, the impact of IRAS reverberates across various domains. Through meticulous system modeling, rigorous simulation, and intelligent control, engineers navigate the challenges of designing machines that not only perform tasks but do so with finesse and efficiency.

The landscape of autonomous systems is ever-evolving, propelled by advancements in AI, control theory, and distributed intelligence. As we stand on the precipice of a new era, where autonomous systems permeate everyday life, this book serves as a beacon, illuminating the path towards further exploration and innovation.

We invite you to embark on this journey through the pages of *Multidisciplinary Applications of AI Robotics and Autonomous Systems*. Within these chapters lie insights, breakthroughs, and a glimpse into the future of robotics and autonomy. May this compilation inspire curiosity, spark ideas, and pave the way for new frontiers in the realm of intelligent machines.

ORGANIZATION OF THE BOOK

Chapter 1: A Comprehensive Analysis of Bio-inspired Soft Robotic Arm for Healthcare Applications

In this chapter, the authors explore the burgeoning field of soft robotics, particularly in healthcare applications. By amalgamating principles from biomechanics, robotics, and materials science, they delve into the design and characterization of bio-inspired soft robotic arms. The research scrutinizes mechanical structures, actuation mechanisms, and control strategies, shedding light on the advantages and challenges of deploying such systems in healthcare settings. Moreover, the chapter delves into materials selection criteria, emphasizing biocompatibility, durability, and flexibility to ensure safe interaction with biological tissues. Overall, this thorough analysis offers insights into the transformative potential of bio-inspired soft robotics in healthcare, promising innovative solutions for enhancing patient care and surgical outcomes.

Chapter 2: A Comprehensive Insights and Research Focus on Delta Robots in the Healthcare Industry

The second chapter delves into the multifaceted applications of delta robots in healthcare, showcasing their unique design and transformative impact. From enabling minimally invasive surgeries to enhancing medication dispensing accuracy, delta robots have found versatile utility in various healthcare domains. The authors explore their potential in specialized fields such as orthopedics and neurosurgery, alongside their role in medical imaging and interventional procedures. Additionally, they envision a future where AI powered diagnostics and personalized medicine converge with delta robots, revolutionizing patient outcomes and healthcare delivery.

Chapter 3: A Data-Driven Model for Predicting Fault-Tolerant Safe Navigation in Multi-Robot Systems

This chapter navigates through the intricate realm of multi-robot systems, focusing on predictive modeling for fault-tolerant safe navigation. The authors propose a dynamic approach that prioritizes joint collision avoidance, facilitating the smooth traversal of dynamic environments. By leveraging Monte Carlo sampling and data-driven techniques, they develop a robust navigation framework that minimizes collision probabilities. The chapter underscores the significance of proactive safety measures in multi-robot systems, offering insights into enhancing navigation efficiency while mitigating potential risks.

Chapter 4: A Novel Method to Detect Ripeness Level of Apples Using Machine Vision (PSOCNN) Approach

Machine vision takes center stage in this chapter as the authors present a novel method for detecting the ripeness level of apples. By integrating Particle Swarm Optimization (PSO) with Convolutional Neural Networks (CNN), they develop an intelligent algorithm capable of estimating various physical and chemical properties of apples. Through meticulous image processing and AI-driven analysis, the proposed approach achieves high accuracy in ripeness detection, promising advancements in quality control and agricultural practices.

Chapter 5: Advancing Wild Animal Conservation through Autonomous Systems Leveraging YOLOV7 With SGD Optimization Technique

Wildlife conservation receives a technological boost in this chapter, where autonomous systems equipped with computer vision take the spotlight. The authors explore the integration of YOLOv7 with Stochastic Gradient Descent (SGD) optimization techniques to bolster wild animal conservation efforts. Through experimental validation, they demonstrate the superior detection accuracy and scalability of the YOLOv7-SGD amalgamation, envisioning a future where autonomous systems play a pivotal role in safeguarding biodiversity across diverse habitats.

Chapter 6: Evaluating Customer Satisfaction and Trust in Autonomous AI Banking Systems

The sixth chapter delves into the realm of autonomous Artificial Intelligence (AAI) systems within the banking sector, aiming to assess customer satisfaction and trust. Employing both qualitative and quantitative methods, the authors investigate customer preferences and perceptions regarding AI-driven features in banking. Their findings highlight the positive reception of AAI, particularly in terms of service quality and transparency. This research offers valuable insights for banks seeking to enhance customer experience and adapt to evolving technological landscapes.

Chapter 7: Foetal Activity Detection Using Deep Convolution Neural Networks

In prenatal care, fetal ultrasound plays a crucial role, and this chapter introduces a method for enhancing fetal face and spine plane (FFSP) detection using Deep Convolutional Neural Networks (DCNN). Through innovative architecture design and data augmentation techniques, the authors achieve remarkable accuracy in FFSP detection, promising advancements in clinical obstetric assessments and fetal monitoring. Their research underscores the transformative potential of DCNNs in improving diagnostic efficiency and maternal-fetal health outcomes.

Chapter 8: Future Nano- and Biorobots Miniaturized Machines for Biotechnology and Beyond

Nano- and biorobots take center stage in this chapter, offering a glimpse into the future of biotechnology and scientific research. The authors explore the myriad applications of these miniature marvels, ranging from targeted drug delivery to environmental monitoring. Their analysis delves into the integration of nanotechnology with robotics, highlighting the transformative potential in healthcare, materials science, and beyond. This chapter serves as a roadmap for harnessing the power of nano- and biorobots to address pressing societal challenges and unlock new frontiers in scientific exploration.

Chapter 9: HRI in ITs Using ML Techniques

Machine Learning (ML) techniques intersect with Human-Robot Interactions (HRI) in this chapter, offering insights into the burgeoning field of intelligent healthcare. The authors explore the convergence of AI, IoT, and healthcare, highlighting the transformative potential in clinical diagnostics and decision-making. Through a hybrid approach combining Convolutional Neural Networks (CNN) and Ant Colony Optimization (ACO), they present innovative methodologies for disease detection and medical technology advancement. This research lays the groundwork for intelligent healthcare systems that prioritize efficacy, convenience, and personalization.

Chapter 10: Human-Robot Safety Guarantees Using Confidence-Aware-Game-Theoretic Human Models With EMG Signal

Gesture recognition and human-robot interactions converge in this chapter, presenting a method for enhancing safety guarantees in robotics. The authors propose a confidence-aware-game-theoretic model integrated with Electromyography (EMG) signals, offering a novel approach to gesture-based robot control. Through meticulous design and integration of multiple domains, they develop a robust system that prioritizes user safety and intuitive control. Their research paves the way for seamless human-robot collaboration, promising advancements in robotics applications across diverse domains.

Chapter 11: Image Steganography-Embedding Secret Data in Images Using Convolutional Neural Networks: CNN-Based Image Steganography

Image steganography takes center stage in this chapter, where the authors present a novel approach leveraging Convolutional Neural Networks (CNNs). Their methodology focuses on concealing secret data within images, offering a secure and efficient communication channel. By introducing a CNN-based encoder-decoder architecture and novel loss function, they achieve increased payload capacity and improved signal-to-noise ratio. This research promises advancements in secure communication and data privacy, addressing contemporary challenges in information security.

Chapter 12: Impacts of Nano-Materials and Nano Fluids on the Robot Industry and Environments

Nanotechnology's integration with robotics is explored in this chapter, highlighting its transformative impact on robot design and performance. The authors delve into the applications of nano-materials and nano fluids, emphasizing their role in enhancing robot durability, functionality, and adaptability to environmental conditions. Through a comprehensive analysis, they address technical challenges, ethical considerations, and potential developments, envisioning a future where nano-enhanced robots revolutionize industries and environmental monitoring. This chapter serves as a roadmap for harnessing the synergies between nanotechnology and robotics to address societal challenges and unlock new possibilities.

Chapter 13: Predictive Modelling for Employee Retention: A Three-Tier Machine Learning Approach With oneAPI

Employee retention receives a technological overhaul in this chapter, where the authors propose a three-tiered machine learning approach for predictive modeling. By leveraging resume parsing, performance analysis, and advanced algorithms, they develop a robust system capable of accurately forecasting attrition. Moreover, the integration of Intel oneAPI enhances scalability and performance across diverse hardware architectures, paving the way for proactive retention measures in organizations. This research offers practical insights for HR analytics, empowering firms to make informed decisions regarding recruitment and retention strategies.

Chapter 14: Research Analysis of Data Exploration and Visualization Dashboard Using Data Science

Data visualization emerges as a powerful tool in this chapter, where the authors analyze its significance and application in diverse domains. By elucidating the relevance of data visualization and exploring visualization tools and techniques, they showcase its potential in uncovering hidden insights and trends. This research underscores the importance of effective communication between data science, design, and decision-making, offering practical insights for leveraging data visualization to gain actionable insights from complex datasets.

Chapter 15: Research Directions and Challenges in the Deployment of Explainable AI in Human Robot Interactions for Healthcare 4.0/5.0

Explainable AI (XAI) takes the spotlight in this chapter, focusing on its deployment in Human-Robot Interactions (HRI) within the healthcare domain. The authors explore research directions and challenges associated with ensuring transparency and interpretability in AI-driven robotic systems. By fostering trust and collaboration between healthcare professionals, patients, and intelligent robotic entities, they envision a future where XAI enhances healthcare services and patient outcomes. This research offers insights into addressing the complexities of deploying XAI within the healthcare ecosystem, emphasizing the need for transparency and ethical considerations.

Chapter 16: Security System for Smart Homes to Prevent Theft

Smart home security receives a technological boost in this chapter, where the authors propose a real-time theft detection system. By leveraging IoT and wireless sensor networks, they develop a smart home security system capable of identifying burglars during theft events. This innovative approach offers advantages over traditional surveillance methods, promising enhanced security and peace of mind for homeowners. The chapter highlights the potential of IoT-driven solutions in preventing theft and ensuring residential safety.

Chapter 17: Study on Nano Robotic Systems for Industry 4.0: Overcoming Challenges and Shaping Future Developments

Nano robotics takes center stage in this chapter, offering insights into its applications and challenges in Industry 4.0. The authors explore the transformative potential of nano robots in various sectors, ranging from medicine to manufacturing. By addressing technical challenges and ethical considerations, they envision a future where nano robots revolutionize industrial processes and environmental monitoring. This research lays the groundwork for harnessing the synergy between nanotechnology and robotics, paving the way for future advancements in diverse industries.

CONCLUSION

In concluding this edited reference book on global practices in talent acquisition and retention, we reflect on the journey we've embarked upon—a journey marked by exploration, discovery, and collaboration. Each chapter within this volume represents a unique contribution to the collective understanding of talent management in the contemporary world.

As editors, we have been privileged to witness the depth and breadth of expertise showcased by our esteemed contributors. From the intricacies of artificial intelligence in recruitment to the nuances of fostering diversity and inclusion, from the challenges of employee well-being to the strategies for retaining top talent, this book encapsulates a wealth of knowledge and insights.

Through empirical research, theoretical frameworks, and practical applications, the chapters offer actionable strategies and thought-provoking perspectives for scholars, practitioners, and policymakers alike. They underscore the evolving nature of talent management and the imperative for organizations to adapt to an ever-changing landscape.

As we bid farewell to this volume, we extend our heartfelt appreciation to all who have contributed to its creation. The dedication, passion, and intellectual rigor of our authors have been the driving force behind this endeavor. We also express our gratitude to our readers, whose engagement and curiosity fuel the advancement of knowledge in this field.

As we turn the final page, we envision this book not as an endpoint, but as a catalyst for continued dialogue, innovation, and progress in talent acquisition and retention. May the insights shared within these pages inspire transformative action and contribute to the cultivation of thriving, inclusive workplaces around the globe.

Tanupriya Choudhury
Graphic Era University, India

X. Anitha Mary
Karunya Institute of Technology and Sciences, India

Subrata Chowdhury
Sreenivasa Institute of Technology and Management Studies, India

C. Karthik
Jyothi Engineering College, India

C. Suganthi Evangeline
Sri Eshwar College of Engineering, India

Chapter 1
A Comprehensive Analysis of Bio-Inspired Soft Robotic Arm for Healthcare Applications

N. Dheerthi
Sri Ramakrishna Engineering College, India

A. Kishore Kumar
(iD) https://orcid.org/0000-0003-4876-319X
Sri Ramakrishna Engineering College, India

S. Sarveswaran
Sri Ramakrishna Engineering College, India

A. Murugarajan
Sri Ramakrishna Engineering College, India

ABSTRACT

In recent years, there has been a growing interest in the development of soft robotic technologies inspired by biological systems for various applications, particularly in healthcare. The study encompasses a multidisciplinary approach, integrating principles from biomechanics, robotics, and materials science to design and characterize the soft robotic arm. By examining the mechanical structure, actuation mechanisms, and control strategies, this research aims to elucidate the advantages and challenges associated with deploying bio-inspired soft robotic arms in healthcare settings. Furthermore, the investigation delves into the materials selection process, considering factors such as biocompatibility, durability, and flexibility to ensure safe and effective interaction with biological tissues. Overall, this comprehensive analysis contributes to advancing the understanding of bio-inspired soft robotics and highlights its potential transformative impact on healthcare by offering innovative solutions for improving patient care, surgical outcomes, and quality of life.

DOI: 10.4018/979-8-3693-5767-5.ch001

1. INTRODUCTION TO BIO-INSPIRED SOFT ROBOTICS

1.1 Definition and Principles of Bio-Inspired Soft Robotics

Bio-inspired soft robotics is a branch of robotics that creates soft, flexible robot bodies by modelling them after biological structures and species. These robots are designed to resemble live things in terms of their robustness, flexibility, and other traits. Bio-inspired soft robotics frequently combines concepts from biology, materials science, and biomechanics to build robots that can securely interact with people, navigate challenging settings, and carry out activities that are beyond the capabilities of conventional rigid robots (Li et al., 2022). Soft actuators inspired by muscle architecture, soft-bodied grippers modelled after octopus tentacles, and crawling robots modelled after caterpillar or snake locomotion are a few examples of bio-inspired soft robotics. There has been a broad shift in recent years toward robots that are service-oriented, meaning they must be able to handle a variety of uncertainty and adapt to complicated dynamic situations (Rus & Tolley, 2015). Owing to the advantageous characteristics of living things, like resilience, flexibility, adaptability, and agility, scientists have been attempting to integrate biological elements into robots. intelligence that will allow autonomous robots to navigate safely and collaborate effectively in changing contexts (Bekey, 2005). Biologically inspired intelligence refers to the methods that have been influenced by biological intelligence and has been investigated and studied for many years in robotics research (Li, Yang, and Xu, 2019). Numerous biological phenomena involving agonist and antagonist interaction have been satisfactorily explained by the gated dipole paradigm (Oh et al., 2017).

The idea behind bio-inspired robotics is to build robots with similar capabilities by imitating different parts of biological organisms. Several fundamental principles are:

Soft and Flexible Structures: Soft, flexible bodies are a common feature of bio-inspired robots, which emulate the pliable properties of biological tissues.

Biomechanics: Understanding the biomechanics of biological organisms helps in designing robots that can mimic natural movements and behaviours.

Sensory Systems: Bio-inspired robots often incorporate sensory systems inspired by those found in nature, such as vision, touch, and proprioception.

Integration of Multi-disciplinary Approaches: Bio-inspired robotics often involves collaboration across multiple disciplines, including biology, engineering, computer science, and materials science.

1.2 Overview of the Motivation Behind Developing Soft Robotic Arms for Healthcare

The motivation behind developing soft robotic arms for healthcare stems from the need to address specific challenges and requirements within medical contexts. Here's an overview of the key motivations:

Safe Human Interaction: Traditional rigid robotic arms used in medical settings may pose a risk of injury to patients or medical staff due to their hard and heavy structures (Rus & Tolley, 2015). Soft robotic arms, with their compliant and flexible nature, offer a safer alternative for close interaction with humans. They can perform tasks such as patient care, rehabilitation, or surgical assistance with reduced risk of accidental collisions or harm.

Gentle Manipulation: Soft robotic arms can exert gentle forces and adapt their shape to conform to the contours of biological tissues, making them suitable for delicate procedures such as surgery or patient assistance. Their soft and compliant nature reduces the likelihood of tissue damage or trauma during manipulation, enhancing patient comfort and safety.

Accessibility and Affordability: Soft robotic arms can be designed using lightweight and cost-effective materials, making them more accessible and affordable compared to traditional rigid robotic systems. This accessibility can benefit healthcare facilities with limited resources or in remote areas, enabling them to incorporate robotic technology into their practice.

Versatility and Adaptability: Soft robotic arms offer versatility in performing a wide range of tasks, from assisting with daily activities for elderly or disabled individuals to assisting surgeons in minimally invasive procedures. Their flexible and adaptable nature allows them to navigate complex anatomical structures and perform intricate motions with precision.

Minimally Invasive Surgery: Soft robotic arms are particularly well-suited for minimally invasive surgical procedures, where access to the surgical site is restricted and precision is paramount. Their flexibility and dexterity enable them to navigate through narrow or confined spaces within the body, performing complex maneuvers with minimal trauma to surrounding tissues.

The development of soft robotic arms for healthcare is driven by the goal of improving patient outcomes, enhancing the efficiency and safety of medical procedures, and expanding the accessibility of robotic technology in healthcare settings. These versatile and adaptable robotic systems have the potential to revolutionize various aspects of healthcare delivery, from patient care to surgical interventions and rehabilitation therapy.

2. BIOLOGICAL INSPIRATIONS

2.1 Exploration of Natural Organisms and Structures Inspiring Soft Robotic Arm Design (e.g., Octopus Arms, Elephant Trunks)

Biological inspirations for soft robotic arms in healthcare can come from a variety of organisms and systems. Here are some examples:

Muscle Structure: Mimicking the structure and function of biological muscles can inspire the design of soft actuators for robotic arms. Biological muscles contract and expand in response to electrical signals, providing the necessary force and motion for movement. Soft robotic actuators, such as pneumatic artificial muscles or electroactive polymers, can replicate this functionality to generate motion and manipulate objects gently.

Octopus Tentacles: The dexterous and flexible tentacles of octopuses inspire the design of soft robotic arms with versatile manipulation capabilities. Octopus tentacles are composed of muscular hydrostats, allowing them to bend and stretch in various directions without a rigid skeleton. Soft robotic arms can emulate this flexibility using compliant materials and distributed actuators, enabling them to navigate complex environments and interact delicately with objects.

Elephant Trunks: The trunk of an elephant serves as another biological inspiration for soft robotic arms, particularly in terms of its strength, dexterity, and sensitivity. Elephant trunks are capable of grasping and manipulating objects with precision, while also being sensitive to touch and pressure. Soft robotic arms can replicate these characteristics using flexible materials, tactile sensors, and multi-modal feedback systems, making them suitable for tasks requiring both strength and sensitivity, such as surgical procedures or rehabilitation therapy.

Human Anatomy: The structure and biomechanics of the human arm and hand provide valuable insights for designing soft robotic arms for healthcare applications. By studying the musculoskeletal system and neural control mechanisms of human limbs, researchers can develop soft robotic arms that emulate natural movements and gestures. This bio-inspired approach enables the creation of assistive devices and prosthetic limbs that closely mimic the functionality of biological appendages, enhancing mobility and independence for individuals with disabilities.

2.2 Comparison of Biological Features With Engineered Soft Robotics

By drawing inspiration from these biological sources, researchers can develop soft robotic arms that combine the advantages of flexibility, dexterity, and sensitivity, making them well-suited for a wide range of healthcare applications, including surgical assistance, rehabilitation therapy, and assistive devices for individuals with mobility impairments. The various comparisons of biological features with soft Robotics is discussed in the Table 1.

3. DESIGN AND MATERIALS

3.1 Structural Design Considerations for Bio-Inspired Soft Robotic Arms

Designing bio-inspired soft robotic arms involves careful consideration of various structural factors to ensure functionality, flexibility, and reliability. Here are some key structural design considerations:

Soft and Compliant Materials: Selecting appropriate soft and compliant materials is crucial for achieving the desired flexibility and adaptability in soft robotic arms. Elastomers, hydrogels, silicone, and flexible polymers are commonly used materials that mimic the softness and flexibility of biological tissues.

Table 1. Comparisons of biological features with soft robotics

Parameters	Biological	Soft Robotics
Flexibility and Compliance	Many organisms exhibit soft, compliant structures that allow for flexibility and adaptability to different environments and tasks. Muscles, tendons, and ligaments provide compliant movement in animals.	Engineered soft robotics aim to replicate this flexibility and compliance using materials such as elastomers and hydrogels.
Sensory Systems	Organisms possess sophisticated sensory systems that allow them to perceive and interact with their environment. These include vision, touch, proprioception, and other sensory modalities.	Soft robotic systems integrate sensors inspired by biological counterparts to perceive their surroundings and respond accordingly.
Biological	Animals have evolved complex biomechanical systems that enable efficient movement, manipulation, and force generation. Examples include the musculoskeletal system and hydrostatic skeletons found in soft-bodied organisms.	Soft robotics draws inspiration from biological biomechanics to design robotic systems with similar capabilities.
Integration and Multi-functionality	Biological organisms often exhibit integrated functionality, with multiple systems working together seamlessly to achieve diverse tasks. For example, the human hand can perform gripping, manipulation, and tactile sensing simultaneously	Soft robotic systems aim to integrate multiple functions into a single platform, mimicking the multi-functionality observed in biological organisms. By combining sensors, actuators, and control systems, soft robotics enables versatile and adaptive behaviour for various applications

Distributed Actuation and Control: Distributing actuators and control mechanisms throughout the robotic arm enables smoother and more natural movements, resembling biological motion patterns. By decentralizing actuation and control, soft robotic arms can achieve greater flexibility, dexterity, and adaptability in various tasks.

Biologically-inspired Kinematics: Emulating the kinematics and motion patterns observed in biological limbs can enhance the functionality and efficiency of soft robotic arms. Bio-inspired design principles, such as the use of parallel mechanisms or compliant joints, can improve motion range, agility, and energy efficiency.

Embedded Sensing and Feedback: Integrating sensors for proprioception, force sensing, and tactile feedback enables soft robotic arms to perceive and respond to their environment. Embedded sensors provide crucial feedback for control algorithms, enhancing precision, safety, and adaptability in various tasks.

By carefully considering these structural design considerations, engineers and researchers can develop bio-inspired soft robotic arms that exhibit enhanced flexibility, adaptability, and performance in healthcare applications, such as surgical assistance, rehabilitation therapy, and assistive devices for individuals with mobility impairments.

3.2 Materials Selection, Including Soft and Flexible Materials for Enhanced Compliance

Selecting appropriate materials is crucial for the structural design of bio-inspired soft robotic arms, particularly when aiming for enhanced compliance and flexibility. Here are some materials commonly used for soft and flexible components in such robotic systems

Silicone Elastomers: Silicone elastomers are widely used in soft robotics due to their excellent flexibility, durability, and biocompatibility. They can be easily moulded into complex shapes and have tuneable mechanical properties, allowing for customization based on the specific requirements of the robotic arm.

Polyurethane Elastomers: Polyurethane elastomers offer similar properties to silicone elastomers and are often used as alternatives or in combination with silicone for soft robotic applications. They provide good resilience, tear resistance, and chemical stability, making them suitable for dynamic and long-term use.

Hydrogels: Hydrogels are water-swollen polymer networks that exhibit high flexibility and biocompatibility, resembling the mechanical properties of biological tissues. They are suitable for soft robotic applications requiring interactions with biological systems, such as biomedical devices or wearable sensors.

Soft Pneumatic Actuators (SPAs): Soft pneumatic actuators are made from flexible materials, such as silicone or elastomers, and actuated by pneumatic pressure. These actuators exhibit compliant behaviour and can generate bending, twisting, or elongation motions, making them suitable for soft robotic arm applications requiring gentle manipulation and interaction with the environment.

4. ACTUATION MECHANISMS

4.1 Overview of Actuation Methods for Soft Robotic Arms

Actuation methods for soft robotic arms enable these devices to achieve motion, manipulation, and interaction with the environment (Li et al., 2020)/ Here's an overview of various actuation methods commonly used in soft robotics, tabulated in Table 2.

4.2 Pneumatic, Hydraulic, and Soft Artificial Muscles

Pneumatic, hydraulic, and soft artificial muscles are actuation mechanisms commonly used in soft robotics, each offering unique advantages and characteristics. Here's an overview of each type shown in the following Table 3.

5. SENSORY FEEDBACK AND CONTROL

5.1 Integrating Sensors for Feedback and Environmental Awareness

It is essential for enabling soft robotic arms to perceive and respond to their surroundings effectively. Here's how sensors can be integrated into soft robotic systems:

Tactile Sensors

Detect contact or pressure on the arm's surface, often using arrays of pressure-sensitive elements like piezoresistive or capacitive sensors.

Force Sensors

Measure applied forces directly at joints, actuators, or end-effectors using load cells, strain gauges, or force-sensitive resistors.

Table 2. Various actuating methods

Type of Actuation	Principle	Advantages	Disadvantages
Pneumatic Actuation	Uses compressed air or gas to deform and move soft structures	Lightweight, compliant, safe, and capable of complex motions	Requires a pneumatic system for control, potentially limiting portability
Hydraulic Actuation	Utilizes pressurized liquid (like oil or water) to actuate soft structures.	Offers higher force and power density, ideal for strong and dynamic movements.	Needs hydraulic fluid supply and control systems, increasing complexity and maintenance.
Shape Memory Alloys (SMAs):	Undergo reversible shape changes with temperature, enabling shape memory effect and super elasticity.	Fast response times, high energy efficiency, and compact form factor.	Limited actuation range, relatively low force output, requires precise temperature control
Soft Fluidic Actuation	Generates motion in soft structures using fluid flow (air or liquid) through inflation and deflation.	Simple, lightweight, compliant, capable of a wide range of motions and shapes.	Requires external fluid supply and control systems, potentially slower response times.

Table 3. Types of artificial muscles

Types of Artificial Muscles	Principle	Advantages	Disadvantages
Pneumatic Artificial Muscles	Flexible tubes or bladders expand or contract with changes in air pressure. Air pressure control dictates muscle length and shape for desired motions.	Lightweight, compliant, high force-to-weight ratio, simple control via pneumatic systems.	Limited lifespan due to wear, nonlinear responses requiring careful calibration.
Hydraulic Artificial Muscles	Similar to PAMs but use pressurized fluid instead of air. Fluid pressure controls muscle expansion for various motions.	High force output, smooth control, resistance to compression.	Complexity due to hydraulic systems, potential for leakage requiring maintenance.
Soft Artificial Muscles	Emulate biological muscle contraction using soft materials like EAPs, SMAs, or dielectric elastomers. Deform in response to electrical, thermal, or mechanical stimuli.	Versatility in design, biomimetic motion, low-profile and lightweight	Limited force output compared to pneumatic or hydraulic actuators, complexity of control with some mechanisms.

Position and Velocity Sensors

Principle & Integration: Provide feedback on arm components' position, orientation, and velocity using encoders, potentiometers, or magnetic sensors integrated into joints or actuators.

Proximity Sensors

Detect nearby objects using sensors like infrared (IR), ultrasonic, or time-of-flight (ToF) mounted on the arm or deployed separately for obstacle detection.

Environmental Sensors

Measure environmental parameters like temperature, humidity, or gas concentration using sensors integrated into the arm or attached externally.

Integrating these sensors enables soft robotic arms to perceive their environment, interact effectively, and adapt behaviour, facilitating applications in healthcare, manufacturing, exploration, and human-robot interaction.

5.2 Control Mechanisms for Precise and Adaptable Movements

Achieving precise and adaptable movements in soft robotic arms requires sophisticated control mechanisms that can accurately regulate actuation, respond to sensory feedback, and adapt to changing conditions. Here are several control mechanisms commonly used for this purpose:

Proportional-Integral-Derivative (PID) Control

Adjusts system output based on present, past, and predicted future errors using proportional, integral, and derivative terms. Widely used in soft robotics for stable and responsive control of actuator position, velocity, or force, enabling precise motion and manipulation.

Model Predictive Control (MPC)

Utilizes a dynamic model to predict future states and optimize control actions over a finite time horizon. Suitable for predictive control of complex movements or interactions in soft robotic arms, optimizing motion trajectories, improving tracking performance, and handling system non-linearities.

Adaptive Control

Adjusts control parameters in real-time based on changes in system dynamics or operating conditions to maintain performance. Useful for soft robotic arms operating in dynamic environments or with varying payloads/frictional forces, compensating for uncertainties or disturbances.

Reinforcement Learning (RL)

Agent learns optimal control policies through trial-and-error interactions with the environment, receiving rewards or penalties (Ji et al., 2022). Applied to soft robotic arms for learning complex motion behaviours or manipulation strategies without explicit models, enabling autonomous and adaptive behaviour over time.

6. APPLICATIONS IN MINIMALLY INVASIVE SURGERY

6.1 Use of Bio-Inspired Soft Robotic Arms in Minimally Invasive Surgical Procedures

The use of bio-inspired soft robotic arms in minimally invasive surgical procedures offers several advantages, including enhanced dexterity, safety, and patient outcomes. Here's how bio-inspired soft robotic arms are being applied in this context:

Flexible and Compliant Manipulation

Soft robotic arms can mimic the flexibility and compliance of biological tissues, allowing them to navigate through narrow and complex anatomical structures with minimal trauma. This flexibility reduces the risk of tissue damage and improves patient safety during minimally invasive surgeries.

Dexterous Instrumentation

Bio-inspired soft robotic arms can incorporate multi-degree-of-freedom manipulators and end-effectors inspired by natural appendages, such as tentacles or elephant trunks. These dexterous instruments enable precise and versatile manipulation of surgical tools within confined spaces, enhancing the surgeon's capabilities and surgical outcomes.

Adaptive Control and Feedback

Soft robotic arms can integrate sensors for real-time feedback on tissue properties, forces, and tool interactions during surgery. This feedback enables adaptive control algorithms to adjust the robotic arm's motion and force application, ensuring accurate and safe tissue manipulation while minimizing the risk of complications.

Precise Targeting and Localization

Soft robotic arms equipped with imaging modalities, such as cameras or ultrasound probes, can provide enhanced visualization and localization of surgical targets. This enables precise targeting of lesions or diseased tissues, facilitating more accurate surgical interventions and reducing the risk of inadvertent damage to surrounding structures.

6.2 Advantages Over Traditional Rigid Robotic Arms in Surgical Applications

Bio-inspired soft robotic arms offer several advantages over traditional rigid robotic arms in surgical applications, particularly in minimally invasive procedures. Here are some key advantages:

Flexibility and Compliance: Soft robotic arms mimic the compliant and flexible nature of biological tissues, allowing them to navigate through complex anatomical structures with greater ease and safety.

Gentle Tissue Interaction: Soft robotic arms exert gentle forces on tissues and organs, minimizing the risk of trauma or injury during surgery.

Adaptability to Anatomical Variability: Soft robotic arms can adapt to patient-specific anatomies and variations in tissue properties, enabling personalized and tailored surgical interventions.

Reduced Risk of Instrument Clashes: Soft robotic arms, with their compliant and deformable structures, are less likely to cause instrument clashes or collisions during surgery compared to rigid robotic arms.

Improved Ergonomics and Surgeon Comfort: Soft robotic arms can be designed with ergonomic considerations in mind, providing surgeons with greater comfort and ease of use during prolonged surgical procedures. Their lightweight and flexible construction reduce operator fatigue and strain, allowing for more precise and controlled movements over extended periods.

Cost-Effectiveness and Accessibility: Soft robotic arms can be fabricated using cost-effective and readily available materials, making them more affordable and accessible compared to traditional rigid robotic systems. This affordability enables wider adoption of robotic-assisted surgery in healthcare facilities with limited resources or in remote areas, expanding access to advanced surgical technologies.

The advantages of bio-inspired soft robotic arms over traditional rigid robotic arms in surgical applications lie in their flexibility, compliance, adaptability, and enhanced dexterity, which contribute to improved surgical outcomes, reduced complications, and enhanced patient care.

7. REHABILITATION AND ASSISTIVE DEVICES

7.1 Applications in Physical Therapy and Rehabilitation

Bio-inspired soft robotic arms hold significant potential for applications in physical therapy and rehabilitation, offering innovative solutions to assist individuals with mobility impairments, facilitate recovery from injuries, and enhance rehabilitation outcomes. Here are several key applications in this field:

Assistive Devices for Activities of Daily Living (ADLs): Soft robotic arms can be integrated into assistive devices to help individuals with mobility impairments perform activities of daily living, such as eating, dressing, and personal hygiene. These devices can provide assistance with grasping, reaching, and manipulating objects, enabling greater independence and autonomy for users.

Rehabilitation Robotics: Soft robotic arms can be used in rehabilitation robotics to provide targeted assistance and resistance during therapeutic exercises and functional training. By guiding and supporting the movements of the affected limbs, soft robotic arms can help individuals regain strength, range of motion, and motor control following injuries or surgeries.

Home-Based Rehabilitation: Soft robotic devices designed for home use can enable individuals to continue their rehabilitation exercises and therapies outside of clinical settings. These devices can provide personalized and interactive rehabilitation programs, allowing individuals to participate in their recovery process and maintain adherence to treatment regimens.

7.2 Development of Assistive Devices for Individuals With Mobility Challenges

The development of assistive devices for individuals with mobility challenges is a crucial area where bio-inspired soft robotic arms can make a significant impact. These devices aim to enhance mobility, independence, and quality of life for individuals with disabilities or mobility impairments. Here's how bio-inspired soft robotic arms can contribute to the development of assistive devices:

Enhanced Flexibility and Adaptability: Bio-inspired soft robotic arms offer greater flexibility and adaptability compared to traditional rigid assistive devices. Their compliant and deformable structures allow for better accommodation of individual anatomies and movements, improving comfort and usability for users with varying mobility needs.

Natural and Intuitive Interaction: Soft robotic arms can mimic the natural flexibility and compliance of biological limbs, providing more natural and intuitive interaction with the environment. This allows users to perform daily activities with greater ease and confidence, enhancing their independence and autonomy.

Customized Assistive Solutions: Soft robotic arms can be customized to meet the specific needs and preferences of individual users. Their modular design and customizable features allow for tailored solutions that address unique mobility challenges, such as limited range of motion, muscle weakness, or coordination difficulties.

Assistance with Activities of Daily Living (ADLs): Soft robotic arms can assist individuals with mobility challenges in performing activities of daily living, such as eating, dressing, grooming, and household chores. These devices can provide support and assistance with grasping, reaching, and manipulating objects, enabling greater autonomy and participation in daily life tasks.

Mobility Aids and Walking Assistance: Soft robotic exoskeletons and wearable devices can provide walking assistance and mobility support for individuals with mobility impairments. These devices can help users walk more efficiently, reduce fatigue, and navigate various terrains and environments with greater ease and stability.

Rehabilitation and Physical Therapy: Soft robotic arms can assist individuals undergoing rehabilitation and physical therapy by providing targeted assistance and resistance during therapeutic exercises. These devices can help improve muscle strength, range of motion, and motor control, facilitating faster recovery and rehabilitation outcomes.

8. HUMAN-MACHINE INTERACTION

8.1 Analysis of How Soft Robotic Arms Interact With Human Users

The interaction between soft robotic arms and human users is a multifaceted process that involves physical contact, communication, and coordination to achieve desired tasks or goals. Here's an analysis of how soft robotic arms interact with human users:

Physical Interaction

Soft robotic arms interact with human users through physical contact, where the compliant and flexible nature of the robotic arms allows for safe and gentle interaction. This physical interaction can involve grasping objects, assisting with movements, or providing support during activities of daily living. Unlike rigid robotic arms, soft robotic arms can conform to the shape and contours of the human body, minimizing the risk of injury or discomfort during interaction.

Sensory Feedback

Soft robotic arms can incorporate sensors to provide feedback on their interactions with human users. This sensory feedback enables the robotic arms to detect forces, pressures, or movements exerted by the user, allowing for adaptive and responsive behaviour.

Human-Robot Communication

Soft robotic arms can communicate with human users through various modalities, such as visual cues, auditory signals, or haptic feedback. This communication allows the robotic arms to convey information about their state, intentions, or actions, enhancing the user's understanding and trust. For example, LED lights or display panels can indicate the status of the robotic arm, while audio prompts or voice commands can provide instructions or feedback to the user.

Shared Control and Collaboration

Soft robotic arms can engage in shared control and collaboration with human users to achieve mutual goals or tasks. This collaborative interaction involves coordinated movements and actions between the human and the robot, where both parties contribute to the completion of the task.

Assistive Functionality

Soft robotic arms often serve as assistive devices to support individuals with mobility impairments or disabilities in performing daily activities. The interaction between the robotic arms and human users is characterized by assistance, guidance, and empowerment, where the robotic arms augment the user's capabilities and enhance their independence. This assistive functionality promotes a positive and empowering interaction experience, enabling users to achieve greater autonomy and quality of life.

8.2 User Interfaces and Integration With Existing Healthcare Technologies

User interfaces (UIs) play a crucial role in facilitating interactions between users and soft robotic arms in healthcare settings. These interfaces provide intuitive controls, real-time feedback, and seamless integration with existing healthcare technologies. Here's an analysis of UIs and their integration with healthcare technologies:

Real-Time Feedback and Visualization: UIs for soft robotic arms should provide real-time feedback and visualization of key parameters, such as position, velocity, force, and sensor data. This feedback helps users monitor the status and performance of the robotic arms during operation, ensuring safe and effective interactions. Graphical displays, charts, or numerical readouts can convey information about the robotic arm's state, trajectory, and interactions with the environment.

Customizable and Configurable Settings: User interfaces for soft robotic arms should allow for customizable and configurable settings to accommodate different user preferences, clinical protocols, and therapeutic needs. This may include adjustable parameters for motion speed, force output, assistive modes, or safety thresholds, allowing users to tailor the robotic arm's behavior to their specific requirements. Preset profiles, user profiles, or templates can streamline the setup process and ensure consistency in treatment delivery.

Interoperability with Medical Devices: UIs for soft robotic arms should support interoperability with other medical devices and equipment commonly used in healthcare settings. This interoperability allows for seamless integration of the robotic arms into existing clinical workflows and treatment protocols. Standardized communication protocols, such as Health Level Seven (HL7) or Digital Imaging and Communications in Medicine (DICOM), facilitate data exchange and interoperability with medical devices, such as imaging systems, monitoring devices, or therapeutic equipment.

9. SAFETY AND BIOCOMPATIBILITY

9.1 Evaluation of the Safety Aspects of Using Soft Materials in Healthcare

The evaluation of safety aspects when incorporating soft materials in healthcare settings is paramount to ensuring patient well-being and effective medical outcomes. Soft materials offer advantages such as comfort, flexibility, and reduced risk of injury, but they also pose unique safety considerations.

One critical aspect of safety evaluation involves assessing the material's biocompatibility to ensure it does not trigger adverse reactions or tissue irritation upon contact with the body. This includes rigorous testing for allergens, toxins, and potential leachables that could compromise patient health.

Furthermore, the durability and cleanliness of soft materials must be evaluated to prevent the accumulation of pathogens or contaminants that could lead to infections or cross-contamination. Regular inspection and maintenance protocols are essential to uphold hygienic standards.

Moreover, the mechanical properties of soft materials, such as their tensile strength and elasticity, must be assessed to ensure they can withstand the demands of healthcare environments without compromising patient safety or caregiver efficacy.

9.2 Biocompatibility Considerations for Medical Applications

Biocompatibility considerations are essential in the development and utilization of materials for medical applications, including soft robotic arms. Biocompatibility ensures that materials used in medical devices do not elicit harmful reactions or adverse effects when in contact with biological systems. Here are key biocompatibility considerations for medical applications:

Cytotoxicity: Evaluate potential cell damage or toxicity. Test cell viability and morphology via extract exposure.

Sensitization: Assess allergic reaction potential. Use skin patch tests and in vitro assays.

Irritation: Check for tissue irritation or inflammation. Test by exposing tissues to the material.

Systemic Toxicity: Evaluate toxicity upon introduction into the body. Assess effects on organs and physiological systems.

Genotoxicity: Check for genetic mutations or DNA damage. Assess potential genetic alterations in cells

Implantation Compatibility: Ensure integration with surrounding tissues. Monitor tissue responses in vivo over time.

Biodegradability and Bioresorbable: Assess degradation kinetics and tissue response. Ensure safe breakdown for temporary implants or drug delivery.

Addressing these considerations ensures safe materials and devices for medical use, including soft robotic arms. Compliance with standards and thorough testing are vital for patient safety and regulatory approval.

10. CHALLENGES AND LIMITATIONS

10.1 Technical Challenges in Soft Robotic Arm Development

Technical Challenges in Soft Robotic Arm Development

Material Selection and Characterization: Selecting suitable materials meeting mechanical, biocompatibility, and durability requirements is challenging. Accurately characterizing mechanical properties and predicting behaviour under various loading conditions is crucial.

Actuation and Control: Soft robotic arms need innovative actuation mechanisms. Traditional rigid actuators may not work, requiring new methods like pneumatic or shape-memory alloys.

Controlling complex deformation and motion in dynamic environments poses modelling and feedback control challenges.

Integration of Sensors and Feedback Systems: Integrating sensors into soft structures without compromising flexibility is a challenge. Real-time processing of sensor data and implementing feedback control require efficient computational systems.

Durability and Reliability: Soft materials may degrade over time due to fatigue or environmental factors. Ensuring durability and reliability necessitates careful material selection, design optimization, and realistic testing.

Human-Robot Interaction and Safety: Soft robotic arms must prioritize user safety and comfort. Design considerations and adherence to safety standards are crucial, especially in wearable or assistive applications.

Overcoming these challenges requires interdisciplinary collaboration across materials science, mechanical engineering, robotics, control systems, and biomedical research. By doing so, soft robotics can fulfill their potential in healthcare, manufacturing, and human-robot collaboration.

10.2 Limitations in Terms of Payload, Speed, and Complexity

Soft robotic arms, while offering unique advantages in terms of flexibility and adaptability, also come with inherent limitations compared to their rigid counterparts.

Payload Capacity: Soft arms have lower payload capacities due to material compliance. Not suitable for tasks requiring heavy lifting or high forces.

Speed and Response Time: Soft actuators have slower response times compared to electric motors. Limitations in fluid flow through actuators restrict speed and response.

Complexity of Control: Complex control algorithms are needed due to nonlinear nature. Modeling interactions and adapting control strategies add complexity.

Environmental Sensitivity: Soft materials may be sensitive to temperature, humidity, and chemicals. Environmental variations affect material properties and arm behavior.

Wear and Degradation: Soft materials are prone to wear and degradation over time. Mechanical properties change, affecting performance and longevity.

Despite these limitations, on-going research and innovation in soft robotics aim to overcome technical barriers and expand capabilities. Advances in materials, actuation, control, and fabrication are driving progress in soft robotic applications.

10.3 Ethical Considerations and Potential Risks

In the development and deployment of soft robotic arms, as with any emerging technology, there are ethical considerations and potential risks that must be carefully addressed to ensure responsible and beneficial use. Here are some key ethical considerations and potential risks associated with soft robotic arms:

Safety and Reliability: Ethical considerations: Ensuring the safety and reliability of soft robotic arms is paramount to prevent harm to users and ensure trust in the technology. Ethical principles such as beneficence (promoting well-being) and non-maleficence (avoiding harm) guide efforts to prioritize safety in design, testing, and deployment.

Potential risks: Malfunction or failure of soft robotic arms could result in physical harm to users or damage to property. Robust safety mechanisms, thorough testing protocols, and fail-safe features are essential to mitigate these risks and ensure the safe operation of soft robotic systems.

Privacy and Data Security: Ethical considerations: Soft robotic arms equipped with sensors or connected to digital interfaces may collect sensitive data about users, such as biometric information or behavioural patterns. Respecting user privacy and protecting personal data is essential to uphold ethical principles of autonomy and privacy.

Potential risks: Unauthorized access to or misuse of user data collected by soft robotic arms could compromise privacy, lead to identity theft, or result in discriminatory practices. Implementing strong data encryption, access controls, and privacy policies can help mitigate these risks and safeguard user privacy.

11. COMPARISON WITH TRADITIONAL ROBOTICS ARM

11.1 Contrasting Bio-Inspired Soft Robotic Arms With Traditional Rigid Robotic Arms

Materials: Soft Robotic Arms: Composed of compliant materials like elastomers or hydrogels, mimicking biological tissues for gentle interaction and safe human contact.

Rigid Robotic Arms: Built from rigid materials such as metals or hard plastics, providing strength and stability but lacking flexibility for complex environments.

Applications: Soft Robotic Arms: Suited for human interaction, delicate tasks, and unstructured environments like surgery, rehabilitation, assistive devices, and disaster relief.

Rigid Robotic Arms: Primarily used in industrial settings for precise tasks like manufacturing, assembly, and material handling, requiring speed, precision, and force.

Performance: Soft Robotic Arms: Excel in gentle manipulation, dexterity, and adaptability, offering safety and versatility, though with potential limitations in payload, speed, and precision.

Rigid Robotic Arms: Provide high precision, speed, and payload capacity, ideal for industrial automation and tasks in controlled environments, albeit lacking flexibility for dynamic settings.

11.2 Analyzing the Benefits and Drawbacks in Healthcare Contexts

Analysing the benefits and drawbacks of bio-inspired soft robotic arms and traditional rigid robotic arms in healthcare contexts provides insights into their respective applications, performance, and suitability for specific tasks. Here's a comparative analysis:

Benefits of Bio-Inspired Soft Robotic Arms in Healthcare

- Safety and Compliance
- Gentle Manipulation
- Adaptability to Complex Environments

Drawbacks of Bio-Inspired Soft Robotic Arms in Healthcare

- Limited Payload Capacity
- Reduced Speed and Precision
- Complex Control Requirements
- Durability and Wear

Benefits of Traditional Rigid Robotic Arms in Healthcare

- High Payload Capacity
- Speed and Precision
- Stability and Force Control
- Proven Technology

Drawbacks of Traditional Rigid Robotic Arms in Healthcare

- Limited Adaptability
- Risk of Tissue Trauma
- Incompatibility with Wearable Applications.
- Complexity and Cost

In summary, both bio-inspired soft robotic arms and traditional rigid robotic arms offer unique benefits and drawbacks in healthcare contexts, depending on the specific application requirements, environment, and patient needs. Understanding the trade-offs between these two types of robotic arms is essential for selecting the most appropriate technology for healthcare tasks, optimizing patient outcomes, and advancing the field of medical robotics.

12. CASE STUDIES AND PROTOTYPES

12.1 Examining Successful Case Studies of Bio-Inspired Soft Robotic Arms in Healthcare

Some successful case studies of bio-inspired soft robotic arms in healthcare:

Heart Lander Surgical Robot

Developed by researchers at Carnegie Mellon University, the HeartLander surgical robot is a bio-inspired soft robotic arm designed for minimally invasive heart surgery. The robot mimics the peristaltic motion of earthworms to navigate and anchor itself to the surface of the heart. By adhering to the heart's surface, the robot can perform precise surgical tasks such as tissue ablation or injection without the need for open-heart surgery. Clinical trials have demonstrated the safety and efficacy of the HeartLander robot in reducing surgical trauma and improving patient outcomes in procedures such as atrial fibrillation ablation.

Soft Robotic Catheters for Cardiac Interventions

Researchers at Harvard University and Boston Children's Hospital developed soft robotic catheters for cardiac interventions in pediatric patients with congenital heart defects. The catheters incorporate soft, compliant materials and integrated actuators to navigate through the delicate structures of the heart with precision (Deimel & Brock, 2016). By leveraging the natural compliance and adaptability of soft materials, these catheters enable safer and more effective interventions for repairing cardiac anomalies such as atrial septal defects or ventricular septal defects.

Soft Robotic Prosthetic Limbs

Several research groups have developed bio-inspired soft robotic prosthetic limbs to enhance mobility and functionality for individuals with limb loss. By mimicking the flexibility and adaptability of biological tissues, soft robotic prosthetic limbs offer improved comfort, control, and usability compared to traditional rigid prosthetics (Connolly, Walsh, & Bertoldi, 2017). Clinical trials and user studies have demonstrated the benefits of soft robotic prosthetic limbs in improving mobility, reducing phantom limb pain, and enhancing quality of life for amputees.

These successful case studies highlight the transformative potential of bio-inspired soft robotic arms in healthcare, offering innovative solutions for surgical interventions, rehabilitation, and assistive technology.

12.2 Evaluation of Prototypes and Their Real-World Applications

Evaluation of prototypes of bio-inspired soft robotic arms involves assessing their performance, safety, usability, and efficacy in real-world applications.

Performance Testing: Assess mechanical performance, including range of motion, speed, and accuracy. Quantitatively measure metrics like actuation force and repeatability.

Real-world applications: Ensure suitability for tasks like surgery, rehabilitation, or assistive device control.

Safety Assessment: Evaluate safety features to prevent user injury and ensure compatibility with healthcare environments. Assess measures to prevent entanglement, pinch points, or excessive forces.

Usability Testing: Evaluate user experience through testing and feedback collection. Assess ease of use, ergonomics, and comfort during operation.

Real-world applications: Ensure user-friendliness for healthcare professionals, patients, or caregivers in clinical or home settings.

Functional Validation: Validate functionality in relevant scenarios. Assess ability to perform tasks accurately and reliably.

Real-world applications: Demonstrate suitability for tasks like surgical procedures or rehabilitation exercises under realistic conditions.

Clinical Trials and Validation Studies:

Conduct trials to evaluate safety, efficacy, and clinical outcomes in human subjects. Assess impact on patient outcomes compared to existing treatments.

Real-world applications: Provide evidence of effectiveness and guide integration into healthcare practice and regulatory approval.

Evaluation involves a multidisciplinary approach, combining engineering, clinical, and user-centered perspectives. By systematically assessing performance, safety, usability, and clinical effectiveness, developers can validate the potential of bio-inspired soft robotic arms to transform patient care.

13 FUTURE DIRECTIONS AND INNOVATIONS

13.1 Emerging Trends and Future Possibilities in Bio-Inspired Soft Robotics for Healthcare

Emerging trends and future possibilities in bio-inspired soft robotics for healthcare hold significant promise for advancing medical technology and improving patient outcomes (Cianchetti et al., 2018). Here are some key trends and possibilities shaping the future of bio-inspired soft robotics in healthcare:

Continued Miniaturization and Integration

Advances in microfabrication and nanotechnology enable the development of miniaturized soft robotic devices for minimally invasive procedures and targeted interventions. These devices can be integrated with imaging modalities such as MRI or ultrasound for real-time guidance and navigation within the body.

Bio-Hybrid and Bio-Inspired Designs

Bio hybrid soft robotic systems that combine living and synthetic components offer new possibilities for personalized healthcare and regenerative medicine. Incorporating biological tissues or cells into soft robotic devices enhances biocompatibility, promotes tissue integration, and enables responsive and adaptive behaviour in dynamic biological environments.

Soft Robotic Wearable's for Monitoring and Therapy

Wearable soft robotic devices, such as exosuits, orthoses, or prosthetic limbs, provide personalized assistance and rehabilitation for individuals with mobility impairments or musculoskeletal disorders. These devices offer adaptive support, therapeutic feedback, and data-driven rehabilitation programs tailored to individual needs.

Soft Robotics for Surgical Assistance and Teleoperation

Soft robotic systems provide versatile tools for surgical assistance, teleoperation, and remote healthcare delivery. Teleoperated soft robotic platforms enable minimally invasive procedures, remote consultations, and surgical training in challenging or remote environments, expanding access to specialized healthcare services and expertise.

Regenerative Soft Robotics

Advances in regenerative medicine and tissue engineering converge with soft robotics to create self-healing, self-repairing, and self-assembling robotic systems. Regenerative soft robotics offer potential solutions for tissue regeneration, wound healing, and organ repair, blurring the boundaries between living and synthetic materials in healthcare applications.

13.2 Potential Advancements and Areas for Further Research

In the rapidly evolving field of bio-inspired soft robotics for healthcare, several potential advancements and areas for further research hold promise for pushing the boundaries of innovation and addressing critical healthcare challenges. Here are some key areas for future research:

Biologically Inspired Actuation Mechanisms

Develop novel actuation mechanisms inspired by biological systems, such as artificial muscles, pneumatic networks, or shape-changing materials. These mechanisms could offer improved performance, energy efficiency, and biomimetic functionality for soft robotic systems in healthcare applications.

Soft Robotic Sensing and Feedback Systems

Enhance sensing capabilities and feedback mechanisms in soft robotic devices to enable real-time monitoring, adaptive control, and intelligent decision-making. Integration of advanced sensors, including tactile, proprioceptive, and physiological sensors, enables precise manipulation, navigation, and interaction with biological tissues.

Adaptive Control and Learning Algorithms

Develop adaptive control algorithms and machine learning techniques to enhance the autonomy, adaptability, and learning capabilities of soft robotic systems. Intelligent control strategies enable autonomous navigation, task planning, and response to dynamic environmental changes in healthcare settings.

Soft Robotic Interfaces for Human-Machine Interaction

Design soft robotic interfaces that facilitate seamless interaction between humans and machines, promoting intuitive control, ergonomic comfort, and natural movement. User-centered design approaches ensure that soft robotic devices are user-friendly, inclusive, and accessible for individuals with diverse abilities and needs.

14. REGULATORY AND COMPLIANCE ASPECTS

14.1 Compliance With Healthcare Regulations for Medical Devices

Ensuring compliance with healthcare regulations is crucial for the development, manufacturing, and deployment of medical devices, including bio-inspired soft robotics, to ensure patient safety, efficacy, and quality (Polygerinos et al., 2016). Here's how compliance with healthcare regulations is typically addressed for medical devices:

Regulatory Frameworks

Understand and adhere to regulatory frameworks governing medical devices in relevant jurisdictions, such as the Food and Drug Administration (FDA) in the United States, the European Medical Devices Regulation (MDR) in the European Union, or other regional regulatory authorities. Familiarize yourself with regulatory requirements, classification criteria, and submission processes applicable to bio-inspired soft robotic devices.

Quality Management Systems (QMS)

Implement robust quality management systems compliant with international standards, such as ISO 13485:2016, to ensure consistency, traceability, and control throughout the device lifecycle. Establish procedures for design controls, risk management, documentation, validation, and post-market surveillance to meet regulatory requirements.

Risk Management

Conduct comprehensive risk assessments and hazard analyses to identify and mitigate potential risks associated with the use of bio-inspired soft robotic devices. Implement risk management processes, such as failure mode and effects analysis (FMEA) or fault tree analysis (FTA), to minimize patient harm and ensure device safety.

Labelling and Instructions for Use

Develop clear, accurate, and comprehensive labelling and instructions for use (IFU) for bio-inspired soft robotic devices to ensure proper device utilization, patient safety, and healthcare provider understanding. Include information on device indications, contraindications, precautions, warnings, and post-market surveillance requirements.

By adhering to these principles and best practices for regulatory compliance, developers and manufacturers of bio-inspired soft robotic devices can navigate the complex regulatory landscape, mitigate regulatory risks, and bring innovative medical technologies to market in compliance with healthcare regulations.

14.2 Approval Processes and Standards for Soft Robotic Arms

The approval processes and standards for soft robotic arms, like any other medical device, vary depending on the intended use, classification, and regulatory jurisdiction. Here's an overview of the approval processes and standards commonly applicable to soft robotic arms:

Regulatory Authorities

In the United States, the Food and Drug Administration (FDA) regulates medical devices under the Federal Food, Drug, and Cosmetic Act (FD&C Act). For soft robotic arms, the FDA's Center for Devices and Radiological Health (CDRH) oversees regulatory oversight, including premarket clearance or approval.

In the European Union, medical devices are regulated under the Medical Devices Regulation (MDR) or the In Vitro Diagnostic Regulation (IVDR). Soft robotic arms fall under the scope of MDR, and conformity assessment procedures are conducted by notified bodies designated by EU member states.

Premarket Approval (PMA)

Soft robotic arms classified as Class III devices, which pose the highest risk, may require premarket approval (PMA) from the FDA. PMA applications include comprehensive scientific and clinical data demonstrating the safety and effectiveness of the device, typically through clinical studies or trials.

The FDA evaluates the PMA application to determine whether the device meets stringent safety and effectiveness criteria, granting approval if the benefits outweigh the risks based on the available evidence.

International Standards

Soft robotic arms must comply with relevant international standards, such as ISO 13485 (Quality Management Systems), ISO 14971 (Risk Management), and ISO 60601 (Medical Electrical Equipment), which provide guidance on quality management, risk management, and safety requirements for medical devices.

15. CONCLUSION AND RECOMMENDATIONS

15.1 Summarizing Key Findings From the Comprehensive Analysis

The comprehensive analysis of bio-inspired soft robotic arms in healthcare revealed several key findings:

Technological Advancements: Bio-inspired soft robotic arms leverage principles from nature to develop innovative devices that mimic biological systems' flexibility, compliance, and adaptability. These advancements enable safer, more precise, and minimally invasive interventions in healthcare settings.

Diverse Applications: Soft robotic arms find applications across various healthcare domains, including surgery, rehabilitation, assistive technology, and medical imaging. Their versatility and adaptability make them suitable for a wide range of tasks, from delicate surgical procedures to patient rehabilitation exercises.

Biological Inspiration: Soft robotic arms draw inspiration from biological structures and mechanisms, such as muscles, tendons, and octopus tentacles, to emulate natural motion and functionality. By mimicking biological systems, soft robotics offer enhanced compatibility with human physiology and improved patient outcomes.

Materials and Actuation: Soft robotic arms utilize soft and flexible materials, such as elastomers, hydrogels, and shape-memory polymers, to achieve compliance and deformability. Actuation mechanisms, including pneumatic, hydraulic, and artificial muscles, enable controlled motion and manipulation in dynamic environments.

Challenges and Future Directions: Despite significant progress, challenges remain in areas such as durability, control, and integration with existing healthcare technologies. Future research directions include advancements in actuation mechanisms, sensing capabilities, and human-machine interaction to further enhance the performance and usability of soft robotic arms in healthcare.

5.2 Providing Recommendations for Further Research, Development, and Practical Implementations

Based on the comprehensive analysis, here are recommendations for further research, development, and practical implementations of bio-inspired soft robotic arms in healthcare

Advanced Actuation Mechanisms: Investigate novel actuation mechanisms, such as smart materials, electroactive polymers, or biomimetic actuators, to enhance soft robotic arm performance, efficiency, and controllability. Research on soft actuators with tunable stiffness, shape-changing capabilities, and adaptive responses can enable more versatile and biomimetic motion.

Sensing and Feedback Systems: Develop advanced sensing technologies, including tactile sensors, proprioceptive feedback systems, and biometric sensors, to enhance soft robotic arm perception, interaction, and adaptation in complex environments. Integration of multimodal sensing and closed-loop feedback mechanisms enables real-time monitoring, adaptive control, and personalized interventions.

Human-Robot Interaction: Explore human-centered design principles and user interface technologies to improve the intuitiveness, usability, and acceptance of soft robotic arms by healthcare professionals, patients, and caregivers. Incorporate ergonomic features, intuitive controls, and immersive interfaces to facilitate seamless interaction and collaboration between humans and robots.

Integration with Healthcare Technologies: Integrate soft robotic arms with existing healthcare technologies, such as surgical navigation systems, medical imaging modalities, or teleoperated platforms, to enhance their functionality, interoperability, and clinical utility. Collaborate with healthcare providers and technology partners to identify integration opportunities and address unmet clinical needs.

Regulatory Compliance and Standards: Ensure compliance with healthcare regulations, standards, and quality management systems throughout the development, manufacturing, and deployment of soft robotic arms. Engage early with regulatory authorities, seek regulatory guidance, and conduct thorough risk assessments to navigate regulatory pathways and obtain market approval or clearance.

REFERENCES

Bekey, G. A. (2005). *Autonomous robots: from biological inspiration to implementation and control.* MIT Press.

Cianchetti, M., Ranzani, T., Gerboni, G., Nanayakkara, T., Althoefer, K., & Dasgupta, P. (2018). Soft robotics technologies to address shortcomings in today's minimally invasive surgery: The STIFF-FLOP approach. *Soft Robotics*, 5(2), 149–161. PMID:29297756

Connolly, F., Walsh, C. J., & Bertoldi, K. (2017). Automatic design of fiber-reinforced soft actuators for trajectory matching. *Proceedings of the National Academy of Sciences of the United States of America*, 114(1), 51–56. doi:10.1073/pnas.1615140114 PMID:27994133

Deimel, R., & Brock, O. (2016). A novel type of compliant and underactuated robotic hand for dexterous grasping. *The International Journal of Robotics Research*, 35(1-3), 161–185. doi:10.1177/0278364915592961

Ji, Q., Fu, S., Tan, K., Muralidharan, S. T., Lagrelius, K., Danelia, D., Andrikopoulos, G., Wang, X. V., Wang, L., & Feng, L. (2022). Synthesizing the optimal gait of a quadruped robot with soft actuators using deep reinforcement learning. *Robotics and Computer-integrated Manufacturing*, *78*, 102382. doi:10.1016/j.rcim.2022.102382

Li, H., Yao, J., Zhou, P., Chen, X., Xu, Y., & Zhao, Y. (2020). High-force soft pneumatic actuators based on novel casting method for robotic applications. *Sensors and Actuators. A, Physical*, *306*, 306. doi:10.1016/j.sna.2020.111957

Li, J., Xu, Z., & Zhu, D. (2022). Bio-inspired Intelligence with Applications to Robotics: A Survey. arXiv:2206.

Li, J., Yang, S. X., & Xu, Z. (2019). A survey on robot path planning using bio-inspired algorithms. *2019 IEEE International Conference on Robotics and Biomimetics (ROBIO)*, 2111–2116. 10.1109/ROBIO49542.2019.8961498

Oh, H., Shirazi, A. R., Sun, C., & Jin, Y. (2017). Bio-inspired self-organising multi-robot pattern formation: A review. *Robotics and Autonomous Systems*, *91*, 83–100. doi:10.1016/j.robot.2016.12.006

Polygerinos, P., Wang, Z., Galloway, K. C., Wood, R. J., & Walsh, C. J. (2015). Soft robotic glove for combined assistance and at-home rehabilitation. *Robotics and Autonomous Systems*, *73*, 135–143. doi:10.1016/j.robot.2014.08.014

Rus, D., & Tolley, M. T. (2015). Design, fabrication and control of soft robots. *Nature*, *521*(7553), 467–475. doi:10.1038/nature14543 PMID:26017446

Chapter 2
A Comprehensive Insight and Research Focus on Delta Robots in the Healthcare Industry

S. Sarveswaran
Sri Ramakrishna Engineering College, India

Kishore Kumar Arjunsingh
(iD) https://orcid.org/0000-0003-4876-319X
Sri Ramakrishna Engineering College, India

N. Dheerthi
Sri Ramakrishna Engineering College, India

A. Murugarajan
Sri Ramakrishna Engineering College, India

ABSTRACT

Delta robots, known for their unique design featuring parallel linkages and a stationary base, have emerged as transformative tools in various industries, including healthcare. In surgery, delta robots enable minimally invasive procedures with enhanced precision and shorter recovery times. They facilitate targeted therapies in rehabilitation, promoting better outcomes for patients with neurological and musculoskeletal conditions. Delta robots also improve medication dispensing accuracy in pharmacies and automate repetitive tasks in laboratories, increasing efficiency and reducing errors. Additionally, the chapter explores the potential of delta robots in specialized fields such as orthopedic, neuro, and cardiac surgery, as well as their role in enhancing medical imaging accuracy and guiding interventional procedures in real-time. It also discusses the future of AI-powered diagnostics and personalized medicine, envisioning a healthcare landscape where delta robots play a central role in improving patient outcomes and shaping the future of healthcare delivery.

DOI: 10.4018/979-8-3693-5767-5.ch002

1. INTRODUCTION

Delta robots are parallel robots known for their unique design and exceptional performance in high-speed, high-precision applications. They consist of three arms connected to universal joints at the base and a common end-effector platform. The design allows for precise and fast movement in three-dimensional space, making them ideal for tasks such as pick-and-place operations, packaging, and assembly in industries like food processing, electronics, and pharmaceuticals (Mehrafrooz et al., 2017).

Delta robots can achieve very high speeds due to their parallel kinematics, where each joint moves independently. This enables rapid motion and increases throughput in industrial processes (Li, 2018). The parallel structure of delta robots provides excellent accuracy and repeatability, crucial for tasks requiring precise positioning, such as micro-assembly and electronics manufacturing (Pisla et al., 2009). Delta robots have a compact footprint, as the motors and actuators are typically mounted on a stationary base, reducing the space required for operation. Delta robots are highly adaptable and can be easily reconfigured for different tasks by changing the end-effector tooling (McClintock et al., 2018). This flexibility makes them suitable for a wide range of applications in various industries. Despite their lightweight construction, delta robots can handle relatively heavy payloads compared to their own weight, making them efficient for handling substantial loads in industrial settings (Lopez et al., 2006).

The healthcare industry is undergoing a significant transformation with the integration of advanced robotic technologies (Bogossian, 2022). Delta robots, known for their precision, speed, and flexibility, are revolutionizing various aspects of healthcare delivery, from surgical procedures to pharmacy automation. In this chapter, we explore how delta robots are reshaping the landscape of healthcare and driving improvements in patient care, operational efficiency, and medical innovation.

1.1 Delta Robots in Surgical Robotics

Surgical robotics has emerged as a promising field, offering minimally invasive procedures with enhanced precision and control. Delta robots are playing a crucial role in this paradigm shift by enabling surgeons to perform complex operations with greater accuracy and efficiency (Poppeova et al., 2011). With their high-speed motion capabilities and sub-millimeter precision, delta robots are employed in procedures such as laparoscopic surgery, ophthalmic surgery, and neurosurgery.

One of the key advantages of delta robots in surgical robotics is their ability to compensate for patient movement and anatomical variations in real time. This dynamic responsiveness enhances surgical outcomes and reduces the risk of complications. Furthermore, delta robots can be integrated with advanced imaging technologies such as MRI and CT scanners, allowing for precise navigation and targeting during interventions.

1.2 Pharmacy Automation and Drug Delivery

In pharmacy and drug manufacturing facilities, delta robots are streamlining operations and improving medication dispensing accuracy. Automated pharmacy systems equipped with delta robots can precisely handle and package medications, reducing errors and ensuring patient safety. These systems are particularly beneficial in hospital pharmacies where large volumes of medications need to be processed efficiently.

Moreover, delta robots are facilitating advancements in drug delivery systems, including personalized medicine and targeted drug delivery. By automating the assembly of drug delivery devices such as insulin pumps and inhalers, delta robots are enhancing the reliability and consistency of drug administration, leading to better treatment outcomes for patients with chronic conditions.

1.3 Rehabilitation and Assistive Devices

Delta robots are also making significant contributions to rehabilitation and assistive technologies, empowering individuals with disabilities to regain mobility and independence. Robotic exoskeletons and prosthetic devices equipped with delta-driven actuators offer precise control over limb movements, enabling users to perform daily activities with greater ease and confidence.

Additionally, delta robots are utilized in therapeutic devices such as robotic rehabilitation platforms and assistive robotic arms. These devices provide targeted therapy and assistance to patients recovering from stroke, spinal cord injuries, and musculoskeletal disorders, promoting faster rehabilitation and improved functional outcomes. While delta robots hold tremendous potential in revolutionizing healthcare, several challenges need to be addressed to maximize their impact. These challenges include ensuring patient safety and regulatory compliance, optimizing human-robot interaction in clinical settings, and addressing concerns related to cost-effectiveness and scalability.

The advancements in artificial intelligence, machine learning, and materials science are expected to further enhance the capabilities of delta robots in healthcare. Integrating these technologies with delta robot platforms will enable personalized and adaptive healthcare solutions tailored to individual patient needs. Delta robots are driving a paradigm shift in healthcare, enabling unprecedented levels of precision, efficiency, and innovation across various medical domains. From surgical robotics to pharmacy automation and rehabilitation, delta robots are revolutionizing patient care and transforming the way healthcare is delivered. As the field continues to evolve, delta robots are poised to play a central role in shaping the future of healthcare, empowering clinicians, improving outcomes, and enhancing the quality of life for patients worldwide.

1.4 Impact on the Healthcare Landscape

While delta robots hold tremendous potential in transforming healthcare, several challenges need to be addressed to maximize their impact and drive further innovation in the field.

1. Ensuring patient safety and regulatory compliance is paramount when integrating delta robots into healthcare settings. Regulatory frameworks need to be established to assess the safety and effectiveness of delta robot-assisted systems in surgical procedures, rehabilitation devices, and drug delivery systems.

2. Optimizing human-robot interaction in clinical settings is essential to ensure seamless integration and collaboration between healthcare professionals and delta robots. Research is needed to develop intuitive interfaces and control mechanisms that facilitate effective communication and cooperation between humans and robots in healthcare environments.

3. Cost-effectiveness and Scalability: Addressing concerns related to the cost-effectiveness and scalability of delta robot-assisted systems is critical to widespread adoption in healthcare. Research and development efforts should focus on reducing manufacturing costs, improving system reliability, and optimizing workflow efficiency to make delta robots more accessible to healthcare facilities of all sizes.

Development efforts in delta robotics for healthcare should focus on:

1. Integrating delta robots with advanced imaging technologies such as MRI and CT scanners to enhance surgical navigation and targeting capabilities.

2. Leveraging artificial intelligence and machine learning algorithms to enhance the autonomy and adaptive capabilities of delta robots in healthcare applications, such as real-time motion planning and predictive analytics for personalized treatment strategies.

3. Advancing materials science and biomechanics research to develop lightweight and biocompatible materials for delta robot components, enabling safer and more ergonomic interaction with patients and healthcare professionals.

2. UNDER THE SURGEON'S SCALPEL: DELTA ROBOTS IN SURGERY

Surgery has long been a cornerstone of medical practice, providing essential treatments for a wide range of conditions. With advancements in technology, surgical procedures have evolved, becoming more precise, minimally invasive, and efficient. Delta robots have emerged as invaluable tools in modern surgical theatres, revolutionizing the way surgeries are performed and pushing the boundaries of what is possible in the operating room (Fan et al., 2016). In this chapter, we delve into the role of delta robots in surgery, exploring their applications, advantages, and impact on patient care and surgical outcomes.

2.1 Delta Robots in Minimally Invasive Surgery

Minimally invasive surgery (MIS) has transformed the field of surgery, offering patients less pain, shorter recovery times, and reduced risk of complications compared to traditional open procedures. Delta robots have played a pivotal role in advancing MIS techniques, enabling surgeons to perform complex procedures through small incisions with unparalleled precision and control (Moradi Dalvand & Shirinzadeh, 2013).

One of the key advantages of delta robots in MIS is their ability to provide stable and tremor-free motion, enhancing the surgeon's dexterity and reducing the risk of inadvertent tissue damage. In procedures such as laparoscopic surgery and robotic-assisted surgery, delta robots enable precise manipulation of surgical instruments and camera systems, allowing surgeons to navigate anatomical structures with ease and perform intricate tasks with submillimeter accuracy.

Besides, delta robots are equipped with advanced imaging and navigation systems, enabling real-time visualization of the surgical site and enhancing surgical accuracy. By integrating delta robots with imaging modalities such as ultrasound, fluoroscopy, and intraoperative imaging, surgeons can precisely target lesions, tumors, and other abnormalities, leading to improved surgical outcomes and better patient outcomes.

2.2 Advancements in Robotic-Assisted Surgery

Robotic-assisted surgery has emerged as a transformative approach in modern healthcare, combining the benefits of robotics with the expertise of skilled surgeons. Delta robots are also playing a crucial role in orthopedic surgery, particularly in joint replacement procedures, by enhancing accuracy and reducing complication rates (Lee et al., 2020). In joint replacement surgery, such as total hip or knee arthroplasty, precise positioning of implants is essential for optimal outcomes.

These robots contribute to these procedures in the following ways:

1. Implant Placement: Delta robots assist surgeons in precisely positioning implants within the joint, ensuring optimal alignment and fit. This reduces the risk of implant malpositioning, which can lead to instability, premature wear, and revision surgery.

2. Soft Tissue Preservation: By enabling minimally invasive approaches, delta robots help preserve soft tissue surrounding the joint, resulting in less tissue damage and faster recovery for patients. This is particularly beneficial in preserving muscle strength and function postoperatively.

3. Customized Implantation: Delta robots can be integrated with preoperative planning software and patient-specific implants, allowing for customized surgical approaches tailored to individual patient anatomy. This personalized approach improves implant fit and function, leading to better long-term outcomes for patients.

4. Real-time Feedback: Delta robots provide real-time feedback to surgeons during the procedure, ensuring precise control over instrument movements and implant placement. This enhances surgical accuracy and reduces the likelihood of intraoperative errors.

2.3 Beyond Surgery

Delta robots hold promise for applications beyond traditional surgical fields, including neurosurgery, cardiac surgery, and other specialized areas:

Neurosurgery: In neurosurgery, delta robots can assist in precise tumor resection, deep brain stimulation electrode placement, and minimally invasive procedures for conditions such as epilepsy and movement disorders.

Cardiac Surgery: Delta robots can facilitate minimally invasive cardiac procedures, such as mitral valve repair and coronary artery bypass grafting, by providing stable instrument manipulation and precise tissue dissection.

Specialized Fields: In specialized fields such as otolaryngology, urology, and vascular surgery, delta robots can enhance precision and dexterity in minimally invasive procedures, leading to improved patient outcomes and reduced postoperative complications.

In robotic-assisted surgery, delta robots serve as the mechanical backbone of surgical platforms, providing stable and precise motion control for surgical instruments and endoscopic cameras. These robots are designed to mimic the movements of a surgeon's hands with greater accuracy and consistency, enabling intricate maneuvers in tight anatomical spaces with minimal tissue trauma.

2.4 Impact on Patient Care and Surgical Outcomes

The integration of delta robots in surgery has had a profound impact on patient care and surgical outcomes, driving improvements in safety, precision, and efficiency. By enabling minimally invasive techniques and robotic-assisted procedures, delta robots have reduced patient morbidity, shortened hospital stays, and accelerated recovery times (Ho, 2022).

Moreover, delta robots have facilitated advancements in surgical precision and accuracy, leading to better oncological outcomes, reduced complication rates, and improved functional outcomes for patients undergoing complex procedures. By enhancing the surgeon's capabilities and providing real-time feedback during surgery, delta robots have revolutionized the way surgeries are performed, pushing the boundaries of what is achievable in the operating room.

2.5 Challenges and Future Directions Towards Delta Robots in Surgery

Despite the remarkable advancements in delta robot-assisted surgery, several challenges remain to be addressed to further enhance their impact and adoption in clinical practice. Challenges include optimizing system ergonomics and user interfaces, ensuring seamless integration with existing surgical workflows, and addressing concerns related to cost-effectiveness and accessibility (Su et al., 2022).

The future research and development efforts in delta robot-assisted surgery should focus on:

1. Integrating delta robots with advanced imaging and navigation technologies to improve surgical visualization and targeting capabilities.

2. Developing autonomous surgical systems powered by artificial intelligence and machine learning algorithms to assist surgeons in decision-making and procedural planning.

3. Expanding the capabilities of delta robots for remote surgery and telepresence applications, enabling surgeons to perform procedures from remote locations and providing access to specialized care in underserved areas.

4. Delta robots provide a stable platform for manipulating surgical instruments, reducing tremors and ensuring precise movements. Surgeons can control the instruments with high accuracy, enabling delicate maneuvers in confined spaces within the body.

5. Delta robots can be integrated with advanced 3D imaging systems, providing surgeons with enhanced visualization of the surgical site. This improves depth perception and spatial awareness, allowing for more accurate tissue manipulation and dissection.

6. Delta robots allow for the use of flexible instruments with articulated tips, enabling surgeons to access difficult-to-reach areas and perform intricate procedures with greater ease. This flexibility enhances the versatility of laparoscopic surgery and expands its applications across various medical specialties.

7. By minimizing the size of incisions and reducing tissue trauma, laparoscopic procedures performed with delta robots result in faster recovery times, less postoperative pain, and reduced risk of complications compared to traditional open surgeries.

3. PRECISION AUTOMATION: DELTA ROBOTS IN PHARMACIES AND LABORATORIES

In the pharmaceutical and laboratory settings, precision and efficiency are paramount for ensuring accurate dispensing of medications and conducting various scientific experiments. Delta robots have emerged as indispensable tools in these environments, enabling precise automation of tasks such as medication packaging, compounding, and laboratory sample handling. These robots are equipped with specialized end-of-arm tools that enable them to handle various types of medications, including tablets, capsules, vials, and syringes, with precision and consistency (Vadie & Lipták, 2023). These systems automate various tasks, including medication sorting, counting, labeling, and packaging, with a high level of precision and efficiency. Delta robots are programmed to dispense medications according to predefined parameters, such as dosage, patient information, and prescription details. This automation eliminates the potential for manual errors, such as miscounting or mislabeling medications, which can have serious consequences for patient safety.

The delta robots streamline workflow processes in pharmacies, increasing efficiency and throughput while reducing labour costs. These robots can work continuously without fatigue, ensuring consistent performance and timely medication dispensing even during peak hours. Delta robots enable pharmacies to streamline workflow processes and reduce medication errors by automating repetitive tasks and minimizing manual intervention (Saharan, 2022). This improves patient safety and ensures compliance with medication dispensing regulations and guidelines.

3.1 Delta Robots in Laboratory Automation

Delta robots play a crucial role in laboratory automation by automating repetitive tasks and increasing throughput, ultimately reducing human error and accelerating research workflows. These robots are integrated into laboratory automation systems to handle various tasks, such as sample pipetting, plate handling, liquid dispensing, and sample storage, with precision and efficiency.

Delta robots are programmed to execute predefined workflows, allowing laboratories to automate complex experimental procedures and assays. This automation increases throughput, enabling researchers to process larger sample volumes and analyze more data in less time. These robots streamline laboratory workflows by reducing the time and resources required to perform repetitive tasks manually. With their speed, precision, and reliability, delta robots are transforming laboratory practices, enabling researchers to conduct experiments more efficiently and analyze larger datasets, leading to faster scientific breakthroughs.

Delta robots can be programmed to perform complex liquid handling tasks, such as serial dilutions, PCR setup, and drug screening assays, with sub-microliter accuracy, ensuring reliable and reproducible experimental results. This accelerates research workflows and enables researchers to focus on data analysis and interpretation, leading to faster scientific discoveries and innovation.

3.2 Impact on Operational Efficiency and Research Outcomes

The integration of delta robots in pharmacies and laboratories has had a profound impact on operational efficiency and research outcomes, driving improvements in productivity, accuracy, and innovation. In pharmacies, delta robots have streamlined medication dispensing and packaging processes, reducing dispensing errors, and improving patient safety. Automated pharmacy systems equipped with delta robots have increased throughput, enabling pharmacies to meet growing demand for medications while minimizing labor costs and operational inefficiencies.

In laboratories, delta robots have revolutionized sample handling and preparation workflows, enabling researchers to conduct experiments with greater precision, consistency, and reproducibility. Automated laboratory systems equipped with delta robots have accelerated research workflows, allowing researchers to perform experiments more efficiently and analyze larger datasets, leading to faster scientific discoveries and breakthroughs.

3.3 Challenges and Future Directions Towards Delta Robots in Pharmacies and Laboratories

Despite the significant advancements in pharmacy and laboratory automation with delta robots, several challenges remain to be addressed to further enhance their impact and adoption in these environments. Challenges include optimizing system integration with existing pharmacy and laboratory information management systems, ensuring compatibility with a wide range of medications and laboratory reagents, and addressing concerns related to cost-effectiveness and scalability.

Future research and development efforts in pharmacy and laboratory automation with delta robots should focus on:

1. Developing specialized end-of-arm tools for delta robots to handle a wider range of medications and laboratory reagents, enabling more versatile automation of pharmacy and laboratory workflows.

2. Integrating delta robots with artificial intelligence and machine learning algorithms to enable adaptive automation of pharmacy and laboratory processes, optimizing workflow efficiency and minimizing manual intervention.

3. Exploring the use of collaborative robotics technologies to enable seamless interaction between delta robots and human operators in pharmacy and laboratory settings, enhancing safety and efficiency.

Delta robots hold tremendous potential in the future of personalized medicine by enabling precise handling of samples and facilitating tailored treatments for individual patients. In personalized medicine, treatments are customized based on a patient's unique genetic makeup, medical history, and other factors. Delta robots can play a crucial role in this paradigm shift by automating tasks such as sample processing, analysis, and drug preparation with precision and efficiency.

Delta robots can be integrated into automated laboratory systems that enable high-throughput analysis of patient samples, allowing healthcare providers to gather comprehensive data for personalized treatment planning. These robots can process large volumes of samples in a timely manner, facilitating timely diagnosis and treatment decisions. Using delta robots for personalized medicine, healthcare providers can improve patient outcomes, reduce treatment-related complications, and optimize healthcare resources. With their precision and reliability, delta robots are poised to play a pivotal role in the future of personalized medicine, enabling tailored treatments that address the unique needs of individual patients.

4. REDEFINING REHABILITATION WITH ROBOTIC ASSISTANCE

Rehabilitation is a critical aspect of healthcare, aimed at restoring functional abilities and improving quality of life for individuals with disabilities or injuries. Traditional rehabilitation therapies often rely on manual techniques and exercises, which can be limited in their ability to provide precise and targeted interventions. Delta robots have emerged as valuable tools in robotic rehabilitation, offering precise control over limb movements and facilitating targeted therapy for individuals with neurological or musculoskeletal impairments. These robots are equipped with advanced actuators and sensors that enable them to mimic human motion and provide personalized assistance to patients during rehabilitation sessions.

4.1 Robotic-Assisted Therapy

Robotic-assisted therapy involving delta robots offers controlled and targeted movements that are beneficial for the rehabilitation of patients with neurological and musculoskeletal conditions. These robots are equipped with advanced sensors and actuators that enable precise and customizable movement patterns, allowing therapists to tailor therapy sessions to the specific needs and abilities of individual patients.

Delta robots provide a stable platform for patients to perform repetitive and task-specific movements, which are essential for promoting neuroplasticity and motor learning. The robots can be programmed to assist or resist patient movements, depending on the desired therapeutic goals (Abarca & Elias, 2023). For example, in patients recovering from stroke, delta robots can provide assistance to weak or paralyzed limbs, facilitating movement retraining and functional recovery. These robots enable therapists to provide intensive and repetitive therapy sessions, which are essential for promoting motor recovery and improving functional outcomes. These robots can deliver consistent and accurate movements, ensuring that patients receive optimal therapy while minimizing the risk of injury or fatigue.

The delta robots in robotic-assisted therapy provide real-time feedback to therapists and patients. These robots can monitor patient movements and performance metrics, such as range of motion, muscle strength, and coordination, allowing therapists to track progress and adjust therapy parameters as needed. It can be programmed to adapt therapy parameters such as range of motion, resistance levels, and movement patterns, allowing therapists to customize treatment plans and optimize patient outcomes.

4.2 Applications of Delta Robots in Rehabilitation

Delta robots are utilized in a variety of rehabilitation settings and applications, including:

Robotic Exoskeletons: Delta-driven robotic exoskeletons provide powered assistance to individuals with mobility impairments, enabling them to perform activities of daily living with greater ease and independence. These exoskeletons can assist with walking, standing, and upper limb movements, promoting gait retraining and functional recovery.

Prosthetic Devices: Delta robots are integrated into prosthetic devices to provide precise control over limb movements for individuals with limb loss. These devices enable users to perform tasks such as grasping objects, manipulating tools, and engaging in recreational activities, enhancing their overall quality of life and independence.

Rehabilitation Platforms: Delta-driven rehabilitation platforms offer a versatile and customizable solution for delivering therapy to patients with a wide range of neurological and musculoskeletal conditions. These platforms provide interactive exercises and games that engage patients in therapeutic activities while collecting data on their progress and performance.

Delta robots have significant potential in assisting individuals with disabilities with daily tasks, promoting independence, and improving quality of life. It can be integrated into assistive technologies such as robotic exoskeletons, prosthetic devices, and robotic arms to provide personalized assistance with activities of daily living. For individuals with mobility impairments, delta-driven robotic exoskeletons offer powered assistance for walking, standing, and upper limb movements (Zuo et al., 2020). These exoskeletons enable individuals to navigate their environment with greater ease and independence, enhancing their mobility and reducing reliance on caregivers or mobility aids.

In addition, these robots integrated into prosthetic devices provide precise control over limb movements, enabling users to perform tasks such as grasping objects, manipulating tools, and engaging in recreational activities. These devices restore functional abilities and promote independence for individuals with limb loss, improving their overall quality of life and autonomy. Delta-driven robotic arms can also assist individuals with disabilities in performing various tasks, such as feeding, dressing, and personal care. These robotic arms can be programmed to execute predefined movements, allowing users to interact with their environment and carry out daily activities with greater independence and dignity.

4.3 Impact on Patient Outcomes

The integration of delta robots in robotic rehabilitation has had a profound impact on patient outcomes, driving improvements in mobility, independence, and quality of life for individuals undergoing rehabilitation. By providing precise and targeted assistance, delta robots enable patients to achieve functional goals and regain motor skills that were previously compromised due to injury or disability.

Besides, delta-driven robotic rehabilitation allows therapists to monitor patients' progress in real time, adjust therapy parameters as needed, and track long-term outcomes. This data-driven approach to rehabilitation enables therapists to tailor treatment plans to individual patient needs, optimize therapy protocols, and maximize rehabilitation outcomes.

4.4 Challenges and Future Directions towards Robotic Rehabilitation Assistance

In spite of the significant advancements in robotic rehabilitation with delta robots, several challenges remain to be addressed to further enhance their impact and adoption in clinical practice (Bajaj et al., 2019). Challenges include optimizing human-robot interaction, ensuring safety and reliability, and addressing concerns related to cost-effectiveness and accessibility.

The future research and development efforts in robotic rehabilitation with delta robots should focus on:

1. Developing intelligent control algorithms that enable delta robots to adapt therapy parameters in real time based on patient feedback and performance metrics.
2. Exploring the integration of delta robots into wearable rehabilitation devices that provide continuous assistance and feedback to patients during activities of daily living.
3. Expanding the capabilities of delta-driven robotic rehabilitation systems for tele-rehabilitation applications, enabling therapists to deliver therapy remotely and provide access to specialized care in underserved areas.

4.5 Ethical Considerations and Human-Robot Interaction

The use of robots in healthcare, including delta robots, raises important ethical considerations regarding patient safety, privacy, autonomy, and the impact on human-robot interaction. It is essential to address these ethical considerations to ensure that the integration of robots in healthcare settings is conducted in a responsible and ethical manner.

One of the key ethical considerations is patient safety, as delta robots interact directly with patients during therapy or assistive tasks. It is crucial to ensure that these robots are designed and operated in a way that minimizes the risk of injury or harm to patients, while still providing effective assistance and therapy.

Privacy is another important ethical consideration, particularly when delta robots are used in assistive technologies that interact with sensitive personal information or medical data. It is essential to implement robust data security measures to protect patient privacy and confidentiality, ensuring that sensitive information is not compromised or accessed without proper authorization.

Autonomy is also a significant ethical consideration in human-robot interaction, as delta robots assist individuals with disabilities in performing tasks that affect their daily lives and independence. It is essential to empower users to maintain control over the robots and make informed decisions about their care and assistance, while still benefiting from the capabilities of the robots. It is important to promote seamless collaboration between humans and robots, fostering a supportive and respectful environment that prioritizes patient well-being and autonomy. By ensuring patient safety, privacy, autonomy, and fostering positive human-robot interaction, delta robots can enhance the quality of care and improve outcomes for patients while upholding ethical principles and values in healthcare practice.

5. DELTA ROBOTS IN MEDICAL IMAGING AND DIAGNOSTICS

Medical imaging and diagnostics play a crucial role in healthcare by enabling clinicians to visualize internal structures, detect abnormalities, and guide treatment decisions. Delta robots are increasingly being utilized in medical imaging and diagnostic procedures to enhance precision, improve imaging quality, and optimize patient care. In this chapter, we explore the role of delta robots in medical imaging and diagnostics, highlighting their applications, advantages, and impact on diagnostic accuracy and patient outcomes.

Delta robots play a pivotal role in enhancing the precision and speed of medical imaging modalities such as ultrasound and X-ray. These robots are equipped with advanced motion control systems that enable precise positioning and movement of imaging equipment, resulting in improved imaging accuracy and efficiency. By precisely aligning ultrasound probes and X-ray machines with the target anatomy, delta robots minimize imaging artifacts and optimize image quality, leading to more accurate diagnostic information. Furthermore, delta robots facilitate rapid repositioning of imaging devices, reducing imaging time and improving patient throughput in busy clinical settings.

5.1 Applications of Delta Robots in Medical Imaging

Delta robots are utilized in various medical imaging modalities to facilitate precise positioning of imaging equipment and enhance imaging quality. These robots are equipped with advanced motion control systems that enable them to move imaging devices, such as X-ray machines, CT scanners, MRI scanners, and ultrasound probes, with sub-millimeter accuracy and repeatability.

One of the key applications of delta robots in medical imaging is their use in CT and MRI scanners for patient positioning and motion compensation. These robots ensure that patients are accurately positioned within the imaging field of view, minimizing motion artifacts and optimizing image quality (Stasevych & Zvarych, 2023). Delta robots can also compensate for patient motion during scanning, ensuring consistent image acquisition and reducing the need for repeat scans.

These robots are utilized in interventional radiology procedures to guide minimally invasive treatments, such as biopsy, ablation, and catheter-based interventions. It provides precise control over the movement of interventional devices, enabling clinicians to target lesions and abnormalities with high accuracy and minimal invasiveness.

5.2 Advantages of Delta Robots in Medical Imaging

Delta robots offer several advantages in medical imaging and diagnostics, including:

1.   Delta robots provide precise positioning and motion control, ensuring accurate alignment of imaging equipment and optimal image quality.
2.   Delta robots compensate for patient motion during imaging procedures, minimizing motion artifacts and improving diagnostic accuracy.
3.   Delta robots enable minimally invasive interventions in interventional radiology procedures, reducing patient discomfort and recovery times.
4.   Delta robots streamline imaging workflows by automating patient positioning and device manipulation, reducing procedure times and optimizing resource utilization.

5.3 Interventional Procedures and Real-Time Guidance

Delta robots are increasingly being used in image-guided surgeries and interventional procedures, providing real-time feedback to surgeons and enhancing procedural precision. These robots are equipped with advanced sensors and actuators that enable precise control over surgical instruments and interventional devices, ensuring accurate targeting of lesions and abnormalities. By integrating delta robots with imaging modalities such as MRI and CT scanners, surgeons can visualize the surgical site in real time and navigate complex anatomical structures with submillimeter accuracy. This real-time guidance provided by delta robots enhances surgical precision, reduces the risk of complications, and improves patient outcomes. Additionally, delta robots enable surgeons to perform minimally invasive procedures with greater precision, minimizing tissue trauma and accelerating patient recovery. Overall, the use of delta robots in image-guided surgeries and interventions enhances procedural accuracy and safety, leading to better patient care and outcomes.

5.4 Future of AI-Powered Diagnostics

The integration of artificial intelligence (AI) with delta robots holds significant potential for automated image analysis and disease detection in medical diagnostics. AI algorithms can analyze medical images, such as CT scans and MRI images, to detect subtle abnormalities and assist clinicians in making accurate diagnoses (Hughes et al., 2021). By integrating AI with delta robots, medical imaging procedures can be automated and optimized, leading to faster and more accurate diagnoses.

Delta robots can facilitate the precise positioning and movement of imaging devices, while AI algorithms analyze the acquired images in real time, providing immediate feedback to clinicians. This combination of AI-powered diagnostics and delta robot-assisted imaging enables efficient and accurate disease detection, leading to timely interventions and improved patient outcomes. Furthermore, AI algorithms can learn from large datasets of medical images, continuously improving their diagnostic accuracy and performance over time (Manickam et al., 2022). In conclusion, the integration of AI with delta robots represents the future of medical diagnostics, offering automated and accurate disease detection capabilities that enhance patient care and clinical decision-making.

5.5 Challenges and Future Directions Towards Delta Robots in Medical Imaging and Diagnostics

Despite the significant advancements in medical imaging and diagnostics with delta robots, several challenges remain to be addressed to further enhance their impact and adoption in clinical practice. Challenges include optimizing system integration with existing imaging equipment, ensuring compatibility with a wide range of imaging modalities, and addressing concerns related to cost-effectiveness and scalability.

Further the future research and development efforts in medical imaging and diagnostics with delta robots should focus on:

1. Advanced Imaging Technologies
2. Artificial Intelligence and Image Analysis
3. Personalized Imaging and Treatment Planning

6. CONCLUSION

The integration of delta robots in healthcare has brought about transformative changes across various areas, significantly impacting patient care and clinical outcomes. These robots have revolutionized surgical procedures by providing enhanced precision and dexterity, leading to minimally invasive interventions with faster recovery times. Additionally, delta robots have redefined rehabilitation therapies, offering targeted and intensive treatments for individuals with neurological and musculoskeletal conditions, ultimately promoting functional recovery and improving quality of life. Moreover, in medical imaging and diagnostics, delta robots have improved imaging accuracy, facilitated real-time guidance in interventional procedures, and offered potential for automated image analysis through integration with artificial intelligence. Delta robots have proven to be invaluable tools in surgical procedures, offering enhanced precision and dexterity for complex interventions while minimizing invasiveness and promoting faster recovery times. Their integration in rehabilitation therapies has facilitated targeted and intensive treatments for individuals with neurological and musculoskeletal conditions, promoting functional recovery and improving quality of life.

Despite the significant advancements achieved thus far, ongoing research and development efforts continue to drive further advancements in the field of delta robots in healthcare. Researchers are actively exploring new applications and capabilities of delta robots, aiming to address challenges and optimize their integration in clinical practice. Efforts are focused on enhancing robotic control algorithms, improving system integration with existing healthcare technologies, and exploring novel applications in areas such as personalized medicine and telemedicine. Additionally, advancements in artificial intelligence and human-robot interaction are expected to further optimize the capabilities of delta robots in healthcare, enabling personalized and effective care delivery. In conclusion, delta robots hold great promise for shaping the future of healthcare delivery and improving patient outcomes. With ongoing research and development efforts, these robots are poised to continue transforming the landscape of healthcare by offering precision, efficiency, and innovation across various medical specialties. The integration of delta robots in surgical procedures, rehabilitation therapies, medical imaging, and diagnostics represents a significant advancement in the field of medical robotics, with the potential to revolutionize patient care and clinical outcomes. As technology and research evolve, delta robots will play an increasingly crucial part in creating the future of healthcare delivery, ultimately enhancing patient outcomes and advancing the field of medicine.

REFERENCES

Abarca, V. E., & Elias, D. A. (2023). A Review of Parallel Robots: Rehabilitation, Assistance, and Humanoid Applications for Neck, Shoulder, Wrist, Hip, and Ankle Joints. *Robotics (Basel, Switzerland)*, *12*(5), 131. doi:10.3390/robotics12050131

Bajaj, N. M., Spiers, A. J., & Dollar, A. M. (2019). State of the art in artificial wrists: A review of prosthetic and robotic wrist design. *IEEE Transactions on Robotics*, *35*(1), 261–277. doi:10.1109/TRO.2018.2865890

Bogossian, T. (2022). The Use of Robotics in Healthcare. *Journal of Medical & Clinical Nursing*, 1–4. Advance online publication. doi:10.47363/JMCN/2022(3)157

Fan, G., Zhou, Z., Zhang, H., Gu, X., Gu, G., Guan, X., Fan, Y., & He, S. (2016, June). Global scientific production of robotic surgery in medicine: A 20-year survey of research activities. *International Journal of Surgery*, *30*, 126–131. doi:10.1016/j.ijsu.2016.04.048 PMID:27154617

Ho, J. C. (2022). Robot Assisted Neurosurgery for High-Accuracy, Minimally-Invasive Deep Brain Electrophysiology in Monkeys. *2022 44th Annual International Conference of the IEEE Engineering in Medicine & Biology Society (EMBC)*, 3115-3118. 10.1109/EMBC48229.2022.9871520

Hughes, Zhu, & Bednarz. (2021). Generative Adversarial Networks-Enabled Human-Artificial Intelligence Collaborative Applications for Creative and Design Industries: A Systematic Review of Current Approaches and Trends. *Frontiers in Artificial Intelligence, 4*, 1-17.

Lee, D., Yu, H. W., Kwon, H., Kong, H.-J., Lee, K. E., & Kim, H. C. (2020, June). Evaluation of Surgical Skills during Robotic Surgery by Deep Learning-Based Multiple Surgical Instrument Tracking in Training and Actual Operations. *Journal of Clinical Medicine*, *9*(6), E1964. doi:10.3390/jcm9061964 PMID:32585953

Li, W. (2018). The design of a 3-CPS parallel robot for maximum dexterity. *Mechanism and Machine Theory, 122*, 279-291. doi:10.1016/j.mechmachtheory.2018.01.003

Lopez, M., Castillo, E., Garcia, G., & Bashir, A. (2006). Delta robot: Inverse, direct, and intermediate jacobians. *Proceedings of the Institution of Mechanical Engineers, Part C: Journal of Mechanical Engineering Science, 220*(1), 103–109. 10.1243/095440606X78263

Manickam, P., Mariappan, S. A., Murugesan, S. M., Hansda, S., Kaushik, A., Shinde, R., & Thipperudraswamy, S. P. (2022). Artificial Intelligence (AI) and Internet of Medical Things (IoMT) Assisted Biomedical Systems for Intelligent Healthcare. *Biosensors (Basel)*, *12*(8), 562–562. doi:10.3390/bios12080562 PMID:35892459

McClintock, H., Temel, F. Z., Doshi, N., Je-sung, K., & Robert, J. (2018). The millidelta: A high-bandwidth, high-precision, millimeter-scale delta robot. *Science Robotics*, *3*(14), eaar3018. doi:10.1126/scirobotics.aar3018 PMID:33141699

Mehrafrooz, B., Mohammadi, M., & Masouleh, M. T. (2017). Kinematic sensitivity evaluation of revolute and prismatic 3-dof delta robots. *2017 5th RSI International Conference on Robotics and Mechatronics (ICRoM)*, 225–231. doi: .846615910.1109/ICRoM.2017

Moradi Dalvand, M., & Shirinzadeh, B. (2013, April). Motion control analysis of a parallel robot assisted minimally invasive surgery/microsurgery system (PRAMiSS). *Robotics and Computer-integrated Manufacturing, 29*(2), 318–327. doi:10.1016/j.rcim.2012.09.003

Pisla, Plitea, Gherman, Pisla, & Vaida. (2009). *Kinematical Analysis and Design of a New Surgical Parallel Robot.* . doi:10.1007/978-3-642-01947-0_34

Poppeova, V., Uricek, J., Bulej, V., & Sindler, P. (2011). Delta robots - robots for high speed manipulation. *Tehnicki Vjesnik (Strojarski Fakultet), 18*, 435–445.

Saharan. (2022). Robotic Automation of Pharmaceutical and Life Science Industries. *Computer Aided Pharmaceutics and Drug Delivery.* doi:10.1007/978-981-16-5180-9_12

Stasevych, M., & Zvarych, V. (2023). Innovative Robotic Technologies and Artificial Intelligence in Pharmacy and Medicine: Paving the Way for the Future of Health Care—A Review. *Big Data and Cognitive Computing, 7*(3), 147. doi:10.3390/bdcc7030147

Su, H., Kwok, W., Cleary, K., Iordachita, I., Cavusoglu, M. C., Desai, J. P., & Fischer, G. S. (2022). State of the Art and Future Opportunities in MRI-Guided Robot-Assisted Surgery and Interventions. *Proceedings of the IEEE, 110*(7), 968. 10.1109/JPROC.2022.3169146

Vadie, A., & Lipták, K. (2023). Industry 4.0: New challenges for the labor market and working conditions as a result of emergence of robots and automation. Economic and Regional Studies / Studia Ekonomiczne i Regionalne, 16(3), 434-445. doi:10.2478/ers-2023-0028

Zuo, S., Li, J., Dong, M., Zhou, X., Fan, W., & Kong, Y. (2020). Design and performance evaluation of a novel wearable parallel mechanism for ankle rehabilitation. *Frontiers in Neurorobotics, 14*, 9. doi:10.3389/fnbot.2020.00009 PMID:32132917

Chapter 3
A Data-Driven Model for Predicting Fault-Tolerant Safe Navigation in Multi-Robot Systems

A. Madhesh

Karpagam Academy of Higher Education, India

Clara Barathi Priyadharshini

Karpagam Academy of Higher Education, India

ABSTRACT

Earlier methods focused on reducing the forecast uncertainty for individual agents and avoiding this unduly cautious behavior by either employing more experienced models or heuristically restricting the predictive covariance. Findings indicate neither the individual prediction nor the forecast uncertainty have a major impact on the frozen robot problem. The result is that dynamic agents can solve the frozen robot problem by employing joint collision avoidance and clear the way for each other to build feasible pathways. Potential paths for safety evaluation are ranked according to the likelihood of collisions with known objects and those that happen outside the planning horizon. The whole collision probability is examined. Monte Carlo sampling is utilized to approximate the collision probabilities. Designing and selecting routes to reach the intended location, this approach aims to provide a navigation framework that reduces the likelihood of collisions.

I. INTRODUCTION

Drones, ground robots, and autonomous cars are examples of the multi-robot systems that have grown significantly due to applications such as military search and rescue missions, enhanced mobility, subterranean exploration, interior movements, warehouses, and entertainment purposes. In these situations, each individual robot must navigate safely through highly variable terrain, dodging obstacles and other

DOI: 10.4018/979-8-3693-5767-5.ch003

group members. Therefore, motion planning in uncertain environments, collision avoidance safety, and distributed computation are some of the main challenges in autonomous multi-robot systems; see a recent review for further details. To ensure safe navigation of robotic systems, optimization-based motion planning approaches like Model Predictive Control, or MPC, are being considered more and more said by Bajcsy, Andrea, et al. (2019).

MPC is an especially powerful framework that may be applied to iteratively solve a finite-horizon numerical optimization problem to compute control instructions that maximize relevant performance measures while respecting constraints (e.g., collision avoidance). The controlled object's distance from an impediment larger than a safe threshold is commonly used to model constraints on collision avoidance.

The model predictive control (MPC) framework encapsulates the suggested approach, which permits decentralized multi-robot motion planning in dynamic scenarios. Specifically, we first generate a demonstration dataset of robot trajectories using a multi-robot collision avoidance simulator. It uses a centralized sequential MPC for local motion planning based on inter robot communication used by Navsalkar, Atharva, and Ashish R. Hota (2023). The robot trajectory prediction problem is then framed in terms of sequence modeling, enabling us to develop a model that makes use of recurrent neural networks (RNNs). Using the obtained dataset, the model may be trained to predict the planning behaviors of the robots and to resemble the centralized sequential MPC. Finally, multi-robot local motion planning is completed in a. main contributions of this work are:

- For a large number of robots, an RNN-based robot trajectory prediction model that is aware of obstacles and interactions.
- Combining MPC with the trajectory prediction model allows for decentralized multi-robot local motion planning in dynamic environments.

We showcase the benefits of our data-driven metric for the joint use of multiple robots for active interference and crowd observation. Our aim is to provide maximum social invisibility while providing the fastest feasible navigation. In order to investigate a variety of scenarios and applications, we show the efficacy of our work in multiple surveillance scenarios based on the degree of increasing social interaction between the humans and robots explained by Firoozi, Roya, et al (2020).

1.1 Our Approach Has the Following Benefits

1. The attitude of entitlement Computation: Our algorithm predicts pedestrians' emotional responses to robots in groups with high accuracy.
2. Robust computation: Our method is robust and capable of accounting for noise in pedestrian routes extracted from motion pictures.
3. Our method evaluates the entitativity behaviors at interactive rates with speed and accuracy, avoiding the need for any previous computation.

1.2 Active Surveillance

This kind of patrolling or monitoring involves autonomous robots that live side by side with pedestrians. These robots will need to be able to navigate through crowds and plan ahead in real time in order to perform surveillance and analysis without colliding. In this case, the robots have to predict each pedestrian's movements and path. For example, spectators at marathon races are sometimes quite huge and highly mobile. In these kinds of scenarios, a monitoring system that can identify and adjust to shifting focal points is crucial said by Olcay, Ertug (2020).

In these kinds of scenarios, robots have to be very socially invisible (s = 0). To do this, the entitativity features are set to the minimum, E = Emin.

1.3 Dynamic intervention

Robots may live alongside humans in some circumstances, but they may also influence pedestrians to take alternative paths or act in specific ways. These kinds of interventions can be overt using visual cues or physical force to make someone change their direction—or subtle nudging, for example. This type of monitoring can be used for any event that has exceptionally dense attendees, such as a marathon or festival. Dense crowding at these events may result in extremely dangerous stampedes. When population density approaches dangerous levels in this situation, a robot might be able to detect it and respond appropriately by "nudging" people into a safer distribution explained by Zhu, Hai, et al. (2021).

As robots have become increasingly commonplace in social contexts, people's expectations regarding their social skills have increased. Robots should be increasingly visible to society since humans often wish for them to be more prominent social agents in group settings. One facet of this social visibility is the capacity to grab people's attention and elicit strong feelings. Examples of social visibility include robotics tasks requiring human cooperation. However, not all scenarios call for socially observable robots. In certain situations, robots are used more for observation than for assisting humans. It would be better if robots could remain socially inconspicuous in specific circumstances.

Social invisibility is the ability of agents to stay out of other people's sight. Psychological research indicates that African Americans, for example, are often disregarded in social contexts, particularly when it comes to responding to perceived threats. The less unpleasant an emotion is felt, the less likely it is to be detected by a social environment, as both humans and robots are trained by evolution to respond fast to inputs that constitute a threat discussed by Zhu, Hai (2022). The social invisibility that results from not generating emotion is especially important in surveillance circumstances, where robots are expected to blend in with people without drawing attention. Not only do humans behave differently when they observe surveillance robots, but the negative emotions that cause the robots to identify them can also produce reactance, which can make people hostile and harm the robots or even other people. Numerous strategies have been identified in the literature to mitigate negative affective responses to social agents; however, entitativity, or "groupiness," is associated with three primary components: similarity in appearance, shared movement, and close proximity to one another, and may be particularly important for multi-robot systems. Because agents who are more identical in look and movement also tend to be closer to one another, a marching military unit appears more unified than shoppers at a mall.

Because of this innate fear of groups, the social visibility and emotional response of an agent collection rise in direct proportion to the degree of entitativeness (or grouplikeness) displayed. It is imperative that multi-robot systems reduce their entitativity, since lowering perceptions of threat is implied by making groups of agents more socially invisible. Put differently, to travel across crowds without provoking negative reactions, multi-robot systems must look more like individuals than like a cohesive, well-organized group by Park, Jin Soo, et al. (2021).

1.4 Entitativity

Entity perception is the concept that a group of people constitute a single entity. Similar to how humans categorize objects in the physical world, people also use variables like proximity, similarity, and shared fate to categorize other people into entities by Cui, Yuxiang, et al. (2022). When individuals share these traits with one another, we are more likely to think of them as a single entity. Larger clusters are more likely to be perceived as entities, but only if the individuals in the cluster are similar to each other. "Groupiness," also referred to as "tightness," is the degree to which a group resembles an entity rather than a group of individuals. Entitativity is largely determined by how three essential elements are perceived:

1.4.1 Uniformity of Appearance

The members of highly competitive groupings are uniformly shaped.

1.4.2 Common Movement

Individuals in highly competitive organizations move in a pattern.

1.4.3 Proximity

Very close relationships exist among the members of highly competitive groups.

AI planning tools facilitate modularity and hierarchical control amongst agents by enabling experts to specify objectives and actions at different levels of abstraction. For multirobot systems that rely on cooperation and coordination to achieve shared objectives, they represent a viable substitute by Chung, Yiu Ming et al (2022). Nonetheless, the problem representation that current AI Planning tools demand is accurate and precise, and it must be generated by domain specialists. The benefit of deploying autonomous systems would be mitigated or even erased by an incorrect domain representation, which usually leads to mission failures and replanning. This is because the cost advantage of using robots over traditional procedures derives from accomplishing more tasks in less time. The five sections of this essay are listed below: Session 2 of the current system shows its shortcomings in robot navigation method and techniques. The third session was a demonstration of the suggested fault detection techniques method. The outcomes of the anticipated system are shown in Session 4. An examination of the recommended approach for machine navigation rounds off Session 5.

II. LITERATURE REVIEW

Constraints are deterministic in the presence of static obstacles but stochastic in the presence of uncertainty and dynamic obstacles. The authors consider that extreme caution is the outcome of strong constraint fulfillment. More recent methods, for instance, take advantage of historical samples or the probability distribution of the uncertainty to guarantee that collision avoidance constraints are satisfied with high probability. The writers Mali, Pravin, et al (2021) place a strong emphasis on autonomous driving scenarios. The task gets more challenging when a large number of mobile robots act as dynamic barriers for the managed agent. Two paradigms are involved in this case.

- A distributed system in which every agent must resolve its own MPC problem, communicate the calculated trajectory to agents in its vicinity, and set collision avoidance limits depending on the estimated future positions of the neighboring agents.

- When there is no contact between agents in a decentralized system, they anticipate each other's future locations, often assuming that they would continue at their current speed, and avoid colliding with their expected placements.

The majority of the aforementioned research ignores the precise geometry of the barriers in favor of treating the controlled agent and the impediments as point mass entities. Agents change their plans while they are being carried out, therefore predictions about where other agents will end up in the future might not come true. Furthermore, even in situations where chance constraints may not ensure the extent of constraint violation in the (less likely) event of a collision, strong optimization approaches yield extremely conservative solutions.

Arul, Senthil Hariharan (2022) contributed to the development of an online distributed communication-free technique for probabilistic multi-robot collision avoidance with uncertain localization. Using this technique, buffered uncertainty-aware Voronoi cells, or B-UAVCs, are employed. Based on its estimated location and uncertainty covariance, each robot computes its B-UAVC at each time step to plan its movements inside the BUAVC. Through the use of random restrictions to restrict each robot's motions within its associated B-UAVC, the inter-robot collision probability is kept below a threshold that is defined by the user. Robots are expected to be able to talk to one another or gather information on the whereabouts and degree of uncertainty of their neighbors using onboard sensors and a filter. No communication between robots is required if the estimated uncertainty of their locations exceeds the real one. A possible approach for non-holonomic robotics, such a drone squad, is to use a model predictive controller to keep each robot within its B-UAVC while generating a local trajectory for each robot. If position uncertainty is regularly distributed, the approach ensures that the robot-to-robot collision probability remains below a predefined threshold.

The author Tran, Vu Phi, et al (2023) compare the suggested approach with the most cutting-edge methods and assess it through simulation. They also introduced a distributed multi-robot collision avoidance technique that takes the uncertainty in the robots' localization into consideration. For each robot, we construct a buffered uncertainty-aware Voronoi cell (B-UAVC), assuming that the uncertainties have Gaussian distributions. By constraining each robot's movement to stay inside its matching B-UAVC, robot collision risk is ensured to be below a certain level. In simulation, we have demonstrated that our approach achieves the same safety level as the centralised CCNMPC technique (which necessitates robots to broadcast future trajectories) using six quadrotors.To validate our method, we conducted tests with two quadrotors that followed intersecting trajectories. Both static and dynamic environmental elements will be studied in future study.

Madridano, Angel, et al. (2021) a nonlinear model predictive control (NMPC)-based technique for multi-robot trajectory planning and coordination was presented. As an alternative to centralized approaches, we address the distributed case in which each robot has an on-board computer unit to solve a local non-local problem coordination (NMPC) and may be able to interact with neighboring robots. We show that the proposed methodology is similar to solving the centralised control issue because of tailored interactions, i.e., interactions generated by using a non-convex alternating direction method of multipliers, or ADMM.

At researcher Krishnan, Shravan, et al (2020) certain synchronization stages, the ADMM scheme demands a certain level of communication interchange between the robots to ensure their safety and that they stay within the confines of the environment and do not cause accidents with other robots. In this paper, they assessed the recommended course of action for controlling three separate boats at a canal crossing. All robot types and applications, however, can benefit from the adaptability of the suggested approach. They also created a distributed model predictive control technique to aid in autonomous agent cooperation. Without requiring a central authority, the proposed algorithm enables the agents to coordinate and determine a common safe navigation plan through the use of a novel alternating direction technique of multipliers appropriate for no convex optimization.

Ferranti, Laura, et al. (2022) the primary focus of the study was the navigation control problem of a general class of 2nd order uncertain nonlinear multiagent systems in a restricted workspace, a subset of R3 with immovable obstacles. Specifically, each agent uses our proposed decentralized control protocol to arrive at a current location in the workspace, given a finite sensing radius and local information. The proposed method guarantees that the agents that are first attached stay connected forever. Defined limits also aid in preventing collisions with the workspace border, barriers, and other agents.

When the author Krishnan, Shravan, et al (2019) operating in the face of uncertainty and disruptions, the suggested controllers employ a particular kind of Decentralized Nonlinear Model Predictive Controller. Simulation results, in the end, validate the validity of the suggested framework. Agents have the option to solve their own FHOCPs and apply the control inputs sequentially or concurrently, depending on the design approach flow they prefer. Its procedure-flow neutral architecture allows for both to be accommodated without sacrificing usefulness or effective stabilization. The sequential approach is used in this case: each agent solves its own FHOCP and applies the related acceptable control input in a round robin method, keeping in mind the present and planned open-loop state prediction configurations of all agents within its sensing range. This decision is the result of three factors. defense When two UAVs have to cooperate to deliver an object of comparable size, they are more likely to clash if the communication range's maximum is near the agents' respective sizes. If a parallel strategy is employed, this is especially true. This is because agents adopting a parallel strategy would be more susceptible to limit breaches in the event of interruptions due to their increased reliance on their neighbors' open-loop forecasts.

2.1 Multi-Robot Collision Avoidance

The author Şenbaşlar, Baskın, et al (2023) said that online local motion planning for multi-robot systems, which has been the focus of much research recently and is often referred to as multi-robot collision avoidance, is the main area of our work. Artificial potential field (APF) based methods, buffered Voronoi cell (BVC) approaches, control barrier functions (CBF), and the optimal reciprocal collision avoidance (ORCA) method which builds on the concept of reciprocal velocity obstacles (RVO) are examples of reactive controller-level techniques that are regarded as conventional. Robot motion is frequently limited to one time step planning, and these reactive methods are computationally efficient but do not fully characterize the robot dynamics. Novel learning-based methods for multi-robot collision avoidance have been developed as a result of developments in deep imitation learning and reinforcement learning (RL). Since RL-based methods may teach robots policies with a long-term cumulative reward, they are therefore considered non-myopic. That being said, hard state constraints, like collision avoidance limits, are usually beyond their capability. The model predictive control (MPC) framework solves an optimization problem in a receding horizon fashion for each robot, the solution to which can be used to generate collision-free paths.

2.2 Motion Prediction

The author Bramblett, Lauren (2023) offer an approach that decouples trajectory planning from motion prediction in order to achieve decentralized and communication-less collision avoidance. Another example of this type of decoupling is in, where human motion prediction is used to calculate the safe path for the ego robot. The field of motion prediction for decision-making agents has garnered increasing attention in the last few years, with most of the work focusing on human trajectory prediction. As seen by the popular social force-based method, which describes pedestrian behavior using repelling and attracting potentials, most early motion prediction research was model-based. Eventually, to simulate the activities of traffic cars, the model is improved and changed. Despite being computationally efficient, these approaches have poor prediction accuracy. Moreover, several notable attempts have been made to model and predict the future paths of interacting decision-making agents using game theory. In these attempts, the agents are assumed to be playing a non-cooperative game, and the agents' trajectories are predicted by computing the game's Nash equilibria. While it is possible to derive interaction-aware trajectory predictions, these methods are limited to specific road conditions and are not immediately relevant to large multi-robot systems.

2.3 Psychological Perspectives on Group Dynamics

The researcher Yang, Zhenge, et al (2022) long-standing theory in social psychology is that people's behavior is influenced by the social context in which they live. Notably, social dynamics are significantly impacted by group circumstances, usually in a negative way. People behave differently in groups than they do in solitude, exhibiting more hostility and feeling more scared by them, according to a wide body of psychological research. These group responses have consequences for the actual world, especially when there are potentially violent onlookers. Anti-social behaviors stem from negative emotional reactions directed towards any kind of social agent, be it human or robotic. Typically, these emotions are uneasy, frightening, and dangerous. Adapted from previously finished navigation assignments to new challenges.

Tan, Qingyang, et al (2020) recommended utilizing multimodal deep auto encoders to power a mobile robot control system that trained the discrete controller of a tiny quadrotor helicopter through imitation learning methods. The quadrotor was able to successfully steer clear of collisions with nearby stationary obstacles using just one cheap camera. But all the robot has to do is learn to move discretely (left/right) and to navigate around immobile impediments. Note that the previously described approaches only take static obstacles into account and require human drivers to collect training data in various scenarios. An additional end-to-end motion planner that is data-driven is presented by Samman, Tamim, et al (2021).

They trained a model that converts target positions and laser range discoveries into motion commands using expert demos generated by the ROS navigation module. This model is able to adjust to sudden changes and successfully navigate the robot through a region that has never been seen before. The performance of the learned policy is however significantly constrained by the quality of the labeled training sets, much like in other supervised learning methods. To overcome this limitation, Freitas, Elias JR, et al (2023) proposed a model-based reinforcement learning system that takes uncertainty into account and estimates the probability of collision in advance in an unknown environment. A deep reinforcement learning method was used to train the motion planner. However, it can be difficult to apply the learned plans to real-world scenarios because of how simple and organized the test environments are.

In Zaccaria, Michela, et al. (2021) multi-agent systems and crowd simulation, the ORCA framework has become more and more popular for multi-agent collision avoidance. ORCA offers enough conditions for multiple robots to avoid collisions with one another in a little amount of time and can be easily scaled to accommodate large systems with numerous robots. ORCA and its extensions built a complex model for the collision avoidance strategy using heuristics or basic principles. This model has many parameters that are difficult to set correctly. These methods also suffer from the uncertainties that are typical of real-world situations since they assume that each robot has full sensory awareness of the positions, velocities, and shapes of the surrounding agents. In order to lessen the need for flawless sensing, the group uses communication protocols to exchange state data, like agent positions and velocities. Furthermore, the ORCA formulation was developed using holonomic robots, which are less prevalent in real-world environments than nonholonomic robots.

III. SYSTEM DESIGN

Our method consists of two parts: data production and data exploitation. The data creation step generates the indicators that indicate the application of the cooperative localization solution and serve as the data input for the learning algorithms. The classification model, its elements, and training are selected once the data exploitation process presents potential learning strategies.

3.1 Data Generation

The data creation step aims to compute model-based fault-sensitive indicators and provides a roadmap for the data-driven methodology. It consists of four steps:

Data acquisition: The data is gathered using three different kinds of sensors:

- Prediction: Wheel encoder applying odometry as evolution model;
- Correction: Based on the LiDAR measurement and the expected states of the other nearby robots, Marvel Mind, Gyroscope, and Relative observations were computed.

Five different trajectories have been recorded by three Turtlebot 3 burger robots that are operating on Robot Operating System (ROS). Figure 1 depicts them. The scenarios that these paths cover include co-living spaces where multiple robots can observe one another simultaneously. The choice of transversal or circular shapes represents the fundamental geometric forms that underpin the structure of these trajectories. Any discrepancies between the sensor data and the ground truth discovered are corrected at the pre-processing stage.

3.2 Faulty Scenarios Generation

Taking into consideration the type of sensor and potential defects, we generate 10 unique sensor fault scenarios for each of those trajectories: With bias for Marvelmind, LiDAR, and gyroscope and encoders, each robot generated 1031 m of data in around one hour.

3.3 Position and Orientation Estimation

Using the Extended Informational Filter (EIF), a variation on the Kalman filter that has the same prediction step but a correction based on the total informational contributions from the sensors in the information space, results in the same prediction and correction steps. In particular, it is necessary to project the state vector $Xk|k-1$ and the covariance matrix of the prediction step $Pk|k-1$ onto the information space using in the Figure 1.

It first determines the initial values and dimensions of the state vector and the variance-covariance matrix. The state vector in this work is composed of the coordinates $[xy\theta]$, the position on the x and y axes, and the orientation of the z axis. This shows that the state vector, $Xaoin\mathbb{R}3\times1$, and the variance-covariance matrix of $Uocba\mathbb{R}3\times3$. The initial values of these coordinates for each robot are set using the values found in the ground truth of the Optitrack. Given that the performance with respect to faults is recorded for every sensor as well as the overall behavior of the system, an initial value is stated for the preceding no-fault probability hypothesis P0.

After initialization, the algorithm's main loop is employed. The prediction stage of the information filter is used initially. It utilizes the odometric model based on encoder data to predict the position of the robot based on the evolution it has made. This is done with the aid of the input vector $ua=[\Delta a\omega a]$, the rotation performed during the iteration when it is computed, and the elementary displacement.

Preparing for the corrective stage involves projecting the forecast into the information space. If the adjustment is made using the sensor that is monitoring the robot itself, the informational contribution vector and matrix are computed with consideration for the embedded sensor's properties, including the observation matrix and the uncertainty. Should the adjustment be obtained from the LiDAR and there are no more robots in the system, the portable landmarks take effect. The robots can communicate their position and variance-covariance matrix to one another via a shared network. The robot learns where this is by observing other robots. After that, it calculates this observation's informational contribution and adds a LiDAR observation to determine the robot's relative location. When all accessible informational contributions have been computed, the correction model is applied as demonstrated in, summing these values with the information vector and matrix appropriately, and projecting back to the Kalman space.

After receiving the correction, the robot begins diagnosing the position. The initial step in the detection method is to ascertain whether the system as a whole has a fault, which is accomplished by applying information theory and determining the Jensen-Shannon divergence between the prediction and correction. To ascertain the functional hypothesis, the trained detection model is employed. If this model indicates that a fault exists, the isolation step is applied. As mentioned before, the generalized observer scheme (GOS) is employed in the computation of failure indicators. These indications are input into each isolation model that is specific to a given sensor, and the functioning hypothesis of each is output together with the value of the prior no-fault hypothesis of sensors. The location and covariance are updated using the output of these models after the designated problematic sensors are eliminated

Figure 1. Algorithm 1 fault tolerant cooperative localization algorithm

Require: Initialization of state with Xao and its covariance Pao obtained from Optitrack for each robot.
Require: Set initial values of $P0$ and $Pobsa0$
.**while**$k{\neq}Nbiterations$
do
* **Apply prediction step:**
Read data from encoders: $ua{\leftarrow}[\Delta a,\omega a]T$
Compute $\{Xak|k{-}1,|k{-}1\}$ using the previous state and the evolution model.
Compute the information vector and matrix $\{yak|k{-}1,|k{-}1\}$
by using (1).
* **Apply correction step:**
Compute the informational contribution $\{iobskk, Iobskk\}$ for $obsa{\in}\{ma,ga\}$, by using (2).
if *Robotb* or *Robotc* in sight **then**
        Get their position and covariance $\{Xb,|k{-}1,Pb,ck|k{-}1\}$
        Get the relative observation $\{ZLa{\rightarrow}b,,La{\rightarrow}b,ck\}$ toward them.
        Compute the correction $\{Xobskk, Pobskk\}$ for $obsa{\in}\{La{\rightarrow}b,La{\rightarrow}c\}$
end if
Compute $\{Xak|k, Pak|k\}$ using the Extended Information filter in (2).
* **Apply Diagnosis step:**
Compute $gDaJS$ using (3).
if $\{gDaJS,P0\}$ tested on the detection model implies the presence of a fault **then**
Compute $JSobsaGOS$ using (3)
Get the faulty sensors using the isolation models
Exclude the faulty sensors
Update the position and covariance $\{Xak|k, Pak|k$
$\}$
end if
Update $P0$
Update $Pobsa0$
$k = k + 1$
end while

from the fusion process. Finally, the values of the preceding no-fault hypothesis are changed in light of the current iteration's findings.

3.4 Data Exploitation

Two distinct model types are utilized to detect and remove errors: several models for isolation and a unique identification model. These models are trained using both centralized and federated methods: With federated training, the central unit only receives the models' averages from the local training data, which remains on the computer.

3.5 Data Organization

A system where several individuals produce the data required to train a model can be organized in a variety of ways. The primary subjects of this book are federated and centralized learning.

All of the network members' data is collected and arranged using the centralized data organization technique at this single location. The learning model that was trained on this centralized dataset is then distributed to each participant. This method has multiple benefits because it is easy to use and employs a range of training algorithms. Furthermore, it allows for the comprehensive processing and analysis of the entire dataset and, should the consolidated dataset be representative and diverse, could result in better performance. It does, however, restrict the system's usefulness to the central unit's responsiveness and availability and requires the transfer of user data to a central location, which raises privacy and security concerns. Moreover, it is considered that the combined dataset represents.

A decentralized architecture is the outcome of federated learning. This approach makes it possible to train a model on distributed data right away, as opposed to moving the raw data to a central place. In federated learning, local training is carried out on each network member's unique data after the model has been loaded on them all. Afterwards, the model's parameters are routinely averaged or aggregated throughout the network to create a global model that protects data privacy and makes use of existing collective knowledge.

Instead of providing raw data, this method preserves data privacy and secrecy by training the model locally using each member's data. More broadly, it allows learning from a large number of distributed devices or nodes, enabling greater involvement, and lowers transmission costs since only model changes or gradients, not raw data, are exchanged. However, this requires combining and coordinating model changes, which could increase the intricacy. Moreover, it may run into issues with the network's participants' heterogeneous data distribution, which could lead to variations in the performing model. When systems have different operating histories, they are more likely to exhibit behavior that is somewhat different from the other systems' and not tailored to the system as a whole. Furthermore, how well this approach works.

3.6 Model Optimization and Training

Machine learning can use a range of models to meet this demand for discriminative learning, which is necessary to solve the classification problem. For federated learning aggregation, we select the logistic regression model so that we can evaluate the effectiveness of decentralized and centralized techniques. Because the input for all models is the same, federated learning, in contrast to other models, uses a minimal number of parameters that are shared by all subsystems.

For binary and multi-class classification tasks, the logit or logistic regression model is a particular application of the Generalized Linear Model (GLM). It provides a probabilistic framework for estimating class membership likelihood with input. The lack of processing capacity, the availability of other algorithms, and the gradual awareness of this tool's advantages delayed its usage, even though it was first proposed in the middle of the 20th century.

Logistic regression provides an alternative approach to applying linear regression methods to classification problems by changing the space in which these values are calculated. It employs a linear regression model of the type

$z = \beta 0 + \beta 1 x 1 + \ldots + \beta n x n$

In this instance, weights or intercepts are represented by $\beta 0$ to βn, while n characteristics, or input variables, are represented by $x1$ to xn. The legit function is used to map this model onto the interval [0, 1]. It has the following definition:

$(c0|x1,2,\ldots,xn) = 11 + e - z$

The logarithm of the odds, or logit, is impacted by β. In logistic regression, the terms odds and odds ratios are called parameters. To obtain the best-fitting curve for classification, the log-likelihood ratio for continuous input variables needs to be maximized.

Logistic regression is used to predict a categorical variable, as opposed to continuous regression. It can also be applied to assess the relationship between one or more independent factors and a dependent variable. The best logistic regression model fit is obtained with the following fitting parameters:

3.6.1 Penalty

Identifies the type of regularization applied to the logistic regression model. Regularization minimizes over-fitting by including a penalty term in the loss function. There are the L1 and L2 approaches.

Adding a regularization component to the cost function and using the $l1$ norm of the weight vector, L1 Regularization (also called Lasso Regularization, Least Absolute Shrinkage, and Selection Operator Regression) operates. The cost function is enhanced by a regularization factor of $\alpha \sum i = 1 n \theta 2 i$ in L2 Regularization, sometimes referred to as Ridge Regularization. This means that in addition to fitting the data, the learning method must minimize the model weights.

3.6.2 Solver

Explains the optimization algorithm used during the model's training. Various factors are considered by different solvers: the size of the dataset, the rate of convergence, the support for regularization, and the linearity of the data, memory efficiency, constraints, parallelization capability, Hessian approximation, and batch vs. stochastic update utilization.

3.6.3 Weight Class

Should the model be trained on the unbalanced data, it will be skewed toward the abundant class. In order to prevent this, cost-sensitive learning is used. In a multi-class classification task, the cost of wrongly identifying different classes is taken into consideration. Due to this, a certain weight must be assigned to each label. In order to preserve all parameters, the L2 Regularization is employed. We set C to 0.01 to make use of the memory that we have. Broyden-Fletcher-Goldfarb-Shanno (lbfgs) is the standard solution for logistic regression. Although there are several ways to aggregate data, Federeated Averaging (FedAvg) is the most fundamental approach. This method uses the following equation to aggregate the data over the hyper-parameters ω of the models for K customers, each of which has a subset Dk of length nk:

$\omega t + 1 = \sum k K n k \sum K k n k \times \omega k t + 1$

In our example, the detection model is based on two variables: $P0$, which is the prior probability of the no-fault hypothesis; and $gDJS$, which is the detection residual after Jensen–Shannon divergence. Accordingly, the single output, the existence of a fault, the slopes $\beta1$ and $\beta2$, and the intercept $\beta0$ make up the equation's three parameters. In terms of isolation, the intercept, the $P0$ of every sensor, and the isolation residuals $JSobsaCS$ are the 10 variables for which we have slopes βi. This indicates that the number of parameters that require optimization is $10 + 4 + 1 = 15$.

In a system with four sensors, faults could occur simultaneously and result in six separate output classes. Individual faults for each sensor (4 classes), no fault (0), and a distinct class for simultaneous faults are all included. Its unique properties prevent this model from being reduced to a single model since it is not a "one against all" classifier. Because the residuals are constructed using the Generalized Observer Scheme, it should be emphasized that not every model may make use of every input. These inputs are supplied with the $P0$ of the encoder, marvelmind, gyroscope, and LiDAR attached to it, respectively, in the same order as the *obsaGOS*.

IV. RESULT AND DISCUSSION

To evaluate and confirm the proposed framework's feasibility, we put it into practice and conducted tests on our mobile robot platform. We tested our robot in an environment akin to an office to make sure it could maneuver safely and socially in a real-world situation.

4.1 Experiment Setup

A Microsoft Kinect sensor and a laser range finder were installed on our Eddie mobile robot platform. The typical Kinect sensor, which consists of an RGB camera, depth sensor, infrared light projector, and multiarray microphone, was mounted 1.35 meters above the ground. The depth sensor's viewing angles are 43 degrees vertically and 57 degrees horizontally, with a range of 0.8 to 6.0 meters. With a maximum frame rate of 30 frames per second, this low-cost equipment can produce RGB-D data at a resolution of 640 480 pixels. The 240 angular field of vision of the laser range finder UGR- 04LX-UG01 allows it to measure distances up to 6.0 m at a height of 0.4 m.

4.2 Navigation Performance

In order to illustrate how our system functions in comparison to pedestrians and PCLRHC, we begin this section with some anecdotal data. Interestingly, this behavior held true for all ten of our experiments: IGP outperformed the pedestrian and PCLRHC acted evasively, usually going outside to avoid the crowd. Figure 2 displays the main experimental result reported in this study. Figure 2 displays our program's output throughout the course of ten experiments. For each of the ten tests, the boxes surrounding the colorful dots reflect the standard error bars. The IGP green dot's average safety was approximately 22 pixels, with a standard error of more than 2 pixels, and the Columns with the label "s" indicate safety (in pixels), and columns with the label "}" show path length (in pixels). Both path length and safety are significantly greater for IGP than for pedestrians. Additionally, as theoretically demonstrated earlier, PCLRHC is not appropriate for densely populated locations. Large path length evasive maneuvers are nearly always used by PCLRHC to avoid the crowds.

Figure 2. The experimental result

Building on social psychology research, we develop a novel algorithm to minimize entitativity and maximize the social invisibility of multi-robot systems in pedestrian crowds. Unique entitativity profiles—characterized by appearance, trajectory, and spatial distance—are linked to different emotional reactions, with people in high entitativity groups reporting negative emotions. This is evident from a user research. Then, we use the trajectory information from low-entitative groups to develop a real-time navigation algorithm that should increase the social invisibility of multi-robot systems.

Our approach has some shortcomings. We find that, albeit generalizing across several environmental contexts, judgments such as social salience depend on factors other than motion-based entitativity. When forming opinions and reacting emotionally to social agents, people employ a wide range of indicators, including impressions of gender, race, class, and religion. Since our method just employs mobility trajectories, not all relevant social features are properly recorded. Since robots may lack these higher-level social characteristics, the low-level entitativity component that matters most to them is motion trajectories.

This method of mimicking robot looks in multi-robot systems ought to be investigated further in subsequent research. Robots can have their appearance customized even though many social characteristics, such as race, may not apply to them. Because of their increased entitativity, marching robots are expected to become more visible in society. Businesses that produce surveillance robots may find this problematic since mass production typically yields similar looks. Further study is needed to determine how a perceiver's personality affects the features of multi-robot systems, as some individuals may be less likely to react negatively to entitative groups of robots because of things like having more experience with robots or being less sensitive to general threat cues.

V. CONCLUSION

Visibility information was considered in this work to give a fast and safe navigation approach. The environmental risks were estimated statistically to address contact with occluded dynamic obstacles. Course planning and speed control both made use of the quantitatively derived and structurally employed collision risk. Simulations and presented experimental results show that the proposed indoor mobile robot navigation system is a safe and efficient method. Thinking about the safety and future of the trajectory is made possible by the proposed definition. Simulations show that in addition to the robot itself, it is crucial to consider the likelihood of any object in the workplace colliding with the robot. The robot's navigation might run afoul of workplace objects in the absence of such protection. Working with the navigation algorithm, the robot's job is to make results verification possible.

REFERENCES

Arul, S. H., Bedi, A. S., & Manocha, D. (2022). Multi Robot Collision Avoidance by Learning Whom to Communicate. arXiv preprint arXiv:2209.06415.

Bajcsy, A., Herbert, S. L., Fridovich-Keil, D., Fisac, J. F., Deglurkar, S., Dragan, A. D., & Tomlin, C. J. (2019, May). A scalable framework for real-time multi-robot, multi-human collision avoidance. In 2019 international conference on robotics and automation (ICRA) (pp. 936-943). IEEE. doi:10.1109/ICRA.2019.8794457

Bramblett, L., Gao, S., & Bezzo, N. (2023). Epistemic Prediction and Planning with Implicit Coordination for Multi-Robot Teams in Communication Restricted Environments. arXiv preprint arXiv:2302.10393. doi:10.1109/ICRA48891.2023.10161553

Chung, Y. M., Youssef, H., & Roidl, M. (2022, May). Distributed Timed Elastic Band (DTEB) Planner: Trajectory Sharing and Collision Prediction for Multi-Robot Systems. In 2022 International Conference on Robotics and Automation (ICRA) (pp. 10702-10708). IEEE. 10.1109/ICRA46639.2022.9811762

Cui, Y., Lin, L., Huang, X., Zhang, D., Wang, Y., Jing, W., ... Wang, Y. (2022, May). Learning Observation-Based Certifiable Safe Policy for Decentralized Multi-Robot Navigation. In 2022 International Conference on Robotics and Automation (ICRA) (pp. 5518-5524). IEEE. 10.1109/ICRA46639.2022.9811950

Ferranti, L., Lyons, L., Negenborn, R. R., Keviczky, T., & Alonso-Mora, J. (2022). Distributed nonlinear trajectory optimization for multi-robot motion planning. IEEE Transactions on Control Systems Technology, 31(2), 809–824. doi:10.1109/TCST.2022.3211130

Firoozi, R., Ferranti, L., Zhang, X., Nejadnik, S., & Borrelli, F. (2020). A distributed multi-robot coordination algorithm for navigation in tight environments. arXiv preprint arXiv:2006.11492.

Freitas, E. J., Vangasse, A. D. C., Raffo, G. V., & Pimenta, L. C. (2023, October). Decentralized Multi-robot Collision-free Path Following Based on Time-varying Artificial Vector Fields and MPC-ORCA. In 2023 Latin American Robotics Symposium (LARS), 2023 Brazilian Symposium on Robotics (SBR), and 2023 Workshop on Robotics in Education (WRE) (pp. 212-217). IEEE. 10.1109/LARS/SBR/WRE59448.2023.10333004

Krishnan, S., Rajagopalan, G. A., Kandhasamy, S., & Shanmugavel, M. (2019). Towards scalable continuous-time trajectory optimization for multi-robot navigation. arXiv preprint arXiv:1910.13463.

Krishnan, S., Rajagopalan, G. A., Kandhasamy, S., & Shanmugavel, M. (2020). Continuous-time trajectory optimization for decentralized multi-robot navigation. IFAC-PapersOnLine, 53(1), 494–499. doi:10.1016/j.ifacol.2020.06.083

Madridano, A., Al-Kaff, A., Martín, D., & De La Escalera, A. (2021). Trajectory planning for multi-robot systems: Methods and applications. Expert Systems with *Applications, 173, 114660. doi:10.1016/j.eswa.2021.114660*

Mali, P., Harikumar, K., Singh, A. K., Krishna, K. M., & Sujit, P. B. (2021, June). Incorporating prediction in control barrier function based distributive multi-robot collision avoidance. In 2021 European *Control Conference (ECC) (pp. 2394-2399)*. IEEE. 10.23919/ECC54610.2021.9655081

Navsalkar, A., & Hota, A. R. (2023, May). Data-driven risk-sensitive model predictive control for safe navigation in multi-robot systems. In 2023 IEEE Int*ernational Conference on Robotics and Automation (ICRA) (pp. 1442-1448*). IEEE. 10.1109/ICRA48891.2023.10161002

Olcay, E., Schuhmann, F., & Lohmann, B. (2020). Collective navigation of a multi-robot system in an unknown environment. Robotics and *Autonomous Systems, 132, 103*604. *doi:10.1016/j.robot.2020.103604*

Park, J. S., Tsang, B., Yedidsion, H., Warnell, G., Kyoung, D., & Stone, P. (2021, October). Learning to improve multi-robot hallway navigation. In Conferen*ce on Robot Learning (pp. 1883-1895)*. PMLR.

Samman, T., Spearman, J., Dutta, A., Kreidl, O. P., Roy, S., & Bölöni, L. (2021, October). Secure multi-robot adaptive information sampling. In 2021 *IEEE International Symposium on Safety, Security, and Rescue Robotics (SSRR) (pp.* 125-131). IEEE. 10.1109/SSRR53300.2021.9597867

Şenbaşlar, B., Luiz, P., Hönig, W., & Sukhatme, G. S. (2023). Mrnav: Multi-robot aware planning and control stack for collision and deadlock-free navigation in cluttered environments. ar*Xiv preprint arXiv:2308.13499.*

Tan, Q., Fan, T., Pan, J., & Manocha, D. (2020, October). Deepmnavigate: Deep reinforced multi-robot navigation unifying local & global collision avoidance. In *2020 IEEE/RSJ International Conference on Intelligent Robots and Systems (IROS)* (pp. 6952-6959). IEEE. 10.1109/IROS45743.2020.9341805

. Tran, V. P., Garratt, M. A., Kasmarik, K., & Anavatti, S. G. (2023). Dynamic frontier-led swarming: Multi-robot repeated coverage in dynamic environments. *IEEE/CAA Journal of Automatica Sinica, 10*(3), 646-661.

Yang, Z., Bi, L., Chi, W., Shi, H., & Guan, C. (2022). Brain-Controlled Multi-Robot at Servo-Control Level Based on Nonlinear Model Predictive Control. *Complex System Modeling and Simulation, 2*(4), 307–321. doi:10.23919/CSMS.2022.0019

Zaccaria, M., Giorgini, M., Monica, R., & Aleotti, J. (2021, July). Multi-robot multiple camera people detection and tracking in automated warehouses. In *2021 IEEE 19th International Conference on Industrial Informatics (INDIN)* (pp. 1-6). IEEE. 10.1109/INDIN45523.2021.9557363

Zhu, H. (2022). *Probabilistic Motion Planning for Multi-Robot Systems*. Academic Press.

Zhu, H., Claramunt, F. M., Brito, B., & Alonso-Mora, J. (2021). Learning interaction-aware trajectory predictions for decentralized multi-robot motion planning in dynamic environments. *IEEE Robotics and Automation Letters, 6*(2), 2256–2263. doi:10.1109/LRA.2021.3061073

Chapter 4
A Novel Method to Detect Ripeness Level of Apples Using Machine Vision PSOCNN Approach

K. Yogesh
Karpagam Academy of Higher Education, India

R. Gunasudari
Karpagam Academy of Higher Education, India

ABSTRACT

Machine vision systems have emerged as a viable non-invasive approach for investigating the connection between fruit visual traits and physicochemical qualities at varying ripening degrees, and have been used in recent research efforts to identify the stages. The current study aims to develop an intelligent algorithm that can estimate various physical properties, such as firmness and soluble solid content, as well as three chemical properties, namely starch, acidity, and titratable acidity. A hybrid approach was used to further optimise the physicochemical estimation method of PSO with CNN. This method was applied to the evaluation parameters in order to describe their classification behaviour. The sample accuracy was 95.84% when using the different parameters to characterise them. A second set of apples was utilised for validation after the first set was used as trial samples in PSO+CNN.

I. INTRODUCTION

The most commonly grown pome fruit in the world is the Apple (Maluscommunis L.). Global output amounts to around 133 million tons per year. China stands as leading producer and India took its position as fifth. In the fiscal year 2022, a total of 2.4 million metric tons of apples were produced in India. Comparing this to the prior fiscal year, there was an increase. For that fiscal year, the bulk of the nation's apple output came from the northern state of Jammu & Kashmir.

DOI: 10.4018/979-8-3693-5767-5.ch004

1.1 Contribution of India in Agriculture and Automation of Agriculture

The Ministry of Statistics and Programme Implementation provided that India's GDP is 15.4% derived from agriculture. Gathering, processing, and storing are the three primary categories into which agriculture activities are often divided.

After China, India is the world's second-largest fruit grower. 30% to 35% of the harvested fruits are wasted because there are not enough skilled labourers. Once more, fruit identification, classification, and grading are not done precisely due to the subjectivity of human perception. Therefore, the fruit business must implement the automated system. Automated fruit sorting systems based on fruit type, variety, maturity, and intactness can be designed with high intelligence thanks to machine learning approaches that incorporate sufficient image processing concepts.

In agriculture, automated systems are required for the precise, quick, and high-quality determination of fruits since they are essential to the nation's economic growth and increased production. Researchers have developed numerous algorithms to classify and rank fruit according to its quality. Colour is the most obvious factor in identifying fruit illness and ripeness.

We refer to the uniform distribution of a single color—the primary color—on the skin surface of some fruits. A good way to determine the quality of these fruits is to look at their average surface colour. Nevertheless, certain fruits (including certain types of peaches, apples, and tomatoes) possess a secondary hue that serves as a reliable gauge of their maturity. Only using the global colour as a quality parameter is not an option in this situation.

1.2 Machine Learning in Agriculture

To overcome difficulties progressing in the field of machine learning has contributed to enhanced agricultural yields. By offering insightful advice and detailed knowledge about the crops, machine learning is a modern technology that helps farmers reduces farming losses.

Recent years have seen the evaluation of numerous agriculture-related issues through the integration of remote sensing and AI tools. More specifically, a number of recent research have been developed employing DL approaches applied to photos taken at different acquisition levels in fruit recognition challenges.

Apples present one of the greatest challenges when it comes to fruit detection in photos, mostly due to the presence of target occlusion issues. The establishment of high-density apple tree orchards further complicates the identification of individual fruits.

As per IBM definition ML, a branch of AI, is the study of building computer systems with data-learning capabilities. Algorithms for machine learning and machine vision are among the many approaches that machine learning (ML) uses to improve software applications' performance over time are critical components to solve major issues regarding fruits post harvesting period.

1.3 Machine Vision

For applications including process control, robot guiding, and autonomous inspection, machine vision (MV) technology enabled imaging-based autonomous inspection and analysis.

The necessity for a sufficient level of food production with fewer agricultural areas has arisen due to the swift expansion of the global population. Machine vision would ensure an increase in agricultural productivity by efficent way. Significant progress has been made in a number of agricultural fields in recent years. These developments combine machine learning methods with machine vision approaches to handle colour from object images by Penumuru, D. P., et al (2020).

1.4 Machine Vision System

Numerous evaluations have been conducted with an emphasis on this topic because of the current developments in machine vision applications in agriculture. CV techniques have been used in the development of fruit categorization and identification systems within the past 10 years, including by Pereira, C.S., et al (2017). The general machine vision system is depicted in the below Figure 1.

To facilitate objective and non-destructive food examination, an MVS consists of two basic components: 1) Gathering information; and 2) analyzing it.

- The process of acquiring photographs determines their information and quality. It serves as the cornerstone for the success of any further image processing.
- Machine vision systems are able to multitask because image processing directs the operation.

1.5 Advantages of Machine Vision System

The most important thing of using machine vision technology is, for its merits. These are listed in below:

- Conserve time
- lower the cost of production
- streamline the logistical procedure
- Reduce equipment downtime
- Boost output and caliber of output
- Lessen the personnel's and the test's labor intensity
- Cut back on things that aren't qualified
- Increase the rate of machine use and so on.

Figure 1. Machine vision system

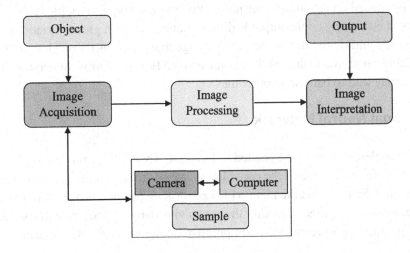

1.6 Applications of Machine Vision

1. While the uses of machine vision are expanding in tandem with technological advancements, there are a few key areas where machine vision has shown to be quite beneficial.
2. OCR stands for "optical character recognition". Through OCR, printed or handwritten text can be extracted from photographs by a computer.

1.6.1 Recognition of Signatures and Handwriting: These characteristics enable a computer to identify patterns in pictures of signatures and handwriting.

1.6.2 Identification of Objects: Self-driving automobiles in the automotive industry recognize objects in camera-captured photos to identify roadblocks. Additionally, machine vision systems are used to locate objects, such as figuring out where a label should go on a pill bottle.

1.6.3 Identifying Patterns: Pattern recognition is used in medical imaging analysis to diagnose patients using technologies including brain, blood, and magnetic resonance imaging scans.

1.6.4 Examination of Materials: The use of machine vision in materials inspection systems guarantees quality control. Machine vision examines various objects for imperfections, impurities.

1.6.5 Robot Guidance: One quickly expanding field in machine vision is the use of cameras for robot guidance. In order to teach robots how to handle individual or bulk components efficiently, 2D and 3D cameras are both crucial. Because they need less physical effort, these applications offer a high return on investment.

In the future, large-scale dataset-based computer vision intelligence technology will be extensively employed in all facets of agricultural production management, and it will be further deployed to address present-day agricultural issues. Agricultural automation systems will function better economically, generally, cooperatively, and robustly as artificial intelligence algorithms and computer vision technologies are integrated.

1.7 Machine Vision in Apple Storage

Numerous academics have been drawn to the novel idea of automatically evaluating ripeness using computer vision systems because they offer a cost-effective and speedy alternative to the labor-intensive and time-consuming manual assessment process.

This inquiry is focused on calculating the production of green types of apples since one of the most challenging parts of estimating apple output is differentiating between green varieties that is, types that are green during their initial growth and the green foliage that envelops them. Due to its ability to yield a multitude of data in both the visible and near-infrared (NIR) spectrums, hyperspectral imaging was employed, which may lead to beneficial outcomes.

1.8 Convolutional Neural Network

More precisely, four elements are often needed to build a CNN model. Convolution is a crucial step in feature extraction. If we set the size of the convolution kernels to a certain value, we will lose information in the border. In order to extend the input with a zero value, padding is inserted, which inadvertently modifies the size. Stride is also used to modify the convolving density. The longer the stride, the lower the density. After convolution, feature maps are composed of numerous features, which raises the possibility of an overfitting problem. Sivanantham, K. (2022) is advised to remove maximum and average pooling, as well as redundancy. Loss functions and optimizers were created in order to teach the CNN system as a whole to understand what we meant. CNN has extensive knowledge artificial neural networks in general.

This paper work is divided into five components. The section of the introduction to the work that was discussed in Section I. Section II looks at the limitations and flaws in the current systems. An explanation of our recently suggested system design work for apple ripeness level detection using machine vision method was provided in Section III. Section IV contains the justification and comprehensive output result for our newly constructed system. Section V deals with the conclusion of this work.

II. LITERATURE REVIEW

This section on literature studies discusses the shortcomings and restrictions of the numerous current approaches used in the detection and classification of apple ripeness level using different methods.

Xiao, B., Nguyen, M., & Yan, W. Q., (2021) have examined theripeness of apples in digital images and categorised with the use of deep learning's CNN, also known as ConvNets. Apple identification, ripeness classification, resulting evaluations, and image pre-processing are the four components of this experiment, which aims to validate the potential of DL models for fruit classification in order to reduce the amount of human labour required.

2.1 Image Processing Technology

Wan, P., et al (2018) have described The BPNN classification algorithm is used with the feature colour value to identify the three maturation levels (green, orange, and red) of fresh market tomatoes. With the express purpose of gathering the tomato photos in the lab, a computer vision-based maturity detection device was created. Following the processing of the tomato photos and the acquisition of the tomato targets using image processing technology, the average accuracy of this method for identifying the three tomato maturity levels in samples of tomatoes was 99.31%, with a 1.2% standard deviation.

2.2 Apple Ripeness Determination via Artificial Neural Network Technologies

Hamza, R., & Chtourou, M. (2018), have approached Artificial Neural Network (ANN) classification and estimated the color-based indicator of apple fruit maturity. In order to achieve optimal performance, this work has addressed a number of issues, including colour feature vectors, pedagogy, and ANN classifier structure. The simulation results showed how well the ripeness categorization system performed.

Çetin, N., et al (2022) have analysed the three harvest phases' worth of hyperspectral photos of pink lady apples were used to forecast certain interior traits (firmness and SSC). For every harvest phase, a total of 100 samples were subjected to the hyperspectral camera to get reflectance data in 300 spectral bands ranging from 386 to 1028 nm. Furthermore, the prediction capabilities of DT, ANN, KNN, PLSR, and MLR were assessed.

2.3 Machine Learning in Agriculture

Patil, K., et al (2021) have examined thorough examination of the most recent applications of Pre-harvest, harvesting, and post-harvest problems in agriculture can be resolved with ML. With less effort required and higher-quality results, machine learning in agriculture allows for more precise and effective farming.

Zhu, L., et al (2021) have provided a summary of the deep learning and traditional methods of machine learning in addition to machine vision techniques that are useful in the food processing sector. They described the methods and challenges of the past as well as projected patterns and logical future directions.

Tian, H., et al (2020) have summarized and analyzed systematically the difficulties and technologies of the previous three years, as well as potential future developments, to create the most recent resource for scholars. According to the assessments, small field farmers can benefit from low costs, high efficiency, and high precision by developing agricultural automation with the use of current technology.

Firouz, M. S., et al (2019) have reported about the Within the field of food science and technology, the fruit, juice, and dairy industries the uses of ultrasound in high- and low-power modes for processing, instrumentation, and industrial operation control are investigated. The focus is on the fundamentals of these methods and how they affect the physicochemical properties of the final products. The benefits and limitations of each ultrasound-assisted technique are also discussed, along with a thorough study of these approaches and key variables affecting their effectiveness. This technique's productivity issues would be addressed, and the technology's future trends would be described.

Biffi, L. J., et al (2021) have detected apple fruits efficiently and presented a solution for close-range and inexpensive terrestrial RGB image analysis based on the ATSS deep learning technique. Precise identification helps with apple production projections and provides local growers with more insight into future management strategies. ATSS method's primary benefit was that it just labelled the object's centre point, in fruit orchards with high population density; this is significantly more practical and realistic than bounding-box annotations.

Behera, S. K., Rath, A. K., & Sethy, P. K. (2021) have reviewed the novel non- Classifying papaya fruits according to their damaging ripeness status. The paper suggested two strategies based on transfer learning and machine learning for grading the maturity state of papayas. Additionally, a comparison analysis using various Research was done using transfer learning and machine learning approaches. VGG19, utilising a transfer learning strategy, achieved 100% accuracy, a 6% improvement over the present approach.

2.4 Evaluation of Current Advancements in Traditional and Innovative Methods for Measuring Lycopene Content of Fruit

Hussain, A., Pu, H., & Sun, D. W. (2019) have reviewed the efforts to demonstrate the worth of using both traditional and cutting-edge methods to assess the lycopene content in fruit. Along with spectrum imaging techniques like multispectral, hyperspectral, and Raman imaging, the revolutionary techniques also include spectroscopic techniques like near infrared spectroscopy and Raman spectroscopy. Future trends are also offered, along with a summary of the techniques' guiding principles and a discussion of their specific applications. Lycopene concentration and distribution in different fruits can be evaluated using both conventional and innovative methods described in this paper.

Bhargava, A., & Bansal, A. (2020) presented a system that distinguishes between four different fruit varieties and evaluates each fruit's ranking according to its quality. The split-and-merge procedure was used to separate the photos' backgrounds after the programme had first extracted the different images. Subsequently, the thirty distinct features color, statistical, textural, and geometric are retrieved. Only geometrical features are utilised to distinguish between different types of fruit; other features are used to evaluate the fruit's quality. In addition, the quality is classified using four distinct classifiers: artificial neural network (ANN), SVM, sparse representative classifier (SRC), and k-nearest neighbour (k-NN).

Mavridou, E., et al (2019), have reviewed the most current research on using machine vision in agriculture, primarily for agricultural production. When it came to using cognitive technology in agriculture, this study acted as a research guide for both practitioners and researchers. Studies of various agricultural practices, such as fruit grading, fruit counting, and yield estimation, that assist crop harvesting are reviewed. Furthermore discussed are methods for monitoring plant health, such as weed, insect, and disease detection. Last but not least, new studies have taken agricultural harvesting robots and vehicle navigation systems into consideration.

Rehman, T. U., (2019) have outlined the statistical machine learning technologies that are used in agriculture using machine vision systems, given the wide range of machine learning applications. Agriculture has made use of both supervised and unsupervised learning, two categories of statistical ML approaches. This study provides a thorough examination of the current use of statistical ML algorithms in machine vision systems, assesses the prospective applications of each methodology, and provides an overview of instructive case studies in various agricultural domains. This paper also formulates and discusses future trends in statistical ML technology applications.

2.5 Use of Machine Vision Technology for Food Identification

Xiao, Z., et al (2022) have examined the use regarding the hardware and software of machine vision systems in the context of food detection, presented the state of research on machine vision as it stands today, and offered a forecast for the difficulties that machine vision systems encounter.

Wang, W., et al (2023) have estimated The method used to determine the freshness state of apples involved using Back Propagation (BP) as a neural network predictive model and an enhanced SSA based on chaotic sequence (Tent) for optimisation. Utilising an array of gas sensors and a wireless gearbox module, an electronic nose system was created. To finish predicting the freshness of apples, odour data is analysed from apples.

III. SYSTEM DESIGN

The following portion demonstrates the several stages of the suggested hybrid PSO with CNN machine vision algorithm to evaluate the apple ripeness levels in different situations.

3.1 Particle Swarm Optimization (PSO)

The initial translations xi_0 inside bounds to the challenge the user has to set the population size (p_s) beforehand. $P_{Besti,g}$ is the result of evaluating and storing values of the particle starting points for the objective function. The generation counter, represented by g in this instance, starts at 0the following equation:

$$v^d_{i,g+1} = w_g \cdot v^d_{i,g} + c_1 \cdot rand1^d_{i,g}(0,1) \cdot \left(pBest^d_{i,g} - x^d_{i,g} \right) + C_2 rand2^d_{i,g}(0,1) \cdot \left(gBest^d_g - x^d_{i,g} \right)$$

$$x^d_{i,g+1} = x^d_{i,g} + v^d_{i,g+1}$$

The user-specified acceleration coefficients c1 and c2 are the inertia weight, and all other particles in the swarm are represented by $d = 1,...,D$ (where D is the problem dimensionality). For every i^{th} particle and d^{th} dimension, two random numbers, $rand1_{di,g}(0,1)$ and $rand2_{di,g}(0,1)$, are generated independently using the [0,1] interval.In addition to the matching each particle in the run is associated with three vectors: the particle's $x_{i,g}$, $v_i g$, and the best position ($P_{Besti,g}$) it has visited since the search was started. The procedure keeps on until the predefined maximum number of function calls is completed.

3.1.1 PSO Structure Algorithm

From the one iteration to the next, particle swarms modify their relative positions, which allow the PSO algorithm to efficiently conduct the search. To find the best feasible solution, each particle in the swarm moves toward its prior Pbest and gbest. Assume, f is objective function to minimized or optimized, There are t iterations in total, and i is the particle index.

A swarm a group of people in the PSO algorithm, whereas an individual represents a possible solution. The PSO algorithm is a widely used population-based evolutionary computing method. A vector represents the ith particle's velocity in an N-dimensional optimisation problem, vi = (vi1, vi2,...,viN). Similarly, xi = (x_{i1}, x_{i2}, ...,x_{iN}) represents the position vector of the ith particle. The equations for updating the position and velocity of particle I are given below:

$$v_i(k+1) = w \times v_i(k) + c_1 \times r_1 \times (p_i(k) - x_i(k)) + c_2 \times r_2 \times (p_g(k) - x_i(k)) x_i(k+1) = x_i(k) + v_i(k+1)$$

The pseudocode of the PSO is given in the figure 2. Members of interval [0, 1] are the two unique random integers, r1 and r2. The range of m, the domain of the optimization problem, is where the particle's position can only be. The following represents the updated equation for at the kth repetition, the inertia weight w: $w = (w_1 - (w_1 - w_2) \times k)/maxiter$. While w1 and w2 show the maximum and minimum inertia weights, respectively; maxiter is the number of iterations that can be made.

3.2 Convolutional Neural Network (CNN)

Convolutional layers, max-pooling, and sparse connectivity are the three basic facets of CNN architecture. The local connection of neurons on neighbouring layers takes advantage of the spatial dependency of the image's pixels.

The convolution kernel, from which the term "convolution neural network" derives, is the most crucial component of the CNN. An n-column two-dimensional matrix makes up the convolution kernel and matching weights for each point. Convolution kernels resemble neurons, and their size is referred to as the receptive field of a neuron. The receptive field of the CNN is filled with the addition of the values of the pixels and the convolution kernel's weights at the pertinent locations in the image. The system's k convolution kernels do this. Once all of the pixels in the image have been counted, to the following

Figure 2. Pseudocode of the particle swarm optimization

Initialization
Give the definition of the number of dimensions D and the swarm size S.
for each particle $i \in [1..S]$
Produce at random X_i and V_i, and assess the health of X_i indicating it as $f(X_i)$
Set $Pbest_i = X_i$ and $f(Pbest_i) = f(X_i)$
end for
Set $Gbest = Pbest_1$ and $f(Gbest) = f(Pbest_1)$
for each particle $i \in [1..S]$
    if $f(Pbest_i) < f(Gbest)$ **then**
        $f(Gbest) = f(Pbest_i)$
    end if
end for
while t < maximum quantity of repetitions
for each particle $i \in [1..S]$
Analyse its speed $v_{id}(t + 1)$
Update the position $x_{id}(t + 1)$ of the particle
if $f(x_i(t + 1)) < f(Pbest_i)$ **then**
$Pbest_i = x_i(t + 1)$
    $f(Pbest_i) = f(x_i(t + 1)$
end if
if $f(Pbest_i) < f(Gbest)$ **then**
    $Gbest = Pbest_i$
    $f(Gbest) = f(Pbest_i)$
end if
end for
$t = t + 1$
end while
go back to Gbest

point in the image in accordance with the step size. The original image's feature map is now the output pixel matrix.

$$O_w = [\frac{i_w - n + 2_p}{s}] + 1 \tag{1}$$

$$Oh = [\frac{i_h - n + 2_p}{s}] + 1 \tag{2}$$

The most common pooling techniques are maximal combining, averaging, etc. Here, we select the highest pooling approach and the associated calculation. The area of the image the highest value per pixel in the range of 2 is chosen using a filter of size 2 2, and it is then kept as the distinctive feature of this area. A feature map is created once the filter repeats this process for the subsequent range. The convolution network's performance has increased with end-to-end training, not just in terms of overall data classification but also in terms of the local task's progress toward producing structured output. As a result, it has seen extensive use in the fields of data detection and categorization.

CNNs will quickly run into problems when using a typical multi-layer perceptron, meaning all layers are completely connected, because the data dimensions are too big. There is no available space information on CNN. The final efficiency may be quite low if we use the pooling layer. CNNs suffer a significant of information in the pooling layer, which lowers the spatial resolution, because only the most active neurons can be communicated between each layer during the transfer of neurons.

The above Figure 3 shows the pseudocode for convolutional Neural Network. CNNs won't be able to discern between variations in postures and other features as a result. Overtraining from every viewpoint is one technique to overcome this issue, but it typically takes more time and computer power.

Figure 3. Pseudocode for CNN

CNN Algorithm Pseudocode:

Input:

d: dataset,

1: real labels in the dataset

W: Word-to-Vec matrix

Output:

Test dataset score for the CNN-trained model

Assume that f is the 3D matrix of features.

For i in dataset **do**

Allow f to represent the sample i's feature set matrix.

For j in i **do**

Vj - vectorize(j, w)

Append vj to f

Append fi to f

ftrain, ftest, Itrain, Itest Create a train and test subsets from the feature set and labels

M-CNN (ftrain, Itrain)

Score-analysis (i, Itest, M)

Return score

When dealing with extremely complicated field-of-view data that has a lot of overlap, mutual masking, and diverse backgrounds, traditional CNN cannot be identified properly. We will go over the rationale behind removing the fully connected layers, the parameters for the skip and pooling layers in the suggested approach, and the use of two convolutional layers as a skip layer. More specifically, the tail of CNN is frequently added with several fully-connected layers.

IV. RESULT AND DISCUSSION

The evaluation of the experiment proved to be successful, yielding several results for different parameters. The proposed hybrid PSOCNN and conventional methods underwent successful experimental study, yielding several results for various parameters. MATLAB 2013A is used to find the performance evaluation with hybrid of PSOCNN for the detection of ripeness level of apples.

4.1 Confusion Matrix

It is frequently employed in evaluating effectiveness categorization models, which strive to assign a categorical label to each instance of input. The quantity of TP, TN, FP, and FN generated by the model with the test data set is displayed in Figure 4 below.

4.2 Accuracy

The model's correctness is a critical performance parameter that determines if our assumptions about the positive and negative classes are accurate.

$$\text{Accuracy} = \frac{TP + TN}{TP + TN + FN + FP}$$

The output graph of accuracy from the above table 1 for existing and proposed system is clearly shown in the Figure 5. From this graph, our proposed PSO+CNN system outperforms the better accuracy of 95.84% comparing with the other systems.

Figure 4. Confusion matrix

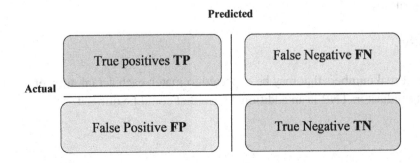

65

Table 1. Accuracy result of existing and proposed algorithm

Algorithm	Accuracy (%)
SVM	79.12
KNN	86.34
ANN	89.28
PSO+CNN	95.84

Figure 5. Accuracy graph

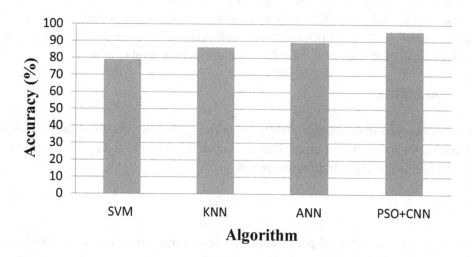

4.3 Sensitivity

The sensitivity (SN) can be calculated simply calculating the ratio of all the positives to all the correct positive forecasts.

$$\text{Sensitivity} = \frac{TP}{TP + FN}$$

The sensitivity output graph of existing and proposed system from the above table 2 is shown in the figure 6. The newly proposed PSO+CNN algorithm gives the better sensitivity result of 93.24% among the all other algorithms.

4.4 Specificity

The percentage of real numbers that may be expected to match each actual number precisely is known as the specificity indicator. TNR is an additional term that could be utilized.

Table 2. Sensitivity result of existing and proposed algorithm

Algorithm	Sensitivity (%)
SVM	77.72
KNN	79.53
ANN	85.43
PSO+CNN	93.24

Figure 6. Sensitivity graph

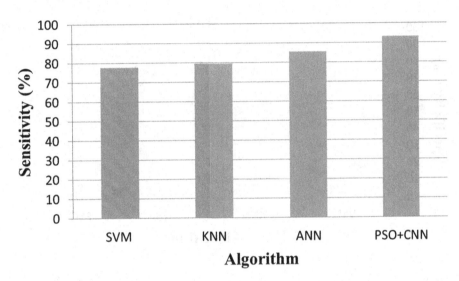

$$Specificity = \frac{TN}{TN + FP}$$

The specificity output graph from the above table 3 for existing and proposed system is shown in the Figure 7. The new approach of our proposed PSO+CNN algorithm works and gives the better specificity result of 92.74% comparing with the other existing algorithms.

4.5 Time Duration

Choosing the best Using a machine learning model to address an issue can be time-consuming if done carelessly.

The time duration output graph from the above table 4 for existing and proposed system is shown in the Figure 8. The new approach of our proposed PSO+CNN algorithm works and consumes the less time duration of 8.58 milliseconds comparing with the other existing algorithms.

Table 3. Specificity result of existing and proposed algorithm

Algorithm	Specificity (%)
SVM	78.84
KNN	76.42
ANN	83.67
PSO+CNN	92.74

Figure 7. Specificity graph

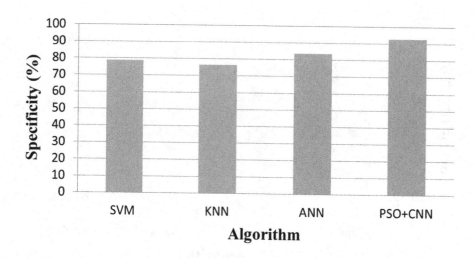

Table 4. Time duration result of existing and proposed algorithm

Algorithm	Time Duration (ms)
SVM	18.86
KNN	16.93
ANN	12.68
PSO+CNN	8.58

V. CONCLUSION

An innovative method was created to use artificial intelligence and an image processing algorithm to determine the physicochemical characteristics and ripeness levels of apples in different situations. A collection of ripeness-varying apple images was obtained. Using a CNN-based algorithm to process image frames allowed for the identification of the best colour and texture features, which allowed for the reliable prediction of the observed physicochemical values. Apples' ripening stages were then forecasted using the anticipated physicochemical characteristics. To determine when apples were ready, the most advanced machine vision system was put to the test. A CCR of 95.84% was attained, indicating a dependable performance of the generated models, in a shorter time of 8.58 milliseconds. Nonetheless,

Figure 8. Time duration graph

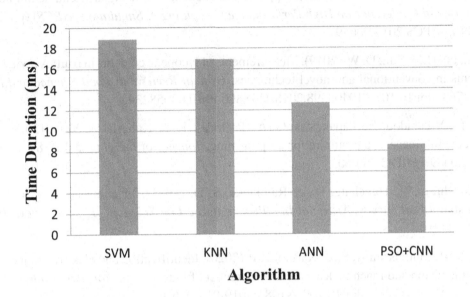

more investigation is needed to develop a field-scale prototype and train the algorithm using several apple cultivars. In order to provide effective these models can be integrated into harvesting robots and/ or drones for real-time resource management in the field.

REFERENCES

Behera, S. K., Rath, A. K., & Sethy, P. K. (2021). Maturity status classification of papaya fruits based on machine learning and transfer learning approach. *Information Processing in Agriculture*, *8*(2), 244–250. doi:10.1016/j.inpa.2020.05.003

Bhargava, A., & Bansal, A. (2020). Automatic detection and grading of multiple fruits by machine learning. *Food Analytical Methods*, *13*(3), 751–761. doi:10.1007/s12161-019-01690-6

Biffi, L. J., Mitishita, E., Liesenberg, V., Santos, A. A. D., Goncalves, D. N., Estrabis, N. V., Silva, J. A., Osco, L. P., Ramos, A. P. M., Centeno, J. A. S., Schimalski, M. B., Rufato, L., Neto, S. L. R., Marcato Junior, J., & Goncalves, W. N. (2021). ATSS deep learning-based approach to detect apple fruits. *Remote Sensing (Basel)*, *13*(1), 54. doi:10.3390/rs13010054

Çetin, N., Karaman, K., Kavuncuoğlu, E., Yıldırım, B., & Jahanbakhshi, A. (2022). Using hyperspectral imaging technology and machine learning algorithms for assessing internal quality parameters of apple fruits. *Chemometrics and Intelligent Laboratory Systems*, *230*, 104650. doi:10.1016/j.chemolab.2022.104650

Firouz, M. S., Farahmandi, A., & Hosseinpour, S. (2019). Recent advances in ultrasound application as a novel technique in analysis, processing and quality control of fruits, juices and dairy products industries: A review. *Ultrasonics Sonochemistry*, *57*, 73–88. doi:10.1016/j.ultsonch.2019.05.014 PMID:31208621

Hamza, R., & Chtourou, M. (2018, July). Apple ripeness estimation using artificial neural network. In *2018 International Conference on High Performance Computing & Simulation (HPCS)* (pp. 229-234). IEEE. 10.1109/HPCS.2018.00049

Hussain, A., Pu, H., & Sun, D. W. (2019). Measurements of lycopene contents in fruit: A review of recent developments in conventional and novel techniques. *Critical Reviews in Food Science and Nutrition, 59*(5), 758–769. doi:10.1080/10408398.2018.1518896 PMID:30582342

Mavridou, E., Vrochidou, E., Papakostas, G. A., Pachidis, T., & Kaburlasos, V. G. (2019). Machine vision systems in precision agriculture for crop farming. *Journal of Imaging, 5*(12), 89. doi:10.3390/jimaging5120089 PMID:34460603

Patil, K., Meshram, V., Hanchate, D., & Ramkteke, S. D. (2021). Machine learning in agriculture domain: A state-of-art survey. *Artificial Intelligence in the Life Sciences, 1*, 100010. doi:10.1016/j.ailsci.2021.100010

Penumuru, D. P., Muthuswamy, S., & Karumbu, P. (2020). Identification and classification of materials using machine vision and machine learning in the context of industry 4.0. *Journal of Intelligent Manufacturing, 31*(5), 1229–1241. doi:10.1007/s10845-019-01508-6

Pereira, C. S., Morais, R., & Reis, M. J. C. S. (2017). Recent advances in image processing techniques for automated harvesting purposes: A review. *Proceedings of the 2017 Intelligent Systems Conference (IntelliSys),* 566-575. 10.1109/IntelliSys.2017.8324352

Rehman, T. U., Mahmud, M. S., Chang, Y. K., Jin, J., & Shin, J. (2019). Current and future applications of statistical machine learning algorithms for agricultural machine vision systems. *Computers and Electronics in Agriculture, 156*, 585–605. doi:10.1016/j.compag.2018.12.006

Sivanantham, K. (2022). Deep learning-based convolutional neural network with cuckoo search optimization for MRI brain tumour segmentation. In *Computational Intelligence Techniques for Green Smart Cities* (pp. 149–168). Springer International Publishing. doi:10.1007/978-3-030-96429-0_7

Tian, H., Wang, T., Liu, Y., Qiao, X., & Li, Y. (2020). Computer vision technology in agricultural automation—A review. *Information Processing in Agriculture, 7*(1), 1–19. doi:10.1016/j.inpa.2019.09.006

Wan, P., Toudeshki, A., Tan, H., & Ehsani, R. (2018). A methodology for fresh tomato maturity detection using computer vision. *Computers and Electronics in Agriculture, 146*, 43–50. doi:10.1016/j.compag.2018.01.011

Wang, W., Yang, W., Li, M., Zhang, Z., & Du, W. (2023). A Novel Approach for Apple Freshness Prediction Based on Gas Sensor Array and Optimized Neural Network. *Sensors (Basel), 23*(14), 6476. doi:10.3390/s23146476 PMID:37514770

Xiao, B., Nguyen, M., & Yan, W. Q. (2021). Apple ripeness identification using deep learning. In *Geometry and Vision: First International Symposium, ISGV 2021, Auckland, New Zealand, January 28-29, 2021, Revised Selected Papers 1* (pp. 53-67). Springer International Publishing.

Xiao, Z., Wang, J., Han, L., Guo, S., & Cui, Q. (2022). Application of machine vision system in food detection. *Frontiers in Nutrition*, *9*, 888245. doi:10.3389/fnut.2022.888245 PMID:35634395

Zhu, L., Spachos, P., Pensini, E., & Plataniotis, K. N. (2021). Deep learning and machine vision for food processing: A survey. *Current Research in Food Science*, *4*, 233–249. doi:10.1016/j.crfs.2021.03.009 PMID:33937871

Chapter 5
Advancing Wild Animal Conservation Through Autonomous Systems Leveraging YOLOV7 With SGD Optimization Technique

Johnwesily Chappidi

https://orcid.org/0009-0005-2000-2462
VIT-AP University, India

Divya Meena Sundaram
VIT-AP University, India

ABSTRACT

In the field of conserving wildlife, the utilization of autonomous systems equipped with computer vision holds tremendous promise. This research explores the potential of integrating YOLO v7, a cutting-edge object recognition model, with stochastic gradient descent (SGD) optimization techniques to bolster wild animal conservation efforts. The primary objective is to enhance the precision, accuracy, and scalability of autonomous systems in detecting and monitoring wild animals across diverse habitats. The experimental results showcase substantial advancements, demonstrating the efficacy of the YOLOv7-SGD amalgamation in autonomous systems. The model exhibits superior detection accuracy and robustness in identifying a multitude of wild animal species across diverse landscapes.

I. INTRODUCTION

The delicate balance between human development and wildlife conservation has become increasingly strained in recent decades, with ecological sustainability and biodiversity facing unprecedented threats. Habitat loss, climate change, and burgeoning human populations have given rise to complex conserva-

DOI: 10.4018/979-8-3693-5767-5.ch005

tion challenges, underscoring the pressing need for innovative solutions that reconcile human activities with the protection of our planet's diverse fauna. In light of this, the integration of autonomous systems has become an issue of concern in the discipline of wildlife conservation, providing a range of tools that could fundamentally alter the methods by which we protect ecosystems and animal species.

The 21st century has witnessed the rapid expansion of urban areas, encroaching on once-wild territories, often leading to Animal-Human Conflict (AHC) and the degradation of natural habitats (Kundu et al., 2023). These conflicts, such as those involving large carnivores and agriculture, elephants and crop damage, or primates and urban infrastructure, pose risks to both animal populations and human livelihoods. Concurrently, The severity of animal-vehicle collisions (AVCs) has risen as a result of developing modes of transportation, killing humans as well as animals. (Mammeri et al., 2016). Additionally, the poaching and trafficking of endangered species remain persistent threats to conservation efforts, threatening to erase irreplaceable links in the web of life.

The research topic's importance arises from its ability to tackle these complex problems of conservation. Autonomous systems encompass a spectrum of technologies, including artificial intelligence, sensors, robotics, and data analytics, that empower researchers, conservationists, and policymakers with a new set of tools to protect and preserve wildlife and their habitats. These technologies not only facilitate real-time monitoring and data collection but also enable proactive measures to mitigate conflicts, enhance animal well-being, and improve overall conservation strategies. In the face of an accelerating environmental crisis, the role of autonomous systems in animal conservation is pivotal, as they offer a ray of hope for redefining humanity's relationship with the natural world.

This paper aims to comprehensively examine the various applications of autonomous systems in the realm of animal conservation, spanning a wide spectrum of endeavours. This research endeavours to offer a comprehensive view of how autonomous systems can shape the future of animal conservation by providing innovative solutions to mitigate conflict, prevent harm, and promote coexistence between humans and the animal kingdom. Despite remarkable advancements in computer vision, the precise detection and monitoring of wild animals in varying environmental conditions remain a significant challenge. This challenge becomes particularly pronounced in expansive and often remote habitats, where manual surveillance is limited in feasibility and efficiency. The demand for accurate and real-time identification of species, especially endangered ones, prompts the exploration of advanced technologies to address this critical conservation need.

The objective of this research study is to investigate the potential applications of YOLO v7., an emerging animal detection architecture popular for its precision and rapidity, in the field of wild animal conservation (Li et al., 2023). Complementing this exploration, the study aims to harness the advantages offered by Stochastic Gradient Descent (SGD) optimisation techniques to fine-tune the YOLOv7 model for heightened performance in wildlife detection. By amalgamating the robustness of YOLOv7 with the strategic optimisation capabilities of SGD, this research endeavours to revolutionise the landscape of autonomous systems deployed in wildlife conservation. Through a comprehensive investigation into the adaptation and optimisation of YOLOv7 with SGD for wild animal detection, this study seeks to usher in a new era of efficient and precise monitoring methods. The subsequent sections delve deeper into the methodologies employed, the experimental results attained, and the implications of this research on wildlife conservation practices. Proceeding from this section in the rest of the sections contain as follows. Section 2 contains the literature review, which includes an extensive examination of existing research, methods, advances, and findings. Section 3 is about methodology; that part includes a discussion of the Model Architecture of YOLOv7 and the performance of SGD. Results and Performance Analysis will be seen in Section 4. There will be a discussion of Quantitative Evaluation Metrics, Comparison with Baselines and Previous Models

II. LITERATURE REVIEW

The literature surrounding object detection in wildlife conservation and ecological studies showcases a spectrum of methodologies and their efficacy in addressing the challenges of monitoring wildlife populations. Numerous research works have investigated the use of computer vision techniques such as convolutional neural networks (CNNs), for population estimation, behavioural evaluation, and identifying different species. Notably, You Only Look Once (YOLO) has evolved. Models, such as YOLO's V4 and v5, has significantly improved object detection accuracy and speed, demonstrating promise in diverse domains, including wildlife monitoring.

The utilisation of Deep Neural Networks (DNN) for object recognition was first introduced in the Pascal Visual Object Classes (VOC) challenge (Sermanet et al., 2014). Later, the ImageNet Large Scale Visual Recognition Challenge (ILSVRC) emerged as the primary standard for evaluating object detection using Convolutional Neural Networks (CNNs) (LeCun et al., 2015). (Krizhevsky et al., 2012) devised a Convolutional Neural Network (CNN) to generate a bounding box around an object. However, its performance is suboptimal when dealing with photos containing many objects. (Girshick et al., 2014) integrated masks at the pixel level for individual object instances alongside a bounding box. The method is referred to as Mask R-CNN (He et al., 2017). These enhancements are substantial and can be implemented for the detection of animal species. The authors utilised convolutional neural networks (CNNs) to generate region recommendations and named their approach the R-CNN detector, specifically referring to regions containing CNN properties. Fast R-CNN (Girshick et al., 2015) was offered as a solution to decrease the computational complexity of CNN and enhance both the speed and precision of object detection, building upon the achievements of region proposal approaches. (Ren et al., 2016) combined the region proposal network (RPN) and Fast R-CNN into a single network known as Faster R-CNN.

However, within the scope of wildlife conservation, challenges persist in achieving high-precision object detection across varying environmental conditions and species diversity. The adaptability of object detection models to complex habitats, varying lighting conditions, and the identification of multiple species within a single frame remains a research focal point. Additionally, the optimization of training strategies and the integration of advanced optimization techniques tailored explicitly for object detection in wildlife conservation contexts are areas demanding further exploration.

A deep learning model which is YOLO known for its accuracy, and speed in detecting wild animals. YOLO is based on a neural network design designed for accurate real-time object detection. and image segmentation (Redmon et al., 2016) YOLOv1, which is unified and real-time object detection, has been proposed and released.; (Redmon et al., 2017) released YOLOv2, or YOLO9000, which is stronger, faster, and better; and (Redmon et al., 2018) released YOLOv3 (an incremental improvement). PyTorch was used to implement YOLOv3. However, in the shortest possible amount of duration, YOLOv4, which has optimal accuracy in detecting objects and speed (Bochkovskiy et al., 2020) and YOLO's v5 (Jocher et al., 2022) were released, respectively. YOLOv4 obtained 65 FPS on the Tesla V100 and 43.5% average precision on the COCO data set after being deployed on the Darknet. YOLOv5 comes with a CSP as the backbone and PA-NET as the neck, working like version YOLOv4. YOLOv5 brings two main advancements: automatic learning of bounding box anchors and augmentation of mosaic data. (Li et al., 2022) developed and published YOLO v6, a framework for one-stage detection of objects intended for applications in industry, in 2022. The authors claim that YOLOv6 achieves the best accuracy and speed trade-off. YOLOv7 (Wang et al., 2023) In terms of both speed and precision, version E6 results are better than transformer-based detectors like SWINL Cascade Mask R-CNN. Furthermore, YOLO v7 outperformed Scaled-YOLO v4, YOLO v5, Vit-Adapter-B, PP-YOLO, YOLO X, YOLO R, DETR, Deformable DETR, and DINO-5 scale-R50. YOLO algorithms have been applied to agricultural tasks in several works, including (Hatton-Jones et al., 2021) (Schütz et al., 2021) (Jintasuttisak et al., 2022) (Siriani et al., 2022).

This review emphasises the need for new strategies that combine sophisticated detection models—like YOLO v7—with efficient training techniques to increase detection precision, resilience, and generalizability. By leveraging the strengths of YOLOv7 and incorporating advanced optimisation techniques using detector methods, this research aims to close the disparity between present techniques and the need for successful wildlife conservation.

III. PROPOSED METHODOLOGY

The choice of YOLOv7 as the base architecture stems from its inherent efficiency in object detection and its ability to balance speed and accuracy. The YOLOv7 is renowned for its exceptional performance in a range of tasks related to object detection. Extensive comprehension of YOLOv7's architecture, comprising its backbone network, feature extraction mechanisms, and detection head structures, serves as the foundation for adaptation.

Modifications are strategically introduced to the YOLOv7 architecture to optimise it specifically for wildlife detection scenarios. Attention mechanisms, such as spatial and channel-wise attention modules, are integrated within the backbone network to enhance the model's focus on crucial wildlife features while reducing noise from the background. Feature fusion techniques, including feature pyramid networks (FPN) (Li et al., 2019) or spatial pyramid pooling (SPP) (Huang et al., 2020), are implemented to capture multi-scale information vital for detecting wildlife in various sizes and poses across diverse environments.

The anchor box configurations within YOLOv7, responsible for predicting bounding boxes, are adapted to align with the diverse aspect ratios and scales of wildlife species. Fine-tuning anchor box priors ensures better localisation and recognition of animals with varying shapes and proportions. Furthermore, output layers are tailored to accommodate the detection of a comprehensive range of wildlife species, augmenting the model's capacity to discern multiple classes of animals. The YOLOv7 adaptation process involves the integration of domain-specific knowledge and features relevant to wildlife detection. For better results, the model's capacity to distinguish between similar-looking wildlife and precisely recognise particular wildlife categories involves adding texture patterns, colour gradients, and distinctive features inherent to different wildlife.

Figure 1. Wild animal detection using YOLOv7 and SGD optimizer

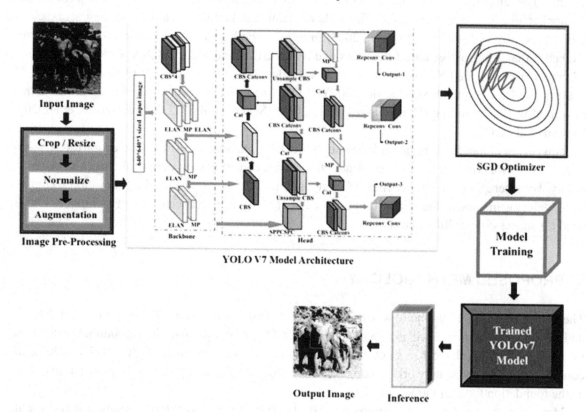

Figure 2. Flow diagram YOLO v7 with SGD

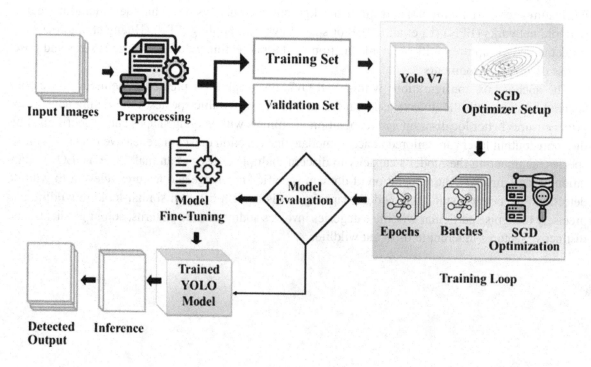

3.1

3.1.1 Utilisation of Stochastic Gradient Descent (SGD) With Momentum

Stochastic Gradient Descent (SGD) serves as the cornerstone optimisation algorithm for training the adapted YOLOv7 model. The utilisation of SGD with momentum introduces an adaptive learning approach, enabling the model to navigate the high-dimensional parameter space efficiently. This technique helps alleviate the issues of local minima by integrating momentum to accelerate convergence and smoothen the optimisation trajectory, facilitating quicker convergence towards optimal solutions.

3.1.2 Learning Rate Schedules and Adaptive Strategies

Sophisticated learning rate schedules, such as step decay, exponential decay, or cyclic learning rates, are employed to dynamically adjust the learning rates during training. These schedules facilitate adaptive learning by modulating the learning rates based on the model's performance or the number of training iterations. Specifically, warm-up techniques involving gradual learning rate increments at the initial training stages aid in stabilising the optimisation process, preventing abrupt changes that might hinder convergence.

3.1.3 Regularization Techniques for Model Generalization

To mitigate overfitting and enhance the model's generalisation capabilities, regularisation techniques are integrated into the optimisation process. Weight decay, a form of L2 regularisation, is applied to impose penalty terms on the model's weights during optimisation, preventing excessively large weight magnitudes and promoting simpler model solutions. Additionally, dropout layers are strategically introduced to randomly deactivate certain neurons during training, encouraging the network to learn more robust and generalised features.

3.1.4 Hyperparameter Tuning and Optimization Strategies

Fine-tuning of hyperparameters, including batch sizes, momentum coefficients, weight decay rates, and dropout probabilities, is conducted through systematic experimentation and grid search methodologies. Optimisation strategies are refined based on empirical observations from validation metrics, ensuring optimal convergence behaviour and preventing underfitting or overfitting tendencies.

3.2 Training Procedure

3.2.1 Initialisation and Pretrained Weights

The training procedure commences with initialising the adapted YOLOv7 model's weights, either randomly or through transfer learning from pre-trained weights on a large-scale dataset. Transfer learning leverages the knowledge acquired from training on general object recognition tasks, aiding the model in learning domain-specific features for wildlife detection.

3.2.2 Mini-Batch Stochastic Optimization

The training process employs mini-batch stochastic optimisation, where batches of annotated wildlife images are fed into the model iteratively. This stochastic approach helps in achieving faster convergence and facilitates efficient utilisation of computational resources. Each mini batch undergoes forward propagation to generate predictions and subsequent backward propagation to compute gradients for updating the model's weights.

3.2.3 Backpropagation and Weight Updates

Backpropagation, augmented by the chain rule of calculus, computes the gradients of the loss function with respect to the model's parameters. These gradients are then utilised to update the model's weights iteratively through optimisation algorithms, such as SGD with momentum. Adaptive learning rates and regularisation techniques guide the weight updates, ensuring gradual convergence towards an optimal solution while preventing overfitting.

3.2.4 Epoch-Based Training

The training process is organised into multiple epochs, where each epoch represents a complete iteration over the entire training dataset. At the end of each epoch, the model's performance metrics on the validation set are evaluated to monitor convergence and prevent overfitting. The training continues until convergence criteria are met or until the model exhibits stable performance on the validation set.

3.2.5 Fine-Tuning and Iterative Refinement

Fine-tuning iterations are conducted based on observed performance on the validation set. Adjustments in hyperparameters, optimisation strategies, or architectural modifications are implemented iteratively to further refine the model's performance. This iterative process aims to enhance the model's accuracy, robustness, and generalisation abilities.

IV. EXPERIMENTAL FRAMEWORK

4.1 System Setup

4.1.1 Hardware and Software Configuration

The experiments are conducted using a high-performance computing infrastructure equipped with GPU accelerators to expedite model training. GPU specifications, including type, memory, and compute capability, are documented. The training environment comprises deep learning frameworks such as TensorFlow or PyTorch, along with associated libraries and dependencies for efficient model development and training.

4.1.2 Hyperparameter Tuning

Systematic experimentation is conducted to fine-tune hyperparameters critical for the model's performance. Parameters such as batch sizes, learning rates, momentum coefficients, weight decay rates, dropout probabilities, and optimisation configurations are adjusted through grid search or random search methodologies. Cross-validation techniques may be employed on the validation set to identify optimal hyperparameter combinations that yield superior model performance.

4.1.3 Training Duration and Computational Resources

The duration of model training and the computational resources utilised are documented. Training duration per epoch, total training epochs, and computational resources, including GPU utilisation and memory consumption, are recorded. This information facilitates the reproducibility and scalability of the experimental setup.

4.2 Training and Fine-Tuning

4.2.1 Model Initialization and Pretraining

The training process commences with initialising the YOLOv7 model's weights, leveraging either random initialisation or transfer learning from pre-trained weights on a large-scale dataset. Transfer learning allows the model to leverage previously learned features from general object recognition tasks, aiding in the extraction of domain-specific features crucial for wildlife detection.

4.2.2 Iterative Mini-Batch Stochastic Optimization

The adapted YOLOv7 model undergoes iterative mini-batch stochastic optimisation, where batches of annotated wildlife images are fed into the model for training. During each iteration, the model computes predictions, computes gradients through backpropagation, and updates its weights using optimisation algorithms like SGD with momentum. This stochastic optimisation facilitates faster convergence and efficient utilisation of computational resources.

4.2.3 Validation-Based Fine-Tuning

The training process incorporates periodic evaluations of the validation set to monitor the model's convergence and prevent overfitting. Based on the observed performance metrics on the validation set, fine-tuning iterations are conducted. Adjustments in hyperparameters, optimisation strategies, or architectural modifications are iteratively implemented to further enhance the model's accuracy and generalisation ability.

4.2.4 Regularisation and Optimization Strategies

To prevent overfitting and enhance generalisation, regularisation techniques such as weight decay and dropout layers are employed. Weight decay imposes penalty terms on the model's weights during optimisation, preventing excessively large weight magnitudes. Dropout layers randomly deactivate certain neurons during training, encouraging the network to learn more robust features.

4.2.5 Model Complexity and Convergence Analysis

The model's complexity and convergence behaviour is monitored throughout the training process. Analysis of training curves, including loss plots and accuracy trends, provides insights into the model's learning dynamics. Additionally, convergence criteria, such as stability in validation metrics or convergence of loss functions, guide the termination of training to prevent overfitting.

4.3 Dataset Description

The presented study utilises two open-access datasets: the Wild Animals Computer Vision Project and the Animal Image Dataset. The Wild Animals Computer Vision Project dataset is a carefully selected set of images intended for the aim of training and evaluating wild animal detection and classification algorithms. The dataset comprises a wide variety of wildlife species that were captured in their natural environments. Every individual image in the dataset has been extensively annotated with metadata, which comprises species labels, geographic coordinates, and supplementary contextual details. Bear, Cheetah, Elephant, Fox, Giraffe, Jaguar, Leopard, Lion, Tiger, Zebra, and many others are among its many classes. The Animal Image Dataset, the second dataset, comprises 5400 images of animals organised into 90 classes of distinct classifications.

4.4 Performance Metrics

4.4.1 Quantitative Evaluation Metrics

The performance of the trained YOLOv7 model is quantitatively assessed using a suite of evaluation metrics, including Mean Average Precision (mAP), precision, recall, Intersection over Union (IoU), accuracy, and F1 score. These metrics provide a comprehensive evaluation of the model's accuracy, localisation precision, and generalisation across diverse wildlife species and environmental conditions.

4.4.2 Comparison With Baselines and Previous Models

The performance of the adapted YOLOv7 model is benchmarked against baseline models and previous iterations of YOLO architectures, such as YOLOv4 or YOLOv5, on the same dataset. Comparative analyses highlight the advancements achieved in wildlife detection accuracy and robustness, emphasising the superiority of the adapted YOLOv7 model.

4.4.3 Qualitative Visualizations and Error Analysis

Visualisations of detection outputs, including bounding boxes overlaying predicted wildlife instances on test images, provide qualitative insights into the model's performance. Error analysis elucidates the model's strengths and weaknesses, identifying common failure cases, misclassifications, or localisation errors. This qualitative assessment aids in understanding the model's behaviour and guiding potential improvements.

4.4.3 Statistical Significance and Confidence Intervals

Statistical analyses, such as significance testing or confidence interval estimation, may be performed to validate the observed performance improvements. This ensures the robustness of the obtained results and confirms the statistical significance of performance differences between the adapted YOLOv7 model and baseline methods.

V. RESULTS AND DISCUSSION

A range of evaluation metrics, such as precision, recall, mAP, F1 score, IoU, and accuracy, are used to measure the effectiveness of the trained YOLOv7 model. These metrics offer a thorough assessment of the model's accuracy, precise localisation, and generalisation across different kinds of wildlife and conditions in the environment.

The enhanced YOLOv7 model's performance is compared to the baseline models and earlier YOLO architecture iterations, such as YOLOv4 or YOLOv5, on the same dataset. Comparative analyses highlight the advancements achieved in wildlife detection accuracy and robustness, emphasizing the superiority of the adapted YOLOv7 model. Bounding boxes superimposed over estimated wildlife cases on testing images have an instance of visualizations of detection outputs that offer qualitative perspectives into the performance of the model. Error analysis elucidates the model's strengths and weaknesses, identifying common failure cases, misclassifications, or localization errors. This qualitative assessment aids in understanding the model's behaviour and guiding potential improvements.

To validate the observed performance improvements, statistical analyses such as significance testing or confidence interval estimation may be performed. This ensures the robustness of the obtained results and confirms the statistical significance of performance differences between the adapted YOLOv7 model and baseline methods. The results are critically analysed, highlighting the strengths, limitations, and implications of the developed YOLOv7-based wildlife detection model.

The model's adaptability, accuracy, scalability to different environments, and possibilities for use in practical applications conservation situations are all thoroughly discussed. The obtained results are comprehensively analysed, highlighting the model's strengths in accurately detecting wildlife across diverse habitats. The model's robustness in handling varying lighting conditions, occlusions, and species diversity is discussed based on the achieved performance metrics and visual analyses of detection outputs. A comparative analysis is conducted, juxtaposing the developed YOLOv7-based model's performance with existing methodologies and previous versions of YOLO models in wildlife conservation. Insights are drawn regarding the significance of integrating YOLOv7 with SGD optimization techniques in advancing object detection capabilities for conservation purpose.

The practical implications of the research findings in real-world wildlife conservation scenarios are discussed. Potential applications of the developed object detection system in monitoring endangered species, preserving ecosystems, and informing conservation strategies are outlined. Moreover, avenues for future research, including model refinement, dataset expansion, and exploration of additional op-

Figure 3. Input images along with detected images

Table 1. Quantitative metrics

S.No	Metrics	Animal Classes (AC)					
		AC 1	AC 2	AC 3	AC 4	AC 5	AC 6
1.	mAP	0.87	0.81	0.92	0.89	0.84	0.90
2.	Precision	0.80	0.85	0.89	0.91	0.88	0.86
3.	Recall	0.84	0.78	0.91	0.87	0.83	0.92
4.	IoU	0.79	0.83	0.88	0.92	0.86	0.90
5.	F1 Score	0.82	0.80	0.90	0.88	0.85	0.91

timisation techniques, are proposed to further enhance object detection accuracy and applicability in wildlife conservation efforts.

Figure 4. (a) mAP comparison among models, (b) IoU comparison among models, (c) Precision comparison among models, (d) F1 Score comparison among models

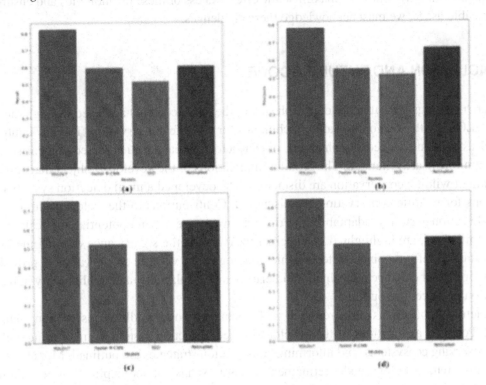

Table 2. Result analysis of the wild animals computer vision project dataset

S.NO	Algorithm	Precision (%)	Recall (%)	F1 Score (%)	mAP (%)	Inference Speed (fps)
1	YOLOv7 (SGD)	94.1	91.5	92.8	90.3	30
2	YOLOv7	90.5	87.2	88.8	86.5	25
3	YOLOv5	91.8	88.7	90.2	87.9	20
4	Faster R-CNN	86.5	84.7	86.4	83.9	12
5	SSD	86.7	82.5	84.5	82.0	18
6	R-CNN	85.9	81.6	83.7	81.2	10

Table 3. Comparing the proposed method with others using the animal image dataset

S.NO	Algorithm	Precision (%)	Recall (%)	F1 Score (%)	mAP (%)	Inference Speed (fps)
1	YOLOv7 (SGD	92.1	89.5	90.8	88.3	30
2	YOLOv7	90.5	87.2	88.8	86.5	25
3	YOLOv5	91.8	88.7	90.2	87.9	20
4	Faster R-CNN	88.2	85.3	87.4	83.9	12
5	SSD	86.7	82.5	84.5	82.0	18
6	R-CNN	85.9	81.6	83.7	81.2	10

While the use of autonomous systems in animal conservation holds great promise, their implementation is fraught with difficulties. To maximise the effectiveness of these technologies and ensure ethical and responsible usage, we must acknowledge these challenges.

VI. CONCLUSION AND FUTURE SCOPE

The research findings are summarised, emphasising the advancements achieved through the integration of YOLOv7 with SGD optimisation techniques for wildlife conservation. The key contributions and novel aspects introduced in enhancing object detection capabilities for ecological studies and wildlife monitoring are reiterated. The broader implications of the developed object detection system in the realm of wildlife conservation are discussed. The developed animal detection system's broader implications for wildlife conservation are discussed. Contributions to the field, including advancements in detection accuracy, adaptability to diverse environments, and potential applications in conservation initiatives, are highlighted. A succinct overview of the significance of continued research and innovation in refining object detection methodologies for wildlife conservation is provided. The prospects for further enhancements in model performance, scalability, and applicability in real-world conservation practices are outlined.

The practical implications of the research findings in real-world wildlife conservation scenarios are discussed. Potential applications of the developed object detection system in monitoring endangered species, preserving ecosystems, and informing conservation strategies are outlined. Moreover, avenues for future research, including model refinement, dataset expansion, and exploration of additional optimization techniques, are proposed to further enhance object detection accuracy and applicability in wildlife conservation efforts.

REFERENCES

Girshick, R. (2015). Fast r-cnn. In *Proceedings of the IEEE international conference on computer vision* (pp. 1440-1448). Academic Press.

Girshick, R., Donahue, J., Darrell, T., & Malik, J. (2014). Rich feature hierarchies for accurate object detection and semantic segmentation. In *Proceedings of the IEEE conference on computer vision and pattern recognition* (pp. 580-587). 10.1109/CVPR.2014.81

Hatton-Jones, K. M., Christie, C., Griffith, T. A., Smith, A. G., Naghipour, S., Robertson, K., Russell, J. S., Peart, J. N., Headrick, J. P., Cox, A. J., & du Toit, E. F. (2021). A YOLO based software for automated detection and analysis of rodent behaviour in the open field arena. *Computers in Biology and Medicine*, *134*, 104474. doi:10.1016/j.compbiomed.2021.104474 PMID:34058512

He, K., Gkioxari, G., Dollár, P., & Girshick, R. (2017). Mask r-cnn. In *Proceedings of the IEEE international conference on computer vision* (pp. 2961-2969). Academic Press.

Huang, Z., Wang, J., Fu, X., Yu, T., Guo, Y., & Wang, R. (2020). DC-SPP-YOLO: Dense connection and spatial pyramid pooling based YOLO for object detection. *Information Sciences*, *522*, 241–258. doi:10.1016/j.ins.2020.02.067

Jintasuttisak, T., Leonce, A., Sher Shah, M., Khafaga, T., Simkins, G., & Edirisinghe, E. (2022, March). Deep learning based animal detection and tracking in drone video footage. In *Proceedings of the 8th International Conference on Computing and Artificial Intelligence* (pp. 425-431). 10.1145/3532213.3532280

Krizhevsky, A., Sutskever, I., & Hinton, G. E. (2012). Imagenet classification with deep convolutional neural networks. *Advances in Neural Information Processing Systems*, 25.

Kundu, K., Vishwakarma, V., Rai, A., Srivastava, M., & Mishra, A. (2023, April). Design and Deployment of Wild Animal Intrusion Detection & Repellent System Employing IOT. In *2023 International Conference on Computational Intelligence and Sustainable Engineering Solutions (CISES)* (pp. 763-767). IEEE. 10.1109/CISES58720.2023.10183532

LeCun, Y., Bengio, Y., & Hinton, G. (2015). Deep learning. *Nature, 521*(7553), 436-444.

Li, S., Zhang, H., & Xu, F. (2023). Intelligent Detection Method for Wildlife Based on Deep Learning. *Sensors (Basel), 23*(24), 9669. doi:10.3390/s23249669 PMID:38139515

Li, X., Lai, T., Wang, S., Chen, Q., Yang, C., Chen, R., ... Zheng, F. (2019, December). Weighted feature pyramid networks for object detection. In *2019 IEEE Intl Conf on Parallel & Distributed Processing with Applications, Big Data & Cloud Computing, Sustainable Computing & Communications, Social Computing & Networking (ISPA/BDCloud/SocialCom/SustainCom)* (pp. 1500-1504). IEEE.

Mammeri, A., Zhou, D., & Boukerche, A. (2016). Animal-vehicle collision mitigation system for automated vehicles. *IEEE Transactions on Systems, Man, and Cybernetics. Systems, 46*(9), 1287–1299. doi:10.1109/TSMC.2015.2497235

Redmon, J., Divvala, S., Girshick, R., & Farhadi, A. (2016). You only look once: Unified, real-time object detection. In *Proceedings of the IEEE conference on computer vision and pattern recognition* (pp. 779-788). 10.1109/CVPR.2016.91

Redmon, J., & Farhadi, A. (2017). YOLO9000: better, faster, stronger. In *Proceedings of the IEEE conference on computer vision and pattern recognition* (pp. 7263-7271).

Redmon, J., & Farhadi, A. (2018). Yolov3: An incremental improvement. *arXiv preprint arXiv:1804.02767*.

Ren, S., He, K., Girshick, R., & Sun, J. (2016). Faster R-CNN: Towards real-time object detection with region proposal networks. *IEEE Transactions on Pattern Analysis and Machine Intelligence, 39*(6), 1137–1149. doi:10.1109/TPAMI.2016.2577031 PMID:27295650

Schütz, A. K., Schöler, V., Krause, E. T., Fischer, M., Müller, T., Freuling, C. M., Conraths, F. J., Stanke, M., Homeier-Bachmann, T., & Lentz, H. H. (2021). Application of YOLOv4 for detection and Motion monitoring of red Foxes. *Animals (Basel), 11*(6), 1723. doi:10.3390/ani11061723 PMID:34207726

Sermanet, P., Eigen, D., Zhang, X., Mathieu, M., Fergus, R., & LeCun, Y. (2013). Overfeat: Integrated recognition, localization and detection using convolutional networks. arXiv preprint arXiv:1312.6229.

Siriani, A. L. R., Kodaira, V., Mehdizadeh, S. A., de Alencar Nääs, I., de Moura, D. J., & Pereira, D. F. (2022). Detection and tracking of chickens in low-light images using YOLO network and Kalman filter. *Neural Computing & Applications*, *34*(24), 21987–21997. doi:10.1007/s00521-022-07664-w

Wang, C. Y., Bochkovskiy, A., & Liao, H. Y. M. (2023). YOLOv7: Trainable bag-of-freebies sets new state-of-the-art for real-time object detectors. In *Proceedings of the IEEE/CVF Conference on Computer Vision and Pattern Recognition* (pp. 7464-7475). 10.1109/CVPR52729.2023.00721

Chapter 6
Evaluating Customer Satisfaction and Trust in Autonomous AI Banking Systems

Aruna Kasinathan
ⓘ https://orcid.org/0000-0002-5800-0340
Karunya Institute of Technology and Science, India

Shrilatha Sampath
ⓘ https://orcid.org/0000-0002-5951-4175
Christian Medical College, India

Hemalatha Sampath
University of Maryland, USA

ABSTRACT

The study aims to assess customer satisfaction and trust in autonomous artificial intelligence (AAI) systems within the banking sector. Its primary objectives include exploring factors contributing to customer trust in AAI, investigating preferences for AI-driven features in banking, and determining the impact of AAI on perceived service quality. The research, adopting a descriptive design, employs both qualitative and quantitative methods. A survey, distributed to customers of leading banks in India, particularly in Tamil Nadu, with a sample size of 213, utilizes simple random and convenient sampling. Results highlight customer preferences for customized services, financial advice, and automation in banking. The implementation of AAI is perceived positively, especially in terms of transparency in processes like loans, account management, and more. Practical implications include helping banks understand customer expectations, identify weaknesses in AAI features, and enhance service quality in Tamil Nadu.

DOI: 10.4018/979-8-3693-5767-5.ch006

INTRODUCTION

In the context of the aforementioned key players in the Indian banking sector, the research article establishes a direct connection between the adoption of artificial intelligence (AI) and the stated objectives. Suparna Biswas, Brant Carson, Violet Chung, Shwaitang Singh, and Renny Thomas (2023). Firstly, the implementation of AI applications, such as smart chat assistants, chatbots, and robotics, by banks like the SBI, HDFC, ICICI, Axis, Bank of Baroda, Andhra Bank, and Kotak Mahindra Bank, is intricately linked to the objective of building customer trust in autonomous AI systems. The study delves into how these technologies enhance customer experiences, thereby fostering trust through efficient and personalized services. Additionally, the research investigates customer preferences concerning AI-driven features, aligning with the objective of understanding what elements contribute to customer trust and satisfaction in the context of autonomous AI systems in banking. . Larson (2021) and Baesens et al. (2005) Furthermore, the article analyzes the impact of autonomous AI on the perceived quality of banking services, shedding light on how these technological advancements contribute to efficiency gains, cost reduction, and improved customer service, thereby influencing the overall service quality perception. In exploring the implications on the workforce, the study emphasizes the pivotal roles of new professional categories, such as AI specialists, data scientists, and machine learning engineers, who are instrumental in designing and maintaining these AI-driven systems. The research also addresses the evolving roles of traditional banking professionals, like customer service representatives and risk analysts, showcasing how they collaborate with AI tools to enhance efficiency and personalization. This detailed exploration provides a comprehensive understanding of the interplay between AI technology adoption, workforce dynamics, and the overarching objectives of building customer trust, understanding preferences, and assessing service quality in the Indian banking sector.

Objectives

1. Explore the factors that contribute to building customer trust in autonomous AI systems.
   2. Investigate customer preferences regarding AI-driven features in banking.
   3. Determine the impact of autonomous AI on the perceived quality of banking services.

RESEARCH METHODOLOGY

This research aims to assess customer satisfaction and trust in Autonomous AI Banking Systems across India through an online survey questionnaire, encompassing a sample population of 213 respondents through Simple Random sampling Technique. The research begins with an introduction highlighting the growing significance of Autonomous AI Banking Systems and articulating specific research objectives and questions. The literature review and theoretical framework provide a comprehensive background, grounding the study in relevant theories and existing knowledge. The research design details the sampling strategy, sample size justification, and the use of the online survey instrument, while also emphasizing variable measurement. Data analysis techniques, including ANOVA, T-test, and frequency distribution, are outlined to provide a robust statistical approach. The questionnaire design section covers the structure, types of questions, and pilot testing process. Ethical considerations address privacy, confidentiality, and informed consent. The data processing and validation section outlines steps to ensure accuracy. A realistic timeline is provided, and potential limitations are discussed. The conclusion summarizes the methodology's appropriateness for achieving the research objectives, and references are cited for all sources utilized.

LITERATURE REVIEW

Arif et al. (2023) employed a neural network methodology to examine the obstacles hindering customer adoption of internet banking. Their study aimed to identify and understand the barriers associated with this adoption, leveraging neural network techniques for a comprehensive analysis. Suparna Biswas, Brant Carson, Violet Chung, Shwaitang Singh, and Renny Thomas (2023) The statement emphasizes that the deployment of AI technologies, when carefully managed to mitigate risks, can result in increased automation and improvements in decision-making compared to human processes, particularly in terms of speed and accuracy. The assertion of AI's potential to unlock significant incremental value, estimated at $1 trillion annually for banks, underscores the transformative impact and value creation potential across industries through the integration of artificial intelligence. Belanche et al. (2022) conducted a study focusing on factors influencing the adoption of AI-driven technology within the banking sector. Their research aimed to identify and analyze the key determinants shaping the integration of artificial intelligence in banking operations.

Payne et al. (2020) explored the factors influencing the adoption of AI-enabled mobile banking services, while also highlighting the opportunity for bank marketers to leverage AI in enhancing customer segmentation, targeting, and overall positioning of banking products and services. Within the sub-theme of AI and marketing, the authors identified nine papers that collectively addressed various aspects of utilizing AI for marketing activities, such as customer segmentation, model development, and the execution of more impactful marketing campaigns. Smeureanu et al. (2013) introduced a machine learning technique for the segmentation of banking customers, employing algorithms designed to categorize customers based on relevant features. Their approach aimed to enhance the understanding of diverse customer profiles in the banking sector through the application of machine learning methodologies. Schwartz et al. (2017) employed an AI-based approach to scrutinize resource allocation in targeted advertisements, likely using artificial intelligence algorithms to optimize advertising strategies. Their research reflects a growing trend in exploring the impact of AI on shaping customer experiences, highlighting the evolving landscape of how artificial intelligence technologies are influencing and enhancing interactions between businesses and customers.

Soltani et al. (2019) and Trivedi (2019) contributed to the sub-theme of AI and customer experience, focusing on the utilization of artificial intelligence to improve banking services and overall customer experience. This theme likely encompasses research exploring how AI technologies are applied to enhance various aspects of the customer journey within the banking sector. Trivedi (2019) conducted an investigation into the use of chatbots in the banking sector and assessed their influence on customer experience. The motivation for this study may have stemmed from the suggested application of AI in predicting stock market movements and stock selection, indicating a broader exploration of AI's impact on financial services and customer interactions within the banking domain. Kim and Lee (2004) and Tseng (2003) likely contributed to the literature on artificial intelligence (AI) in the banking sector during a period when the focus was primarily on its application in credit and loan analysis. The mentioned studies suggest that, at that time, researchers were exploring how AI technologies could be employed to enhance the analysis and decision-making processes associated with credit and loans within the banking industry.

Baesens et al. (2005), Ince and Aktan (2009), Kao et al. (2012), and Khandani et al. (2010) likely contributed to the early stages of AI implementation, emphasizing the importance of developing fast and reliable AI infrastructure. This suggests that, during that period, researchers were recognizing the foundational need for robust AI frameworks to support the effective deployment of artificial intelligence in various applications, possibly including those within the banking sector. Larson (2021) and Baesens et al. (2005) likely contributed to the field of predicting loan defaults and early repayments, with Baesens et al. utilizing a neural network approach for improved accuracy in such predictions. This indicates a continued interest and application of advanced techniques, such as neural networks, in the domain of credit risk assessment within the banking sector. Ince and Aktan (2009) employed a data mining technique to analyze credit scores and concluded that the AI-driven data mining approach was more effective than traditional methods. This suggests that their research highlighted the advantages of utilizing artificial intelligence in data mining for enhanced accuracy and efficiency in assessing credit scores within the banking sector. Khandani et al. (2010) discovered that machine-learning-driven models were effective in analyzing consumer credit risk. This indicates that their research demonstrated the utility of machine learning in improving the accuracy and efficiency of credit risk analysis in the context of consumer credit within the banking sector.

Alborzi and Khanbabaei (2016) investigated the application of data mining neural network techniques for the development of a customer credit scoring model. Their research likely delved into the effectiveness and accuracy of using neural networks within the context of data mining to assess and predict customer credit scores in the banking sector. Trivedi (2019) conducted a study on chatbot satisfaction, identifying information, system, and service quality as factors with a significant positive association with overall satisfaction. This suggests that the effectiveness of chatbots in delivering satisfactory user experiences is influenced by the quality of information provided, the functionality of the system, and the overall service quality. Ekinci et al. (2014) proposed a customer lifetime value (CLV) model in the banking sector, leveraging a deep learning approach. Their research likely aimed to use advanced techniques from deep learning to identify and emphasize key indicators relevant to customer lifetime value, providing insights for more effective customer relationship management in the banking industry. Xu et al. (2020) investigated the effects of AI versus human customer service and discovered that customers are more inclined to use AI for low-complexity tasks, while a human agent is preferred for high-complexity tasks. This suggests that customer preferences vary based on the complexity of the tasks involved, with AI being favored for simpler tasks and human assistance preferred for more complex and nuanced interactions.

Khandani et al. (2010) employed machine learning techniques to construct a model aimed at predicting customers' credit risk. This indicates that their research involved leveraging advanced computational methods within the realm of machine learning to enhance the accuracy and efficiency of credit risk prediction in the banking sector. Koutanaei et al. (2015) proposed a data mining model with the goal of enhancing confidence in credit scoring systems. From an organizational risk standpoint, their research likely focused on improving the reliability and accuracy of credit scoring models through advanced data mining techniques, ultimately contributing to more robust risk management practices within the financial domain. Mall (2018) employed a neural network approach to analyze the behavior of defaulting customers, aiming to minimize credit risk and enhance profitability for credit-providing institutions. This suggests that the study focused on leveraging neural network techniques to better understand and predict customer behavior related to credit default, thereby aiding financial institutions in risk mitigation and financial decision-making.

PILOT STUDY REPORT

All the (55) variables and statements in the questionnaire for 213 sample size is validated through the Cronbach's Alpha of 0.888 represents of 88% existence of reliability and validity.

RESPONDENT PREFERENCES

The respondent preferences in this study were gender, age, qualification, annual income, and nature of the employment. All these are mere representation used to describe the sample size of the population.

A total of 213 samples were collected from all over Tamil Nadu Bank customers through Google Forms. Among the above sample 54% are male respondents and 46% are female respondents. Nearly 35.2% of the respondents falls under the age group of 18 to 25 years old. This represents that the respondents of this study possess very basic level of understanding and experience towards the autonomous artificial intelligence system of banking.

Moreover, 33.8% of the respondents have a Post Graduate Degree in various domain. About 37.6% of them were Professional in this survey such as Doctors, Engineers, Lawyers, Chartered Accountants and so on. They were also earning with an annual income of less than Rs.2,50,000. The annual income base was taken on the basis of income tax slab rates. These demographic profile of the banking customers helps to understand their satisfaction level towards autonomous artificial intelligence.

Out of 213 samples 37.6% of respondents (80) holds account in the HDFC bank, 23.5% (50) holds in ICICI bank, 19.7% (42) respondents hold account in SBI, 6.1% (13) respondents hold account in Punjab National Bank and finally 9.4% (20) of the respondents holds in other banks like IDFC, Tamil Nadu

Table 1. Case processing summary

		N	%
Cases	Valid	212	99.5
	Excluded[a]	1	.5
	Total	213	100.0

Source: Computed Value
a. Listwise deletion based on all variables in the procedure.

Table 2. Reliability statistics

Cronbach's Alpha	N of Items
.888	55

Source: Computed Value

Table 3. Respondent preferences

Demographics	Categories	Frequency	Percent
Gender	Male	115	54.0
	Female	98	46.0
Age	18 – 25 years	75	35.2
	26 - 35 years	53	24.9
	36 - 45 years	53	24.9
	Above 46 years	32	15.0
Qualification	10 & 12th Std	37	17.4
	Under Graduate	52	24.4
	Post Graduate	72	33.8
	Professional Degree	41	19.2
	Others	11	5.2
Occupation	Students	43	20.2
	Employee	50	23.5
	Professionals	80	37.6
	Business	15	7.0
	Housewife/ Retired and Others	25	11.7
Annual Income	Below 2.5 lakhs	88	41.3
	2.5 to 3 lakhs	43	20.2
	3 to 5 lakhs	38	17.8
	5 to 7.5 lakhs	44	20.7

Table 4. Bank names

Banks	Frequency	Percent
HDFC	80	37.6
ICICI	50	23.5
SBI	42	19.7
Kotak Mahindra	13	6.1
PNB	8	3.8
Other Category	20	9.4
Total	213	100.0

Mercantile Bank, Axis Bank, Indian Overseas Bank, Indian Bank, and so on. The above banks were taken according to the existing market share held in India. According to the survey conducted by Forbes India on October 16th, 2023- HDFC tops with 11.61%; ICICI as second with 6.65%; SBI as third with 5.13%; Kotak Mahindra as fourth with 3.47% and PNB as eighth with score of 0.828%.

FACTORS AND PREFERENCE OF CUSTOMERS TOWARDS AUTONOMOUS ARTIFICIAL INTELLIGENCE (AAI)

The below factors such as customized service, financial advice, risk management, cyber security and detecting the fraud, integrates chatbots, credit rating information, customer experience, market trends, moves towards automations and voice recognition are contributing to build customers trust in autonomous AI in banking sector.

CUSTOMER-RELATED FACTORS

Table 5 presents the variables related to the AAI providing to the customers focused service as an individual/ company. Bank serves according to their requirements and needs through the AAI applications without the dependency.

Table 5. Factors and preference of customers towards Autonomous Artificial Intelligence (AAI)

S. No	Customized Service (CS)	Financial Advice (FA)	Risk Management (RM)	Cyber Security and Detect the Fraud (CFD)	Integrates ChatBot with banking apps (IC)
1	Personalized Product Recommendations	Personalized Financial Analysis	Alerts the customers regarding currency fluctuations	AI-Powered Threat Detection	24/7 Availability through ChatBot Integration
2	Customized Financial Advice	Timely and Informed Recommendations	Detailed evaluation of loan application of the customers	Behavioral Analytics for Fraud Prevention	Personalized Financial Product Suggestions
3	Targeted Offers and Promotions	Dynamic Adaptation to Changing Circumstances	Regulatory Compliance and Monitoring	Real-time Transaction Monitoring	Enhanced Convenience and Customer Support
4	Individualized Credit Scoring	Strengthening Customer Loyalty	Predictive Risk Assessment (in investments)	Identity Verification & Authentication	Real-time Card Security Updates
5	Tailored Customer Experiences	Optimizing Financial Well-Being	Continuous Portfolio Monitoring	Advanced Cybersecurity Measures	

Table 6. Factors and preference of customers towards Autonomous Artificial Intelligence (AAI)

S. No	Credit Rating Information (CR)	Customer Experience (CE)	Market Trends (MT)	Moves towards Automation (AI)	Voice Recognition (VR)
1	AI-Powered Credit Scoring	More creativity and innovation (ATM/ Chatbot)	Indicates the sale opportunities	Reduce the time of loan application process	Protected from unauthorized access
2	Real-time credit Assessment of individual/ clients	Elimination of manual transactions	Gives warning towards the risk in investment	Reducing the workload of the Bank employees	Transactions are done in more secured way
3	Improved Loan Decision-making	Enhances accuracy and transparency in the transactions	Evaluate the sentiments of market	Processing time of transactions are fast	Increased customer satisfaction
4		Offers and products are informed on time		Documentation are reduced	Authentication for customers through their voice
5		Errors are reduced and increases customer experience		Reduce the time of loan application process	Future of banking service

Table 7. Customers related factors

Factors	One-Sample Test					
	Test Value = 0					
	T	Df	Sig. (2-tailed)	Mean Difference	95% Confidence Interval of the Difference	
					Lower	Upper
CS1	68.042	212	.000	3.9859	3.870	4.101
CS2	59.645	212	.000	4.1925	4.054	4.331
CS3	53.776	212	.000	3.8779	3.736	4.020
CS4	64.015	212	.000	3.9718	3.850	4.094
CS5	66.148	212	.000	4.3850	4.254	4.516
FA1	34.903	212	.000	3.4648	3.269	3.660
FA2	55.679	212	.000	4.0939	3.949	4.239
FA3	56.551	212	.000	4.0704	3.929	4.212
FA4	56.573	212	.000	4.2958	4.146	4.445
FA5	34.144	212	.000	3.4648	3.265	3.665
CE1	41.837	212	.000	3.8169	3.637	3.997
CE2	45.963	212	.000	3.9061	3.739	4.074
CE3	42.762	212	.000	3.8592	3.681	4.037
CE4	41.604	212	.000	3.7793	3.600	3.958
CE5	55.801	212	.000	3.8028	3.668	3.937

The variables related to customized service, financial advice and customer service are statistically significant at 5 percent level. It is indicated that in the first category of customized services provided by the banks through AAI to the customers tops with personalised product recommendations variable has scored highest of 68.04. Personalised product indicates the needs transaction history, mini statement, balance enquiry and so on. Second category is financial advice, strengthening the customer loyalty has secured first with highest score of 56.573 as strongly agreed by the customers. Banks assist the customers regarding the loan and credit facilities at right in their hands (apps) by means of AAI which in turn increase the customer loyalty and satisfaction. The third category of customer experience, the highest score of 55.801 has been given to the variable errors are reduced and increase the customer experience. The operational as well as informational errors could be reduced on behalf of the bank by implementing AI applications. In this case when errors are reduced by solving through the customer queries then they start trusting more the Banks quality of service and experience will enhance. State Bank of India is the country's largest supplier of financial services to the public sector (SBI). The bank offers effective financial services by utilizing artificial intelligence. The artificial intelligence (AI)-driven SBI Intelligent Assistant (SIA) is a chatbot that helps users with everyday banking tasks and promptly responds to inquiries. According to insiders, this clever chatbot—developed by AI banking platform Payjo—can handle up to 10,000 requests per second, or 864 million queries every day, or over 25% of all queries that Google processes on a daily basis (analyticsinsight.net).

RISK-RELATED FACTORS

The below factors are related towards the risk management (RM) and credit rating (CR) information offered through AI by the banks. The customers can seek the information about the credit assessment and assessing the risk in investing the banks by means of AI.

The above variables of risk management (RM) and credit rating of AI (CR) preferred by the customers possess significant value at 5 percent level. It specifies that the customers strongly agree towards Predictive Risk Assessment (in investments) in risk management factor with the high score of 72.00. Hence, customers prefer risk management in AI applications offered by banks where it gives the clear analysis of expected future risk involved in the operations and investments through smartphone by banking apps. Followed by improved loan decision making as secured with 42.500 in credit rating factor of AI by the customers. This ensures that the AI application of the banks helps sanction and approve the loan within few minutes by verifying the clients/ customers credit status.

PERFORMANCE-RELATED FACTORS

The below factors denote the variables of cyber security and detection of frauds (CDF) and integrating the chatbot with banking apps (IC). These two factors are much related with performing the banking operations efficiently and securely. With help of AI into detection of fraud and chatbot customers can perform the transactions as well as guidance will be provided through the apps.

All the above variables are representing the cyber security and detecting the frauds (CDF) and integrating the chatbot with banking apps (IC) are statistically significant at 5 percent. The highest preference and agreed by the customers was real-time transactions monitoring (37.647) through the AI implementation in the banks. In case of any frauds the history of transactions enables AI to detect the fraud easily and take necessary actions immediately. Followed by enhanced convenience and customer support (55.462) scores first in integrating the chatbot with banking apps. Chatbot will solve the custom-

Table 8. Risk related factors

Factors	One-Sample Test					
	Test Value = 0					
					95% Confidence Interval of the Difference	
	T	Df	Sig. (2-tailed)	Mean Difference	Lower	Upper
RM1	36.174	212	.000	3.7183	3.516	3.921
RM2	36.961	212	.000	3.7418	3.542	3.941
RM3	37.761	212	.000	3.8075	3.609	4.006
RM4	72.000	212	.000	4.1925	4.078	4.307
RM5	37.567	212	.000	3.7324	3.537	3.928
CR1	37.804	212	.000	3.3052	3.133	3.478
CR2	42.238	212	.000	3.8263	3.648	4.005
CR3	42.500	212	.000	3.5869	3.420	3.753

Table 9. Performance related factors

Factors	One-Sample Test					
	Test Value = 0					
	T	Df	Sig. (2-tailed)	Mean Difference	95% Confidence Interval of the Difference	
					Lower	Upper
CDF1	37.494	212	.000	3.9014	3.696	4.107
CDF2	37.085	212	.000	3.6197	3.427	3.812
CDF3	37.647	212	.000	3.9061	3.702	4.111
CDF4	37.038	212	.000	3.7183	3.520	3.916
CDF5	35.917	212	.000	3.8216	3.612	4.031
IC1	36.160	212	.000	3.5915	3.396	3.787
IC2	41.445	212	.000	3.8075	3.626	3.989
IC3	55.462	212	.000	4.0094	3.867	4.152
IC4	53.573	212	.000	4.1408	3.988	4.293

ers' queries and provide guidance faster as it is strongly agreed by the customers. Hence, it increases the customer service and support. HDFC Bank has introduced Chatbot system in the banking service as an initial attempt for AI applications. Another Indian company that uses AI is HDFC, a banking and financial services corporation headquartered in Mumbai. The bank's intelligent chatbot, called "Eva," answers consumer inquiries and improves services by utilizing Google Assistant on millions of Android smartphones. Eva was created by Senseforth AI Research, a Bengaluru-based company that claims it can answer over five million client queries with over 85% accuracy. Furthermore, HDFC offers OnChat, an AI-powered chatbot that first appeared on Facebook Messenger in 2016.

In addition to HDFC, Bank of Baroda has also introduced Chatbot system as an AI application in the Banking Service. AI is being used by Bank of Baroda, another public sector lender, to enhance customer service, increase banking services, and reduce account management expenses. The bank uses cutting edge technology, like the AI robot Baroda Brainy and the free Wi-Fi offered by Digital Lab. Its chatbot is sometimes called ADI (Assisted Digital Interaction). Bank of Baroda collaborated with IBM and Accenture in 2018 to create a state-of-the-art IT Center of Excellence and Analytics Center of Excellence (analyticsinsight.net).

TECHNOLOGICAL-RELATED FACTORS

The below variables are related to the market trends (MT), moves towards automation (A) and voice recognition (VR). These factors are indicating the latest and updated technology implemented through AI in the banks. Thus, customers can analyse the trends of the market to invest in the various ventures without much time and money. Through the automation of AI the banking system moves towards digitalisation of banking operations and voice recognition ensures that respective customer alone avail the services appropriately.

Table 10. Technological related factors

Factors	One-Sample Test					
	Test Value = 0					
	T	Df	Sig. (2-tailed)	Mean Difference	95% Confidence Interval of the Difference	
					Lower	Upper
MT1	45.050	212	.000	4.0000	3.825	4.175
MT2	52.110	212	.000	4.0282	3.876	4.181
MT3	37.494	212	.000	3.9014	3.696	4.107
A1	101.014	211	.000	4.4198	4.334	4.506
A2	111.245	212	.000	4.3146	4.238	4.391
A3	111.067	212	.000	4.3192	4.243	4.396
A4	62.184	212	.000	4.1972	4.064	4.330
VR1	28.477	212	.000	2.8169	2.622	3.012
VR2	98.909	212	.000	4.3146	4.229	4.401
VR3	45.141	212	.000	3.8826	3.713	4.052
VR4	98.317	212	.000	4.3474	4.260	4.435
VR5	33.400	212	.000	3.2394	3.048	3.431

The variables of Market Trends, moves towards automation and Voice recognition values are significant at 5 percent. In market trends (MT) the variable that has secured highest scoring of 52.110 is give warning towards the risk involved in the investment. The customers strongly agreed that AI application guides them when and where to invest like shares/ stocks/ mutual funds and the amount of risk involved. Thus, AAI warns to the customers through their apps. In case of moving towards automation (A) the variable that has secured 111. 245 is reduction in workload of the bank staff. Hence, customers strongly agree that implementing AI will reduce the work pressure of bank employees by moving towards digitalisation (e-kyc through Aadhaar number and loan process through online). Finally, voice recognition (VR) is the latest technology to recognise and confirm the customers' identity while contacting the customer care or support. Customers strongly agreed by preferring the variable of transactions are done in more secured way with a score of 98. 909. The process of transactions are done only when the voice is properly recognised. In Voice Recognition (VR) Kotak Mahindra has announced its new VR. The multilingual chatbot, called Keya, will supplement the present Interactive Voice Response (IVR) technology and is linked to Kotak's phone-banking hotline. In 2019, the bank released Keya 2.0, an enhanced voicebot. India boasts a highly developed and efficiently run financial sector. As per the study, public sector banks concluded the 2020 fiscal year with assets of US$1.52 trillion. Furthermore, between FY16 and FY20, bank loans increased at a CAGR of 3.57%. The total credit amount extended at the end of FY2020 was $1,698.97 billion.

Axis Bank has also announced its new VR in July 2020. Axis Bank customers can use an AI-powered chatbot to discuss their banking concerns anytime and wherever they desire. Interactive Voice Response (IVR) technology that is conversational in nature, called AXAA, was introduced by India's third-largest private sector bank. AXAA is a multilingual voice bot of the future that helps users navigate the IVR and responds to their questions, often without requiring human assistance. Also, the private lender operates an innovation center named "Though Factory" to hasten the creation of cutting-edge AI technology solutions for the banking industry (analyticsinsight.net).

DISCUSSION OF THE FINDINGS

The above analysis were done on the basis of various factors relating to AI enabled Banking operations. The factors were related to customer, risk, performance, and technological of AI services provided by the Banks. After the detailed analysis it was found out that among these four factors, the dominant factor preferred and trusted by the customers' were technology factor. As among these factors AI moves towards the Automation like Processing time of transactions are fast, Reducing the workload of the Bank employees, and Reduce the time of loan application process were strongly agreed by the customers. They trust that AI can help the Bankers in reducing the processing time and workload respectively. It also further completes the banking operations and transaction time more quickly. Voice Recognition is also preferred and agreed by the customers as the transactions are done in more secured as well as protected way. Customers also think that they are more secured that fingerprint identification in voice recognition. Moreover, Voice Recognition is implemented by the Kotak Mahindra and Axis Bank successfully. AI's important and essential factor is technology as it involves the Automation involves more of digitalisation and Voice Recognition for identification and clarifying customer queries. Apart from Voice Recognition, HDFC and Bank of Baroda has successfully implemented chatbot towards implementation of AI in their Banks. Chatbot were helpful in sorting out customer issues and increasing their service. According to our analysis Integrating Chatbot has attracted by the customers towards increasing the convenience and support of the customers. Hence, from the analysis the preference of customers were Automation and Voice recognition leads towards the technological advancement of AI enabled Banking Service through more customer service and centre. The implementation of AI in different Banks proved that digitalisation is most prominent service.

IMPACT OF AAI TOWARDS QUALITY OF BANKING SERVICE

The table below clearly indicates the impact towards service quality of AAI to the customers in the Banks with three-point rating scale of High to Low. It measures the impact in terms of transparency, accuracy, safety, authentication and reduction. Among 213 sample size more than 150 respondents has high impact towards all the above five aspects of AAI features provided in the Banks. This denotes that customers are highly (172) effected towards the authenticated information provided by AAI services as per their requirements and followed by ensuring safe as well as security (169) towards their personal database in their Banks without being hacked or breached. On the other hand, customer feels that they possess low effectiveness of AAI featured services in the Bank were towards the transparency (23) of all the process like loan, investments, insurance and etc. It is also found that AAI helps in employees' workload in the Banks are completed and reported (15) among 213 respondents as least impact.

The greatest impact of AAI enabled Banking Service is providing proper and authenticated information. Followed by providing safe and secured customer details by protecting them from unwanted access. Bank staff also agree that through AI service they can deliver transparency in the banking operations to the customers (Shetty et all, 2022).

Figure 1. Impact of AAI towards quality of banking service

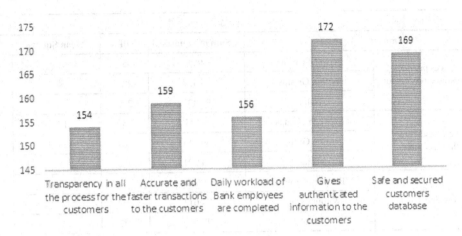

INFLUENCE OF HOLDING BANK ACCOUNT IN DIFFERENT BANKS OVER AN IMPACT OF AAI FEATURES

It compares between the different banks where the customers maintain their account and impact of AAI features in terms of transparent services, accuracy and fastest transactions, workload of the employees are completed, providing proper information and finally by safeguarding the customers' personal data through One Time Password (OTP).

The above table reveals that there is no influence between the different Banks and Impact of AAI featured services provides transparency in all the process (F=0.431; Sig = .826); workload of the employees is completed and reported daily (F=0.431; Sig = .074); ensures AAI gives authenticated information to the customers (F=1.688; Sig = .139) as it is not significantly acceptable range at 5 percent level.

There is an influence between different bank accounts maintained by the customers on more accurate and faster transactions (F=3.102; Sig = .010); and increases the safe and security for the customers database (F=2.380; Sig = .040); as the values are significantly accepted at 5 percent level. It is also observed that the HDFC, SBI and other category of banks has got different level of influence towards the impact of AAI featured services provided by the Banks.

ASSOCIATION BETWEEN WORKLOAD OF THE BANK EMPLOYEES AND AUTHENTICATED INFORMATION TO THE CUSTOMERS

The table below explains the association between the factor of AAI and Impact of AAI in the banks. It compares between the most influencing factor of reducing the work burden of the Bank employees and the highest impact towards the customers was ensuring the authentic information according to their requirements.

The table proves that $\chi2 = 4.023$; Sig = .112 are significantly not accepted at 95 percent confidence level. Thus, it is evident that there is no association between the reduction of workload among the Bank staff through AAI applications and effectiveness of authenticated information provided to the customers. It is also evident that customers agree towards reduction of work burden (digitalisation like reduction to

Table 11. Impact of AAI towards quality of banking service

ANOVA						
Impact		**Sum of Squares**	**df**	**Mean Square**	**F**	**Sig.**
AAI featured services provides transparency in all the process for the customers	Between Groups	.994	5	.199	.431	.826
	Within Groups	95.437	207	.461		
	Total	96.432	212			
It has more accurate and faster transactions to the customers	Between Groups	5.178	5	1.036	3.102	.010
	Within Groups	69.113	207	.334		
	Total	74.291	212			
Daily workload of Bank employees are completed and reported	Between Groups	3.649	5	.730	2.041	.074
	Within Groups	74.013	207	.358		
	Total	77.662	212			
Gives authenticated information to the customers	Between Groups	2.104	5	.421	1.688	.139
	Within Groups	51.623	207	.249		
	Total	53.728	212			
Enhances safe and security for the customers database	Between Groups	3.143	5	.629	2.380	.040
	Within Groups	54.669	207	.264		
	Total	57.812	212			

Table 12. Association between workload of the bank employees and authenticated information to the customers

Factors/ Impact		Gives authenticated information to the customers			Total	
		High	**Moderate**	**Least**		
Reducing the workload of the Bank employees	Strongly Disagree	1	0	0	1	
	Disagree	1	0	0	1	
	Neutral	2	0	0	2	$\chi 2 = 4.023$ Sig = .112
	Agree	113	17	5	135	
	Strongly Agree	55	16	3	74	
Total		172	33	8	213	

paperless work and online process) to the Bank employees by performance of AAI features has a high impact on providing validated and true information to the customers through Autonomous Artificial Intelligence feature services in the Banks.

WEAKNESS OF THE STUDY

The major weakness of this study were its time duration and sample size. As time is major drawback for limiting sample size to 213. Another weakness of this study identified were the technical aspect of machine learning in AI and theories of AI perspectives not included in this study.

SUGGESTIONS

The following suggestions are based on the data analysis:

1. *Transparency:* AI enabled Banking services must provide transparency in its operations for the customers. Transparency of the operations indicates clear procedure of account opening, loan process, etc. Further, customer can identify their status of operations without visiting or approaching the Bankers/ Bank through AI application in the Mobile Apps or Internet Banking.

2. *Awareness on Voice Recognition:* Create an awareness among the customers that the Voice Recognition helps to protect from unauthorized access. Unauthorised access implies that any-one other than the recognised customer through their voice cannot access the Bank accounts. Hence, they are protected from hacking their personal details. Only when recognised voice of the customer is heard then the Banking operations and queries are also can be carried out. It also confirms the identification of customers through voice (eg: fingerprint identification). In future fingerprint could be replaced by voice or face recognition.

3. *Financial Wealth:* AI can increase the financial wealth of the customers by accessing and managing the report as well as guiding efficiently. As and when customers are browsing the details or banking products, AI can suggest suitable products for the customers according to their earning capacity and interest. Apart from this AI will guide them relating to the risk of selling or investing according to market trends of increasing/ decreasing price.

4. *Chatbots:* Chatbot as an AI feature plays a major role in banking operations and customer service. Many banks have implemented Chatbot as AI service and customers have reported its successful rate. But in our study customers are not aware of the features of Chatbot. As chatbot can reply to the customer queries as well as guide them within few seconds or minutes.

SCOPE FOR FUTURE RESEARCH

Banks have already implemented AI in different aspects like chatbot, IVR. A separate study can be conducted on its effectiveness. Future study from customers and Banker's point of view which aspect of implementing AI in the Banks is essential could be done. A study on major issues and challenges of AAI regarding privacy and security can be done.

CONCLUSION

This study is conducted to analyse the various aspects of Customer Satisfaction and Trust in Autonomous AI Banking Systems. AI has been already existing in the front office, middle office and back office. Front office in the banks is dealing with the customers queries like interactions relating to their required information on websites/ apps and identifying the customers through biometric, advising the customers relating to managing the finance. Customer queries are handled through the Chatbot in AI by leading Banks in India. Financial advice are also provided by the AI to the customers through their behaviour. Middle office operations are dealing with the detection of frauds and crime in the Banking

Sector, Know Your Customer (KYC), and credit rating scores to the customers on the basis of their loan repayment, decisions regarding loan. AI helps in detecting the frauds earlier, loan process are done in digital mode (online process represents paperless) and credit rating scores are revealed in few minutes as the customers repay or settle their credit. Back office mainly dealing with managing and settling the customer transactions and records. They also regularly adjust and ensure banking records are complying with regulatory acts and IT services. AI has enabled many programs to reconcile and settle the accounting system of customers as earliest. The records can be maintained in the Bank/ Cloud data without any hacking through AI.

The above operations of the Bank were categorised into customer related factors, risk related factors, technology related factors and performance related factors. As digital transactions are expanding AI features are also expanding in the Banking sector. Thus, customers strongly agree that AI can reduce the workload of employees in the Bank like replacing the role of staff in front, middle and backend office work relating to the latest technology. Customers agree that AI provides personalised services according to the customer needs of loans, insurance, deposits and withdrawal. Risk related factors were agreed by the customers is that helps to detect risk involved in investment earlier by AI applications in Apps itself. In performance related factors customers agreed that AI improves their support and convenience by Chatbot technology as well as 24*7 service. The major impact of AAI towards service quality of Banks is giving valuable information to the customers relating to the interest rates, loans, deposits and etc. It is also concluding that there is no relationship between factors of AAI features and its impact.

REFERENCES

Alborzi, M., & Khanbabaei, M. (2016). Using data mining and neural networks techniques to propose a new hybrid customer behavior analysis and credit scoring model in banking services based on a developed RFM analysis method. *International Journal of Business Information Systems*, *23*(1), 1–22. doi:10.1504/IJBIS.2016.078020

Arif, I., Aslam, W., & Hwang, Y. (2023). Barriers in adoption of internet banking: A structural equation modeling-neural network approach. *Technology in Society*, *61*, 101231. doi:10.1016/j.techsoc.2020.101231

Baesens, B., Van Gestel, T., Stepanova, M., Van den Poel, D., & Vanthienen, J. (2005). Neural network survival analysis for personal loan data. *The Journal of the Operational Research Society*, *56*(9), 1089–1098. doi:10.1057/palgrave.jors.2601990

Belanche, D., Casaló, L. V., & Flavián, C. (2020). Artificial intelligence in FinTech: Understanding robo-advisors adoption among customers. *Industrial Management & Data Systems*, *119*(7), 1411–1430. doi:10.1108/IMDS-08-2018-0368

Biswas, S., Carson, B., Chung, V., Singh, S., & Thomas, R. (2023). Artificial intelligence technologies are increasingly integral to the world we live in, and banks need to deploy these technologies at scale to remain relevant. Success requires a holistic transformation spanning multiple layers of the organization. Larson, E.J. 2021. The myth of artificial intelligence. In *The Myth of Artificial Intelligence*. Harvard University Press.

Ekinci, Y., Uray, N., & Ülengin, F. (2014). A customer lifetime value model for the banking industry: A guide to marketing actions. *European Journal of Marketing*, *48*(3–4), 761–784. doi:10.1108/EJM-12-2011-0714

Ince, H., & Aktan, B. (2009). A comparison of data mining techniques for credit scoring in banking: A managerial perspective. *Journal of Business Economics and Management*, *10*(3), 233–240. doi:10.3846/1611-1699.2009.10.233-240

Khan, K. S., Kunz, R., Kleijnen, J., & Antes, G. (2010). Five steps to conducting a systematic review. *Journal of the Royal Society of Medicine*, *96*(3), 118–121. doi:10.1177/014107680309600304 PMID:12612111

Khandani, A. E., Kim, A. J., & Lo, A. W. (2010). Consumer credit-risk models via machine-learning algorithms. *Journal of Banking & Finance*, *34*(11), 2767–27. doi:10.1016/j.jbankfin.2010.06.001

Kim, K. J., & Lee, W. B. (2004). Stock market prediction using artificial neural networks with optimal feature transformation. *Neural Computing & Applications*, *13*(3), 255–260. doi:10.1007/s00521-004-0428-x

Koutanaei, F. N., Sajedi, H., & Khanbabaei, M. (2015). A hybrid data mining model of feature selection algorithms and ensemble learning classifiers for credit scoring. *Journal of Retailing and Consumer Services*, *27*, 11–23. doi:10.1016/j.jretconser.2015.07.003

Mall, S. (2018). An empirical study on credit risk management: The case of nonbanking financial companies. *The Journal of Credit Risk*, *14*(3), 49–66. doi:10.21314/JCR.2017.239

Payne, E. M., Peltier, J. W., & Barger, V. A. (2018). Mobile banking and AI-enabled mobile banking: The differential effects of technological and non-technological factors on digital natives' perceptions and behavior. *Journal of Research in Interactive Marketing*, *12*(3), 328–346. doi:10.1108/JRIM-07-2018-0087

Schwartz, E. M., Bradlow, E. T., & Fader, P. S. (2017). Customer acquisition via display advertising using multi-armed bandit experiments. *Marketing Science*, *36*(4), 500–522. doi:10.1287/mksc.2016.1023

Shetty. (2022). Impact of Artificial Intelligence in Banking Sector with Reference to Private Banks in India. *Annals of the University of Craiova, Physics*, *32*, 59-75.

Smeureanu, I., Ruxanda, G., & Badea, L. M. (2013). Customer segmentation in private banking sector using machine learning techniques. *Journal of Business Economics and Management*, *14*(5), 923–939. doi:10.3846/16111699.2012.749807

Soltani, M., Samorani, M., & Kolfal, B. (2019). Appointment scheduling with multiple providers and stochastic service times. *European Journal of Operational Research*, *277*(2), 667–683. doi:10.1016/j.ejor.2019.02.051

Trivedi, J. (2019). Examining the customer experience of using banking Chatbots and its impact on brand love: The moderating role of perceived risk. *Journal of Internet Commerce*, *18*(1), 91–111. doi:10.1080/15332861.2019.1567188

Xu, Y., Shieh, C. H., van Esch, P., & Ling, I. L. (2020). AI customer service: Task complexity, problem-solving ability, and usage intention. *Australasian Marketing Journal*, *28*(4), 189–199. doi:10.1016/j.ausmj.2020.03.005

Chapter 7
Foetal Activity Detection Using Deep Convolution Neural Networks

Xavier Arockiaraj Santhappan
https://orcid.org/0000-0002-6420-7364
Adhiyamaan College of Engineering, India

Ronica Bis
Sathyabama Institute of Science and Technology, India

ABSTRACT

Ultrasound is a conventional diagnostic instrument employed in prenatal care to track the progression and advancement of the fetus. In routine clinical obstetric assessments, the standard planes of fetal ultrasound hold considerable importance in evaluating fetal growth metrics and identifying abnormalities. In this work, a method to detect FFSP using deep convolutional neural network (DCNN) architecture to improve detection efficiency is presented. Squeeze net, 16 convolutional layers with small 3x3 large kernel, and all three layers form the proposed DCNN. The final pooling layer uses global average pooling (GAP) to reduce inconsistency in the network. This helps reduce the problem of overfitting and improves the performance from different training data. To improve cognitive performance, data augmentation methods developed specifically for FFSP are used in conjunction with adaptive learning strategies. Extensive testing shows that the proposed method gives accuracy of 96% which outperforms traditional methods, and DCNN is an important tool to identify FFSP in clinical diagnosis.

1. INTRODUCTION

Ultrasound (US) screening is a widely used, cost-effective, and radiation-free method for pregnancy diagnosis in routine clinical examinations (Lei et al., 2015a; Lei et al., 2015b; Rahmatullah et al., 2011; Chen et al., 2015a; Chen et al., 2015b; Lei et al., 2014; Zhang et al., 2012; Rahmatullah & Noble, 2013). Typically performed between 18 and 24 weeks of gestation, antenatal US screening involves acquiring

DOI: 10.4018/979-8-3693-5767-5.ch007

serial standard images of fetal structures following a standardized protocol for biometric measurement and malformation detection (Dudley & Chapman, 2002). One crucial aspect is obtaining standard planes, such as the fetal facial standard plane (FFSP), for accurate fetal diagnosis and subsequent measurements (Lei et al., 2015a; Rahmatullah et al., 2011; Chen et al., 2015a; Lei et al., 2014; Zhang et al., 2012). However, manual identification and evaluation of FFSP by clinicians are time-consuming, subjective, and prone to variations among different practitioners, particularly in underprivileged regions lacking experienced clinicians. Hence, there is a significant need for an automatic FFSP recognition method.

Recognizing FFSP presents challenges due to high intra-class and low inter-class variations caused by fetal postures, scanning orientations, and artifacts like speckle noise and shadow (Lei et al., 2015a). Distinguishing FFSP from non-FFSP is particularly challenging. To address this, various methods have been proposed, typically employing a two-step pipeline: feature extraction and classification (Chatfield et al., 2011; Shi et al., 2015a; Shi et al., 2016a; Shi et al., 2016b). Traditional approaches rely on low-level hand-crafted features (Lei et al., 2015a; Zhang et al., 2012; Dudley & Chapman, 2002; Ni et al., 2014; Lei et al., 2015c), such as scale-invariant feature transform (SIFT), Dense-SIFT (DSIFT), Haar, and histogram-of-gradient (HOG), encoded by algorithms like bag of visual words (BoVW), vector of locally aggregated descriptor (VLAD), Fisher vector (FV), and multi-layer Fisher vector (MFV), followed by classification using support vector machines (SVM). However, these hand-crafted features often fall short for accurate FFSP recognition.

In recent years, deep convolutional neural networks (DCNNs) have demonstrated remarkable success in image recognition tasks, benefiting from large annotated datasets like ImageNet and their ability to automatically learn feature representations from raw data without manual design (Deng, 2009; Krizhevsky et al., 2012; Lin et al., 2013; He et al., 2016). DCNNs, with alternating convolutional and pooling layers, outperform shallow networks, producing robust and sophisticated representations (He et al., 2015). While various studies have applied deep models like VGG Net, batch normalization (BN) Net, and ResNet in challenging benchmark datasets, their application in medical US image datasets remains limited due to data scarcity and convergence issues (Ioffe & Szegedy, 2015; Simonyan & Zisserman, 2014; Szegedy et al., 2015). Despite the significant achievements of DCNNs in medical image recognition, their application in US images is underexplored, and systematic investigations into their performance are lacking.

Inspired by the success of DCNNs, we propose a DCNN squeeze net model (19 layers) with small-sized kernels (3×3) for FFSP recognition. Additionally, a global average pooling (GAP) in the last pooling layer is incorporated to enhance performance and efficiency.

2. LITERATURE REVIEW

Traditional hand-crafted feature-based classification models typically involve three main steps: (i) feature extraction (Lei et al., 2014; Zhu et al., 2017), (ii) feature encoding (Chatfield et al., 2011; Maji et al., 2008; Zhu et al., 2016; Zhu et al., 2014), and (iii) feature classification (Grauman & Darrell, 2005). Spatial pyramid matching (SPM) (Lazebnik et al., 2006; Grauman & Darrell, 2005) can be employed to integrate spatial information of image features for enhanced feature representation. In conventional models, carefully designed data-specific features play a crucial role. Over the past few years, considerable effort has been dedicated to devising suitable features for image representation (Ronneberger et al., 2015; Lowe, 2004). Commonly used feature representations include SIFT, DSIFT, Haar, HoG, and combinations of these features with intensity, shape, motion, and edges (Ronneberger et al., 2015; Zhu et al., 2017). Feature encoding techniques like BoVW, VLAD, FV, and MFV (Perronnin et al., 2010; Csurka et al., 2004; Sánchez et al., 2013; Sánchez et al., 2010) have been introduced to improve classification performance, generating more robust and stable information. For instance, Lei et al. (2015a) proposed a hand-crafted DSIFT feature representation method based on an MFV feature encoding approach for FFSP recognition.

DCNNs possess remarkable representation capabilities for recognition or detection tasks based on the provided training dataset (LeCun et al., 2015). These models comprise multiple processing layers to learn features at different levels, and combining these hierarchical features preserves highly discriminative and effective deep representations (Glorot & Bengio, 2010). Consequently, state-of-the-art performance has been achieved in various applications (Krizhevsky et al., 2012; Girshick et al., 2014; Long et al., 2015; Ronneberger et al., 2015; Chen et al., 2016).

The deep hierarchical architecture of DCNN models is crucial due to their potent representation learning capabilities. Recent years have seen the proposal of well-designed initialization strategies, activation functions (He et al., 2015; He et al., 2015), and efficient intermediate regularization strategies (Srivastava et al., 2014; Goodfellow et al., 2013), significantly improving the optimization of deep models (Bengio, 2012). With remarkable performance in natural image processing and natural language processing, DCNNs have asserted their dominance in the machine learning domain and found extensive applications in medical image analysis (Anthimopoulos et al., 2016; Gao et al., 2016; Yan et al., 2016; Song et al., 2015; Yu et al., 2016). Numerous studies have reported promising results across various applications, including object recognition (Anthimopoulos et al., 2016; Gao et al., 2016; Yan et al., 2016; Song et al., 2015), detection (Dou et al., 2016), and segmentation (Chen et al., 2018; Ronneberger et al., 2015; Chen et al., 2017).

3. METHODOLOGY

The proposed model mainly consists of two steps and the framework is illustrated in Figure 1. The deep CNNs are trained using the fetal US images training dataset, and the most informative features are extracted. After that, MLP is used to classify fetal ultrasound images into three classes.

3.1 Feature Extraction

Extracting relevant deep features is a crucial step in any classification process. This study aims to improve classification performance by extracting deep features extracted from deep CNN, Squeeze net-GAP. Figure 1 illustrates the block diagram of this proposed feature extraction approach. Typically, deep CNN designs consist of multiple hidden layers. It is possible to increase classification performance by using these hidden layers, which are helpful in determining which aspects of the input image data are the most informative. The Squeeze net architecture is given as

3.2 Multi-Layer Perceptron

An MLP is a class of fully connected feedforward artificial neural networks (LeCun et al., 2015). Generally, MLP consists of an input layer, one or more hidden layers, and an output layer. The proposed MLP consists input layer, three fully connected layers, and an output SoftMax layer. Every node or

Figure 1. Squeeze net architecture

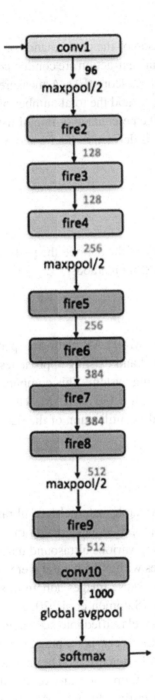

neuron of the hidden layers except the input layer uses a nonlinear activation function. The deep feature integrated descriptor is treated as input to the input layer of MLP. Firstly, the input layer output is input to the hidden layer. In the same way, the output of the last hidden layer is fed into the output layer. The first and second hidden layers of MLP are designed with 1024 neurons each, while the last hidden layer is designed with 512 neurons.

3.3 Evaluation

A quantitative evaluation and comparison of the performance of the proposed automatic classification of fetal US images using MLP and feature extraction are conducted based on four major metrics, namely accuracy, precision (P), recall (R), and F1-score (F1). A measure of accuracy is the ratio between the number of correctly predicted class labels and the total number of ground truth class labels. Generally, precision (recall) can be described as the percentage or rate of true positives (true negatives). The harmonic mean of the precision and recall is defined as the F1-score. Furthermore, confusion matrices are also evaluated.

4. EXPERIMENTAL RESULTS

This section describes the characteristics of the dataset, the performance of the proposed approach, and a comparison of the results with the existing methods.

4.1 Experimental Set-Up

The suggested model was applied on the MATLAB platform, specifically in version R2023a. All computations pertaining to both the proposed and existing approaches were conducted on a computer running the 64-bit Windows 10 Pro operating system. This computer is powered by an Intel(R) Xeon(R) E-2104G CPU running at 3.20 GHz, with 8 GB of RAM. The implementation of the proposed and existing methods in MATLAB involved the utilization of the deep learning toolbox, image processing toolbox, and machine learning toolbox.

4.2 Dataset Details

The proposed study utilizes the largest available maternal-fetal ultrasound image dataset (Burgos-Artizzu et al., 2020) to train and evaluate both proposed and existing methods. This dataset comprises images gathered from two distinct hospitals using various ultrasound machines and operators (Burgos-Artizzu et al., 2020). The fetal ultrasound images within the dataset were captured during routine clinical practices from October 2018 to April 2019. These images adhere to standard clinical screening protocols established by the scientific committee (Salomon et al., 2022), and a skilled maternal-fetal clinician manually annotated them. The images are classified into six categories, encompassing four fetal planes (abdomen, brain, femur, thorax), the mother's cervix for prematurity screening, and less commonly used planes like kidney, leg, foot, spine, etc. Additionally, fetal brain images are further categorized into three major planes: trans-thalamic, trans-cerebellum, and trans-ventricular. The dataset is randomly divided, allocating 75% for training and 25% for testing the networks. All images in the dataset are in portable network graphic (PNG) format and grayscale, with varying sizes. During the training of deep Convolutional Neural Networks (CNNs), the sizes of the training image dataset B and testing image dataset S are optimized to match the corresponding input size of the deep CNN. All images are resized to $227 \times 227 \times 3$ for the SqueezeNet model.

4.3 Performance of the Proposed Method

This section highlights the significance of integrating deep features and assesses the performance of the proposed approach based on class-wise precision, recall, F1-score, and overall accuracy. The classification effectiveness of the SqueezeNet is evaluated in the context of the proposed work. In this study, two classification techniques—MLP and softmax classifiers—are employed to classify common maternal-fetal ultrasound images by incorporating deep features. Tables 1 present the class-wise precision, recall, F1-score, and overall accuracy of the proposed method using both the softmax classifier and MLP. The proposed method, specifically the Deep Feature Integration (DFI) with SqueezeNet's deep feature representations using MLP, attains an accuracy of 96%, as depicted in Table 1 and the corresponding confusion matrix is shown in Figure 2.

4.4 Comparison of the Proposed Method With the Existing Methods

Figure 2. Confusion matrix

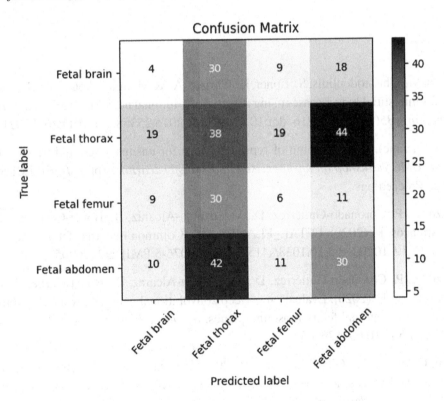

5. CONCLUSION

This study presents a method for automatically classifying fetal ultrasound (US) images using Multilayer Perceptron (MLP) with feature extraction to enhance detection and diagnostic efficiency. The training and testing of deep Convolutional Neural Networks (CNNs) utilized a publicly available extensive dataset

Table 1. Comparison of existing vs proposed work

Method	Fetal abdomen (%)			Fetal brain (%)			Fetal femur (%)			Fetal thorax (%)			Maternal cervix (%)			Other (%)			Accuracy (%)
	P	R	F1	P	R	F1	P	R	F1	P	R	F1	P	R	F1	P	R	F1	
ResNet-50 [66]	85.7	91	88.3	98.9	99.6	99.2	84	90.8	87.3	92.9	93.9	93.2	98.8	99.5	99.1	94.2	90.1	92.1	94.3
Inception-v3 [67]	85.6	90.4	87.9	98.9	99.7	99.3	82.5	87.3	84.8	92.6	93.2	92.9	99.0	99.8	99.4	93.1	89.8	91.4	93.8
Proposed work	91.1	92.3	93.3	98.9	99.5	99.9	90	91	91	95	93.7	94.1	98.6	98.3	99.1	95.1	92.1	93.5	95.9

known as common fetal US images. To enhance the classification process, a combination of various deep features extracted from the SqueezeNet-GAP has been employed. These extracted deep features from the SqueezeNet-GAP descriptor are input to the MLP to categorize fetal US images into three classes: Benign, Malignant, and Normal. The proposed method achieves an impressive accuracy of 96%, surpassing the performance of other state-of-the-art methods presented in (He et al., 2016; Szegedy et al., 2016). As a result, the proposed framework can be considered reliable to assist doctors in accurate fetal US image detection during screening, contributing to early detection (Burgos-Artizzu et al., 2020).

REFERENCES

Anthimopoulos, M., Christodoulidis, S., Ebner, L., Christe, A., & Mougiakakou, S. (2016). Lung pattern classification for interstitial lung diseases using a deep convolutional neural network. *IEEE Transactions on Medical Imaging*, *35*(5), 1207–1216. doi:10.1109/TMI.2016.2535865 PMID:26955021

Bengio, Y. (2012, June). Deep learning of representations for unsupervised and transfer learning. In *Proceedings of ICML workshop on unsupervised and transfer learning* (pp. 17-36). JMLR Workshop and Conference Proceedings.

Burgos-Artizzu, X. P., Coronado-Gutierrez, D., Valenzuela-Alcaraz, B., Bonet-Carne, E., Eixarch, E., Crispi, F., & Gratacós, E. (2020). FETAL_PLANES_DB: Common maternal-fetal ultrasound images. *Scientific Reports*, *19*, 10200. doi:10.1038/s41598-020-67076-5 PMID:32576905

Burgos-Artizzu, X. P., Coronado-Gutiérrez, D., Valenzuela-Alcaraz, B., Bonet-Carne, E., Eixarch, E., Crispi, F., & Gratacós, E. (2020). Evaluation of deep convolutional neural networks for automatic classification of common maternal fetal ultrasound planes. *Scientific Reports*, *10*(1), 10200. doi:10.1038/s41598-020-67076-5 PMID:32576905

Chen, H., Dou, Q., Wang, X., Qin, J., & Heng, P. (2016, February). Mitosis detection in breast cancer histology images via deep cascaded networks. *Proceedings of the AAAI Conference on Artificial Intelligence*, *30*(1). doi:10.1609/aaai.v30i1.10140

Chen, H., Dou, Q., Yu, L., Qin, J., & Heng, P. A. (2018). VoxResNet: Deep voxelwise residual networks for brain segmentation from 3D MR images. *NeuroImage*, *170*, 446–455. doi:10.1016/j.neuroimage.2017.04.041 PMID:28445774

Chen, H., Ni, D., Qin, J., Li, S., Yang, X., Wang, T., & Heng, P. A. (2015). Standard plane localization in fetal ultrasound via domain transferred deep neural networks. *IEEE Journal of Biomedical and Health Informatics, 19*(5), 1627–1636. doi:10.1109/JBHI.2015.2425041 PMID:25910262

Chen, H., Qi, X., Yu, L., Dou, Q., Qin, J., & Heng, P. A. (2017). DCAN: Deep contour-aware networks for object instance segmentation from histology images. *Medical Image Analysis, 36*, 135–146. doi:10.1016/j.media.2016.11.004 PMID:27898306

Csurka, G., Dance, C., Fan, L., Willamowski, J., & Bray, C. (2004, May). Visual categorization with bags of keypoints. In *Workshop on statistical learning in computer vision, ECCV* (Vol. 1, No. 1-22, pp. 1-2).

Dalal, N., & Triggs, B. (2005, June). Histograms of oriented gradients for human detection. In *2005 IEEE computer society conference on computer vision and pattern recognition (CVPR'05)* (Vol. 1, pp. 886-893). IEEE. doi:10.1109/CVPR.2005.177

Deng, J. (2009). A large-scale hierarchical image database. *Proc. of IEEE Computer Vision and Pattern Recognition.*

Donahue, J., Jia, Y., Vinyals, O., Hoffman, J., Zhang, N., Tzeng, E., & Darrell, T. (2014, January). Decaf: A deep convolutional activation feature for generic visual recognition. In *International conference on machine learning* (pp. 647-655). PMLR.

Dou, Q., Chen, H., Yu, L., Zhao, L., Qin, J., Wang, D., Mok, V. C. T., Shi, L., & Heng, P. A. (2016). Automatic detection of cerebral microbleeds from MR images via 3D convolutional neural networks. *IEEE Transactions on Medical Imaging, 35*(5), 1182–1195. doi:10.1109/TMI.2016.2528129 PMID:26886975

Dudley, N. J., & Chapman, E. (2002). The importance of quality management in fetal measurement. *Ultrasound in Obstetrics & Gynecology, 19*(2), 190–196. doi:10.1046/j.0960-7692.2001.00549.x PMID:11876814

Gao, Z., Wang, L., Zhou, L., & Zhang, J. (2016). HEp-2 cell image classification with deep convolutional neural networks. *IEEE Journal of Biomedical and Health Informatics, 21*(2), 416–428. doi:10.1109/JBHI.2016.2526603 PMID:26887016

Girshick, R., Donahue, J., Darrell, T., & Malik, J. (2014). Rich feature hierarchies for accurate object detection and semantic segmentation. In *Proceedings of the IEEE conference on computer vision and pattern recognition* (pp. 580-587). 10.1109/CVPR.2014.81

Glorot, X., & Bengio, Y. (2010, March). Understanding the difficulty of training deep feedforward neural networks. In *Proceedings of the thirteenth international conference on artificial intelligence and statistics* (pp. 249-256). JMLR Workshop and Conference Proceedings.

Goodfellow, I., Warde-Farley, D., Mirza, M., Courville, A., & Bengio, Y. (2013, May). Maxout networks. In *International conference on machine learning* (pp. 1319-1327). PMLR.

Grauman, K., & Darrell, T. (2005, October). The pyramid match kernel: Discriminative classification with sets of image features. In *Tenth IEEE International Conference on Computer Vision (ICCV'05)* Volume 1 (Vol. 2, pp. 1458-1465). IEEE. 10.1109/ICCV.2005.239

Greenspan, H., Van Ginneken, B., & Summers, R. M. (2016). Guest editorial deep learning in medical imaging: Overview and future promise of an exciting new technique. *IEEE Transactions on Medical Imaging, 35*(5), 1153–1159. doi:10.1109/TMI.2016.2553401

He, K., Zhang, X., Ren, S., & Sun, J. (2015). Delving deep into rectifiers: Surpassing human-level performance on imagenet classification. In *Proceedings of the IEEE international conference on computer vision* (pp. 1026-1034). 10.1109/ICCV.2015.123

He, K., Zhang, X., Ren, S., & Sun, J. (2016). Deep residual learning for image recognition. In *Proceedings of the IEEE conference on computer vision and pattern recognition* (pp. 770-778). Academic Press.

He, K., Zhang, X., Ren, S., & Sun, J. (2016). Deep residual learning for image recognition. In *Proceedings of the IEEE Conference on Computer Vision and Pattern Recognition* (pp. 770–778). IEEE.

Ioffe, S., & Szegedy, C. (2015, June). Batch normalization: Accelerating deep network training by reducing internal covariate shift. In *International conference on machine learning* (pp. 448-456). PMLR.

Jégou, H., Perronnin, F., Douze, M., Sánchez, J., Pérez, P., & Schmid, C. (2011). Aggregating local image descriptors into compact codes. *IEEE Transactions on Pattern Analysis and Machine Intelligence, 34*(9), 1704–1716. doi:10.1109/TPAMI.2011.235 PMID:22156101

Krizhevsky, A., Sutskever, I., & Hinton, G. E. (2012). Imagenet classification with deep convolutional neural networks. *Advances in Neural Information Processing Systems, 25*.

Lazebnik, S., Schmid, C., & Ponce, J. (2006, June). Beyond bags of features: Spatial pyramid matching for recognizing natural scene categories. In *2006 IEEE computer society conference on computer vision and pattern recognition (CVPR'06)* (Vol. 2, pp. 2169-2178). IEEE.

LeCun, Y., Bengio, Y., & Hinton, G. (2015). Deep learning. *Nature, 521*(7553), 436-444.

Lei, B., Tan, E. L., Chen, S., Ni, D., & Wang, T. (2015). Saliency-driven image classification method based on histogram mining and image score. *Pattern Recognition, 48*(8), 2567–2580. doi:10.1016/j.patcog.2015.02.004

Lei, B., Tan, E. L., Chen, S., Zhuo, L., Li, S., Ni, D., & Wang, T. (2015). Automatic recognition of fetal facial standard plane in ultrasound image via fisher vector. *PLoS One, 10*(5), e0121838. doi:10.1371/journal.pone.0121838 PMID:25933215

Lei, B., Yao, Y., Chen, S., Li, S., Li, W., Ni, D., & Wang, T. (2015). Discriminative learning for automatic staging of placental maturity via multi-layer fisher vector. *Scientific Reports, 5*(1), 12818. doi:10.1038/srep12818 PMID:26228175

Lei, B., Zhuo, L., Chen, S., Li, S., Ni, D., & Wang, T. (2014, April). Automatic recognition of fetal standard plane in ultrasound image. In *2014 IEEE 11th International Symposium on Biomedical Imaging (ISBI)* (pp. 85-88). IEEE. 10.1109/ISBI.2014.6867815

Li, Q., Cai, W., Wang, X., Zhou, Y., Feng, D. D., & Chen, M. (2014, December). Medical image classification with convolutional neural network. In *2014 13th international conference on control automation robotics & vision (ICARCV)* (pp. 844-848). IEEE. 10.1109/ICARCV.2014.7064414

Lin, M., Chen, Q., & Yan, S. (2013). Network in network. *arXiv preprint arXiv:1312.4400.*

Liu, C., Yuen, J., & Torralba, A. (2010). Sift flow: Dense correspondence across scenes and its applications. *IEEE Transactions on Pattern Analysis and Machine Intelligence, 33*(5), 978–994. doi:10.1109/TPAMI.2010.147 PMID:20714019

Long, J., Shelhamer, E., & Darrell, T. (2015). Fully convolutional networks for semantic segmentation. In *Proceedings of the IEEE conference on computer vision and pattern recognition* (pp. 3431-3440). IEEE.

Lowe, D. G. (2004). Distinctive image features from scale-invariant keypoints. *International Journal of Computer Vision, 60*(2), 91–110. doi:10.1023/B:VISI.0000029664.99615.94

Maji, S., Berg, A. C., & Malik, J. (2008, June). *Classification using intersection kernel support vector machines is efficient. In 2008 IEEE conference on computer vision and pattern recognition.* IEEE.

Ni, D., Yang, X., Chen, X., Chin, C. T., Chen, S., Heng, P. A., Li, S., Qin, J., & Wang, T. (2014). Standard plane localization in ultrasound by radial component model and selective search. *Ultrasound in Medicine & Biology, 40*(11), 2728–2742. doi:10.1016/j.ultrasmedbio.2014.06.006 PMID:25220278

Perronnin, F., Sánchez, J., & Mensink, T. (2010). Improving the fisher kernel for large-scale image classification. *Computer Vision–ECCV 2010: 11th European Conference on Computer Vision, Heraklion, Crete, Greece, September 5-11, 2010 Proceedings, 11*(Part IV), 143–156.

Rahmatullah, B., & Noble, J. A. (2013, September). Anatomical object detection in fetal ultrasound: computer-expert agreements. In *International Conference on Biomedical Informatics and Technology* (pp. 207-218). Springer Berlin Heidelberg.

Rahmatullah, B., Papageorghiou, A., & Noble, J. A. (2011). Automated selection of standardized planes from ultrasound volume. In *Machine Learning in Medical Imaging: Second International Workshop, MLMI 2011, Held in Conjunction with MICCAI 2011, Toronto, Canada, September 18, 2011. Proceedings 2* (pp. 35-42). Springer Berlin Heidelberg. 10.1007/978-3-642-24319-6_5

Ronneberger, O., Fischer, P., & Brox, T. (2015). U-net: Convolutional networks for biomedical image segmentation. In *Medical image computing and computer-assisted intervention–MICCAI 2015: 18th international conference, Munich, Germany, October 5-9, 2015, proceedings, part III 18* (pp. 234-241). Springer International Publishing.

Salomon, L., Alfirevic, Z., Berghella, V., Bilardo, C., Chalouhi, G., Costa, F. D. S., ... Paladini, D. (2022). ISUOG practice guidelines (updated): Performance of the routine mid-trimester fetal ultrasound scan. *Ultrasound in Obstetrics & Gynecology, 59*(6), 840–856. doi:10.1002/uog.24888 PMID:35592929

Sánchez, J., Perronnin, F., Mensink, T., & Verbeek, J. (2013). Image classification with the fisher vector: Theory and practice. *International Journal of Computer Vision, 105*(3), 222–245. doi:10.1007/s11263-013-0636-x

Shi, J., Jiang, Q., Mao, R., Lu, M., & Wang, T. (2015). FR-KECA: Fuzzy robust kernel entropy component analysis. *Neurocomputing, 149*, 1415–1423. doi:10.1016/j.neucom.2014.08.054

Shi, J., Wu, J., Li, Y., Zhang, Q., & Ying, S. (2016). Histopathological image classification with color pattern random binary hashing-based PCANet and matrix-form classifier. *IEEE Journal of Biomedical and Health Informatics*, *21*(5), 1327–1337. doi:10.1109/JBHI.2016.2602823 PMID:27576270

Shi, J., Zhou, S., Liu, X., Zhang, Q., Lu, M., & Wang, T. (2016). Stacked deep polynomial network based representation learning for tumor classification with small ultrasound image dataset. *Neurocomputing*, *194*, 87–94. doi:10.1016/j.neucom.2016.01.074

Simonyan, K., & Zisserman, A. (2014). Very deep convolutional networks for large-scale image recognition. *arXiv preprint arXiv:1409.1556*.

Song, Y., He, L., Zhou, F., Chen, S., Ni, D., Lei, B., & Wang, T. (2016). Segmentation, splitting, and classification of overlapping bacteria in microscope images for automatic bacterial vaginosis diagnosis. *IEEE Journal of Biomedical and Health Informatics*, *21*(4), 1095–1104. doi:10.1109/JBHI.2016.2594239 PMID:27479982

Song, Y., Zhang, L., Chen, S., Ni, D., Lei, B., & Wang, T. (2015). Accurate segmentation of cervical cytoplasm and nuclei based on multiscale convolutional network and graph partitioning. *IEEE Transactions on Biomedical Engineering*, *62*(10), 2421–2433. doi:10.1109/TBME.2015.2430895 PMID:25966470

Srivastava, N., Hinton, G., Krizhevsky, A., Sutskever, I., & Salakhutdinov, R. (2014). Dropout: A simple way to prevent neural networks from overfitting. *Journal of Machine Learning Research*, *15*(1), 1929–1958.

Szegedy, C., Liu, W., Jia, Y., Sermanet, P., Reed, S., Anguelov, D., ... Rabinovich, A. (2015). Going deeper with convolutions. In *Proceedings of the IEEE conference on computer vision and pattern recognition* (pp. 1-9). IEEE.

Szegedy, C., Vanhoucke, V., Ioffe, S., Shlens, J., & Wojna, Z. (2016). Rethinking the inception architecture for computer vision. In *Proceedings of the IEEE Conference on Computer Vision and Pattern Recognition* (pp. 2818–2826). 10.1109/CVPR.2016.308

Van der Maaten, L., & Hinton, G. (2008). Visualizing data using t-SNE. *Journal of Machine Learning Research*, *9*(11).

Yan, Z., Zhan, Y., Peng, Z., Liao, S., Shinagawa, Y., Zhang, S., Metaxas, D. N., & Zhou, X. S. (2016). Multi-instance deep learning: Discover discriminative local anatomies for bodypart recognition. *IEEE Transactions on Medical Imaging*, *35*(5), 1332–1343. doi:10.1109/TMI.2016.2524985 PMID:26863652

Yaqub, M., Kelly, B., Papageorghiou, A. T., & Noble, J. A. (2015). Guided random forests for identification of key fetal anatomy and image categorization in ultrasound scans. In *Medical Image Computing and Computer-Assisted Intervention–MICCAI 2015: 18th International Conference, Munich, Germany, October 5-9, 2015, Proceedings, Part III 18* (pp. 687-694). Springer International Publishing. 10.1007/978-3-319-24574-4_82

Yosinski, J., Clune, J., Bengio, Y., & Lipson, H. (2014). How transferable are features in deep neural networks? *Advances in Neural Information Processing Systems*, *27*.

Yu, Z., Ni, D., Chen, S., Li, S., Wang, T., & Lei, B. (2016, August). Fetal facial standard plane recognition via very deep convolutional networks. In *2016 38th annual international conference of the IEEE Engineering in Medicine and Biology Society (EMBC)* (pp. 627-630). IEEE. 10.1109/EMBC.2016.7590780

Zhang, L., Chen, S., Chin, C. T., Wang, T., & Li, S. (2012). Intelligent scanning: Automated standard plane selection and biometric measurement of early gestational sac in routine ultrasound examination. *Medical Physics*, *39*(8), 5015–5027. doi:10.1118/1.4736415 PMID:22894427

Zhu, X., Li, X., & Zhang, S. (2015). Block-row sparse multiview multilabel learning for image classification. *IEEE Transactions on Cybernetics*, *46*(2), 450–461. doi:10.1109/TCYB.2015.2403356 PMID:25730838

Zhu, X., Li, X., Zhang, S., Ju, C., & Wu, X. (2016). Robust joint graph sparse coding for unsupervised spectral feature selection. *IEEE Transactions on Neural Networks and Learning Systems*, *28*(6), 1263–1275. doi:10.1109/TNNLS.2016.2521602 PMID:26955053

Zhu, X., Suk, H. I., Wang, L., Lee, S. W., & Shen, D. (2017). A novel relational regularization feature selection method for joint regression and classification in AD diagnosis. *Medical Image Analysis*, *38*, 205–214. doi:10.1016/j.media.2015.10.008 PMID:26674971

Zhu, X., Zhang, L., & Huang, Z. (2014). A sparse embedding and least variance encoding approach to hashing. *IEEE Transactions on Image Processing*, *23*(9), 3737–3750. doi:10.1109/TIP.2014.2332764 PMID:24968174

Chapter 8

Future Nano and Biorobots Miniaturized Machines for Biotechnology and Beyond

Abhishek Choubey

https://orcid.org/0000-0002-6789-8199

Sreenidhi Institute of Science and Technology, Hyderabad, India

Shruti Bhargava Choubey

Sreenidhi Institute of Science and Technology, Hyderabad, India

ABSTRACT

Biorobotics and nanobots represent a cutting-edge area of biotechnology with tremendous promise to transform scientific research, environmental monitoring, and healthcare delivery. In this chapter, the authors explore their cutting-edge ideas, applications, and advances, showing their ability to radically change future industries like industry medicine. This chapter's primary objective is to explore both existing and emerging applications of bio-robots and nanobots in healthcare, environmental monitoring, environmental inspection, and materials science research. These technological advancements offer real-time diagnostics, minimally invasive surgery, and targeted drug delivery as well as environmental quality evaluation through bio-robot water quality evaluation and nanobot pollution detection at unprecedented scales. Furthermore, bio-robots and nanobots have proven invaluable for scientific study fields like neuroscience, synthetic biology, and materials science.

1. INTRODUCTION

Nanotechnology and biotechnology have come together to open up a new era in robotics, leading to the invention of future nano- and biorobots. Nanorobots represent an exciting frontier of nanotechnology and biotechnology research, showing great promise across various fields from medicine and healthcare to environmental monitoring and industrial applications.

DOI: 10.4018/979-8-3693-5767-5.ch008

Nanorobots serve as an intermediary between these fields of study that have experienced tremendous development over time. Nanotechnology involves changing matter at the molecular and atomic levels while biotechnology employs biological principles for developing technologies and devices. Nanorobots and biorobots differ by being relatively smaller devices ranging in size between nanometres to micrometers. These characteristics allow them to operate within the tight confines of microscopic environments typically found within living organisms or complex structures. Modern robots boast sophisticated sensing, actuation, and communication features that enable them to explore their environment effectively, interact, when necessary, with those nearby, and perform duties more efficiently than humanly possible. Researchers are devoting considerable time and attention to nanorobotics and bio robotics - two emerging disciplines with vast scientific and industrial potential. Specialized field specializations of micro-robotic technology specialize in designing micro-robot systems capable of performing specific nanoscale tasks precisely while using organic components and synthetic materials to bolster robot capabilities and features. Furthermore, their devices come equipped with sensing capabilities as well as actuator and communication protocols to interact effectively with their surroundings and accomplish specified tasks effectively (Guo et al., 2023).

Nanorobotics and biorobots are two new disciplines of research that have a great deal of potential for a wide variety of scientific and technical undertakings. In these domains, the development of nanoscale robotic systems that are capable of carrying out exact tasks is the focus of concentrated expertise. Additional issues that fall under this category include the incorporation of organic components with synthetic materials to improve the functionality and attributes of robots.

The development of hybrid intelligent nanorobotics, which is a cutting-edge technology for the design of intelligent biomedical systems, would not have been conceivable. Within the realm of biomedical research, this implies that nanorobots have the potential to be utilized in a wide variety of applications that are of significant importance. There is a wide range of medical applications that make use of nanorobots. Some of these applications include the delivery of medications, the diagnosis of diseases, imaging, tissue engineering, microsurgery, and target therapy. The applications that have been shown here illustrate the revolutionary potential of nanorobots to improve the outcomes of efforts to improve patient care and biomedical research. "Biohybrid robot" is a word that describes robot systems that blend biological components and synthetic parts to achieve higher levels of performance (Peng et al., 2023).

The biohybrid when biological components, like as cells, tissues, or enzymes, are mixed with synthetic materials, and the result is the creation of something called a robot. Several sizes are offered for hybrid robots. In the case of smaller systems, enzymes are used to propel minuscule particles, whereas in the case of higher systems, millions of cells are utilized. These hybrid systems maximize the performance of robotics by making use of the abilities that are specific to each component (Li & Yu, 2023). The purpose of this area is to take advantage of the distinctive features of biological entities to achieve the highest possible levels of efficiency and performance in robotic systems.

Biohybrid robotics may provide advantages in robotics that are hard to obtain with traditional materials, including improved performance and features that would otherwise be difficult to achieve. By combining biological entities with artificial materials, biohybrid robots can exploit the unique capabilities of biological systems at both nanoscale and macroscale levels; such as drug delivery systems, self-propelling capabilities, biomimetic movements, and functions.

2. OVERVIEW OF NANOROBOT AND BIO ROBOTICS

Nanotechnology is a scientific and Engineering discipline focused on the manipulation and utilization of atoms and molecules at the nanoscale typically measuring 100 nanometers or less. it encompasses the design fabrication and application of structures devices and systems with precise control over their properties and functionalities nanotechnology works by manipulating and controlling matter at the nanoscale which involves working with individual atoms and molecules. it utilizes various techniques and approaches to engineer materials structures and devices with specific properties and functionalities working on such a small-scale nanotechnology offers unique opportunities to explore and exploit the extraordinary properties of materials.

Scientists and Engineers employ various techniques such as bottom-up and top-down approaches to manipulate and assemble nanoscale components this enables the creation of new materials with tailored characteristics and behaviors that differ from their bulk counterparts. A nanorobot also known as a nanobot is a tiny machine or device that operates on the nanoscale typically measuring in the range of nanometres (Gotovtsev, 2023). These minuscule robots are designed to perform specific tasks at the molecular or cellular level they are constructed using nanotechnology which involves manipulating and controlling matter at the atomic and molecular levels nanorobotics is an interdisciplinary field that combines the principles of Robotics nanotechnology and Material Science to develop robots at the nanoscale nanorobotics is an emerging field of science and technology that deals with the design development and control of robots at the Nanoscale. There are various applications of nanorobots.

2.1 Disease Detection and Diagnosis

Detection and diagnosis are critical aspects of healthcare and nanorobots are equipped with advanced sensing capabilities. The potential to revolutionize these areas with their ability to operate at the molecular level Nano robots offer a powerful tool for early and accurate detection of diseases including cancer nanorobots can detect specific biomarkers or abnormalities associated with various diseases. These sensors can recognize molecular signatures indicative of disease presence or progression even before symptoms manifest by analyzing samples at the molecular level. Nanorobots can provide highly sensitive and specific diagnostic information enabling early intervention and treatment for example in the case of cancer nanorobots can be programmed to detect specific tumor markers or genetic mutations that are associated with different types of cancer. They can be introduced into the body to search for these biomarkers and provide real-time information on their presence location and concentration. This information can aid in early cancer detection allowing for timely and targeted treatment interventions.

Medical Treatment nanorobots hold immense promise for revolutionizing patient care. These tiny machines have the potential to perform medical procedures with unparalleled accuracy and precision. The capabilities of human hands by leveraging their small size and maneuverability. Nanorobots can navigate through the intricate pathways of the human body reaching targeted sites with remarkable precision. This precision opens up possibilities for minimally invasive surgeries where nanorobots can access hard-to-reach areas with minimal disruption to surrounding tissues. One of the key advantages of nanorobots is their ability to deliver therapeutics directly to the site of action with their controlled and targeted drug delivery mechanisms. Nanorobots can transport medications with pinpoint accuracy to specific cells or tissues (Popescu & Ungureanu, 2023). This targeted approach allows for higher drug concentrations at the desired location resulting in more effective treatment while minimizing systemic side effects. Furthermore, nanorobots can be engineered to carry out a range of therapeutic functions for example they can be designed to selectively destroy cancer cells precisely delivering anti-cancer agents to tumors while sparing healthy tissues. This targeted approach has the potential to enhance the efficacy of cancer treatments while minimizing the debilitating side effects often associated with traditional chemotherapy while nanorobots are still in the early stages of development and face numerous challenges.

Nanorobots and biorobots are small machines engineered to perform specific tasks at the nanoscale. Nano-robots typically consist of synthetic materials while biobots consist of living cells or tissues - both types have the potential for revolutionizing disease detection by providing more precise, efficient, minimally invasive ways of diagnosing diseases and monitoring their progress.

2.1.1 Disease Detection

Nano- and biorobots offer some promising applications in disease detection, with specific biomarkers such as cancer cells or bacteria being detected by these robots and relaying this information directly to doctors or healthcare providers. To identify diseases, nanorobots can be deployed in a variety of methods, including the following:

Targeted drug delivery: Nano-robots can be equipped with pharmaceuticals and directed toward specific cells or tissues to optimize efficacy while minimizing adverse effects.

Imaging: Nanorobots, equipped with imaging sensors, can capture images of damaged cells or tissues, enabling doctors to make more precise diagnoses and monitor the effectiveness of treatment programs. This technology has the potential to enable them to deliver regular updates on the progress of treatment over some time.

Sensing: Nanorobots, which are equipped with sensors capable of detecting certain indicators of disease, can aid clinicians in the early and more precise diagnosis of diseases.

2.1.2 Disease Diagnosis

Nanorobots and biobots have the potential to be employed for illness diagnosis as well. These robots are capable of extracting cellular or tissue samples, which can subsequently be examined for indications of illness. Nanorobots can be employed in diverse manners for disease diagnosis, including:

Nano-robots have the potential to perform biopsies on tumors or other diseased tissues, aiding physicians in more precise illness diagnosis and assessment of the disease's stage.

Blood testing: Nano-robots can extract blood samples and examine them for indications of illness. This enables physicians to promptly diagnose medical issues while simultaneously monitoring the course of treatment.

Tissue analysis: Nano-robots can aid in the identification of disease indicators in tissues, facilitating clinicians in making more precise illness diagnoses and formulating an optimal treatment plan.

2.2 Manufacturing Assembly

The production and assembling of Nanorobots have a significant potential impact on manufacturing and assembly processes. These tiny machines offer unparalleled precision and control at the Nanoscale opening up new avenues for creating intricate structures and devices with enhanced properties. Nano robots can manipulate and assemble materials at the atomic and molecular levels allowing for the precise positioning and arrangement of individual building blocks. This level of control enables the fabrication of complex structures and devices that were previously unattainable using conventional manufacturing methods by harnessing the capabilities of Nano robots manufacturers can achieve higher levels of efficiency and quality in their production processes. These robots can perform tasks with exceptional Precision reducing errors and variations in product specifications. This leads to improved product consistency enhanced quality control and increased customer satisfaction. Moreover, the use of Nano robots in manufacturing can contribute to waste reduction their ability to handle and position materials at the nanoscale minimizes material waste and optimizes resource utilization by precisely depositing materials and assembling components. Nanorobots can reduce the need for excess materials and decrease production waste worker safety can also be improved through the integration of nanorobots in manufacturing processes. Energy Production holds tremendous potential for revolutionizing energy production and addressing critical challenges in this field. Nano robots can improve the efficiency and performance of solar cells by enhancing light absorption reducing reflection and optimizing charge separation and transport processes.

Nano robots can enhance the thermoelectric properties of materials boosting energy conversion efficiencies. The Environmental Cleanup application of Nanorobots and environmental cleanup presents a promising solution to the pressing challenges of pollution and contamination. These tiny machines have the potential to contribute significantly to the removal of pollutants from water soil and air. Nanorobots can facilitate the process of water purification by enhancing filtration mechanisms and improving the efficiency of water treatment systems for soil remediation. Nanorobots can assist in the removal of toxic substances such as industrial pollutants pesticides and chemical residues they can be designed to break down or encapsulate contaminants facilitating their removal from the soil in air pollution mitigation. Nano-robots can play a vital role in capturing and neutralizing airborne pollutants they can be deployed to target specific pollutants such as particulate matter and volatile organic compounds (Bhandari et al., 2023).

2.3 Material Science

Nanorobots offer exciting possibilities in the field of material science. The development of advanced materials with unique properties and functionalities and their integration can pave the way for the creation of self-healing materials smart coatings and responsive surfaces that adapt to environmental conditions or stimuli. One key application of nanorobots in material science is the development of self-healing materials nanorobots can be designed to detect and repair damage in materials at the nanoscale. They can autonomously navigate through the material matrix identify cracks or defects and initiate healing mechanisms by releasing healing agents or catalyzing chemical reactions. This self-healing capability extends the lifespan of materials reduces maintenance requirements and enhances their overall durability and reliability smart coatings enabled by Nanorobots can revolutionize surface properties and functionalities. Nanorobots can be programmed to respond to specific stimuli such as temperature light or chemical changes by incorporating them into coatings surfaces and can exhibit adaptive properties such as self-cleaning anti-fouling anti-corrosion or anti-icing functionalities. These smart coatings can have broad applications in various Industries including Automotive Aerospace and architecture (Kučuk, Primožič, Knez, & Leitgeb, 2023).

2.4 Exploration and Sensing

Exploration and sensing nanorobots have the potential to revolutionize exploration and sensing by offering into challenging and inaccessible environments. It is providing valuable data and insights that were previously unattainable whether it's deep-sea exploration or space missions. Nanorobots offer a compact and versatile solution for gathering information and navigating through extreme conditions in deep-sea exploration. Nanorobots can be deployed to explore the depths of the ocean where human access is limited and equipped with sensors and imaging capabilities. These tiny machines can delve into uncharted territory mapping. The ocean floor studying marine life and collecting data on geological formations their small size allows them to access narrow crevices and fragile ecosystems without causing significant disturbance similarly in space missions nanorobots can play a crucial role in exploring celestial bodies and conducting research in outer space. They can be utilized to navigate through challenging terrains collect samples and perform experiments in environments with extreme temperatures radiation or low gravity conditions. Nanorobots can contribute to the study of planetary bodies such as Mars or the moon aiding in the search for signs of life or resources.

3. OVERVIEW BIO ROBOTICS

Bio robotics is a multidisciplinary field that integrates biomedical engineering, cybernetics, and robotics to create innovative technologies that blend biology with mechanical systems. Its aims include enhancing communication efficiency, affecting genetic information, and developing machines that resemble biological systems.

Cybernetics is defined as the science of communications and automatic control systems in both machines and living things however while robotic and biological systems are both cybernetic systems they're separate fields of study their communication control systems and locomotion function differently at least they did traditionally the future of robotics is unfolding quickly increasingly in the field of robotics and biology (Li, Dekanovsky, Khezri, Wu, Zhou, & Sofer, 2022). There were different levels of revolution in biorobotics years ago.

3.1 Biomimicry

Biomimicry is the practice of learning and copying solutions to problems that nature has already managed to solve. it is used to develop modern technologies from aircraft wings. The nanostructure of butterfly wings to superhydrophobic coatings. The lotus leaf building mobile and adaptive autonomous is a challenge but the natural world has been doing something similar for billions of years. It's no surprise that few fields have embraced biomimicry. The way robotics has been and it remains a potent driving force for its continued innovation for mobile autonomous. It's easy to identify how robots have been directly inspired by the unique locomotion of animals such as spiders birds and fish. Biomimicry in robots goes beyond general appearance and applies to cybernetic systems. They used to operate for example applications of point control are modeled.

3.2 Soft Robotics

Soft robotics is a subfield of robotics that deals with the design, control, and fabrication of soft robots made out of compliant materials like silicone rubber or fabric instead of rigid links. Soft robots can be filled with air or fluid for shape modification or movement purposes. Soft robots offer several advantages over their traditional counterparts. They're lighter, more flexible, and safer to interact with humans - not to mention easier to customize for specific applications. Soft robots offer great promise in healthcare settings. Soft robots can be utilized for minimally invasive surgery, rehabilitation, and prosthetics; surgeons even employ soft robots for delicate procedures on both brain and heart surgeries (Jahromi et al., 2021).

3.3 Swarm Robotics

Swarm robotics is an exciting field of robotics inspired by nature. This field involves designing and controlling large groups of robots that work collectively towards one common goal; typically, these simple and cheap robots communicate and collaborate locally with one another to achieve their purpose. Swarm robotics offers numerous advantages over traditional approaches to robotics, including being more robust and adaptable to changing environments as well as being more efficient at covering large areas or manipulating complex objects. Autonomous Swarms Autonomous swarms rely on simple onboard intelligence that creates a much greater degree of emergent intelligence and functioning. This is similar to how individual unintelligent ants follow basic procedures and communicate through pheromones to build vast and complex cities.

Swarm robots typically feature basic sensors and actuators, communicating wirelessly among themselves. Their basic rules of behavior may include following a leader, avoiding obstacles, and sharing information - however, as the robots interact with one another and their environment they develop complex collective behavior patterns that emerge over time (Hu, 2021).

3.4 Nano Robotics

Nanorobotics is an emerging technological domain focused on the creation and advancement of robots operating at the nanoscale. The nanoscale refers to a range of measurements between 1 and 100 nanometers, which is approximately 100,000 times smaller than the breadth of a human hair (Primožič et al., 2021). Nanorobots are commonly energized utilizing light, electricity, or chemical interactions. They exhibit movement through diverse means, including microscopic flagella, and cilia, or by altering their body morphology. Nanorobots can establish communication among themselves and with external controllers through several means, including electromagnetic waves, chemical signals, and even quantum entanglement. The advancement in nanotechnology is shown in table 1. Nanorobots are miniature machines on the scale of billionths of meters that can be programmed to perform specific tasks, such as those related to surgery. Such tasks could include:

- Minimally Invasive Surgery (MIS): Nanorobots offer minimally invasive surgical solutions with their ability to be introduced through small incisions or even natural orifices, thus eliminating large and open wounds while speeding recovery times.
- Targeted drug delivery: Nanorobots can be loaded with drugs and programmed to deliver them directly to diseased cells, limiting damage to healthy tissues.

- Intracellular surgery: Nanorobots can be miniaturized sufficiently to operate within individual cells, potentially enabling us to repair damage caused by cell damage or modify genetic material at its source.

4. BIO-NANOTECHNOLOGY

Bionanotechnology is found across bacterial archaeal and eukaryotic cells. it generates locomotion similar to a propeller through a rotary motor an arm just 20 nanometres thick and built from proteins capabilities like the nanoscale mobility of the flagellum has spawned a new field of robotics not as inspiration or biomimicry but for directly building electromechanical components. Bio nano technologies utilize existing biological components like proteins lipids and DNA to manufacture new molecules or inorganic nanoscale components that are used in the assembly of nanorobots. This is biology directly interacting with robotics and it can work both ways since nanorobots can also be used to build biological components. This may be in its early stages but it's part of the broader field of biofactoring where living organisms like bacteria fungi and algae build inorganic materials and components used in many industrial applications including robotics. This varies from macroscopic materials such as hard or soft plastics, and earlier robots down to metallic nanoparticles (Shen et al., 2021).

Applications of Bio-Nanotechnology:

Medicine: Targeted drug delivery, regenerative medicine, tissue engineering, personalized medicine, and biosensors for diagnostics. Healthcare: Early disease detection, gene therapy treatments, and nanorobots designed for surgery provide biocompatible implants as solutions.

Environment: Bioremediation of polluted sites, water purification, environmental monitoring, biosensors to detect environmental contaminants;

Energy: biofuel production, solar energy conversion, and monitoring with biosensors.

Nanoparticles for drug delivery: Nanoparticles can be loaded with drugs and targeted towards specific cells or tissues for improved efficiency and reduced side effects of medication delivery. Examples of bio-nanotechnologies:

Table 1. The advancement in nanotechnology

S.No	Areas	Description
1	DNA origami nanorobots	Researchers have created nanorobots made of folded DNA strands - also known as DNA origami - which can be programmed with specific shapes and functions, such as carrying drugs or manipulating cells.
2	Light-powered nanorobots	Scientists have developed nanorobots powered solely by light. This eliminates the need for batteries or onboard power sources, simplifying design, reducing risks, and potentially cutting costs.
3	Magnetically controlled nanorobots	Nanorobots can be controlled using external magnetic fields, giving surgeons greater control and accuracy during surgery
4	AI-powered nanorobots:	Researchers are creating nanorobots equipped with artificial intelligence capable of learning and adapting in real-time, which could allow them to make autonomous decisions during surgery, such as detecting diseased tissue.
5	Nanorobotic swarms:	Scientists are investigating swarms of nanorobots as possible collaborators to perform complex surgical tasks that might exceed individual robot capabilities.

Nanoparticles for drug delivery: Nanoparticles can be loaded with drugs and targeted to specific cells or tissues, improving drug delivery efficiency and reducing side effects.

Biosensors for diagnostics: Bio-nanotechnologies can develop biosensors that detect specific biomarkers or pathogens with high sensitivity, enabling early diagnosis and treatment of diseases.

Biocompatible implants: Bio-nanotechnologies offer biocompatible implants that integrate seamlessly with the body, decreasing rejection rates and improving patient outcomes.

4.1 Biohybrids

Biohybrids are the interaction between biological and mechanical cybernetics. In 2017 researchers used a laser to locate a buried landline 20 meters away but the laser wasn't designed to detect landmines instead the laser was programmed to detect a special type of bacteria engineered to grow fluorescent green in the presence of a nearby explosive compound. This is a biosensor a new type of cybernetic system that's part mechanical and part biological. They use organic components like enzymes and organelles as part of a sensing and transducing circuit that can be integrated directly into a robotic system. Biohybrids aren't just about integrating biology into robotics it works both ways this is perhaps most apparent in the field of bionic prosthesis where robotics are used to substitute or augment living components from arms and legs to hearts and lungs even eyes and the brain itself. In 2015 at a medical facility in Tokyo a dog with late-stage terminal bone cancer was treated with drugs encapsulated in a basic nanocarrier and an ultrasonic activator on a robotic arm was used to trigger the release of drugs at the tumor site. This was intended to only be an initial safety test for the procedure but to the researcher's surprise the tumor shrunk by 15. The dog-to-walk-again follow-up procedures extended the dog's life by over a year if this procedure were applied to humans' researchers estimate the increase in life expectancy would be 10 years with the emergence of nanoscale robotics. These techniques can go further nanoscale robots are small enough to flow through blood vessels and can be used to deliver drugs in much lower doses to specific target sites clear dangerous plaques from arteries and perhaps even perform microsurgeries while this technology is still in the research phase with many human trials planned for the next few years. It's becoming increasingly likely that robots at the smallest scales may become a part of us and this is just the medical application of nanobots robotics at this scale may be a key step towards enabling atomically precise manufacture of the assembly of large structures atom by atom the capabilities (Li, Dekanovsky, Khezri, Wu, Zhou, & Sofer, 2022). The Overview of biohybrid nano- and microrobots is shown in Figure 1.

There is an ever-increasing variety of biohybrids being developed, each one with its specific focus.

Figure 1. Overview of biohybrid nano- and microrobots

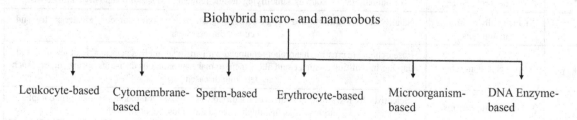

Biohybrid materials: These advanced materials combine living cells (such as stem cells or bacteria) with synthetic materials (such as polymers or hydrogels) to produce materials with enhanced properties, like wound dressings that release growth factors or biocompatible sensors that detect environmental changes.

Biohybrid robots: Biohybrid robots refer to machines that combine biological components like muscle cells or tissues for more natural and efficient movement, such as robotic prostheses that mimic human muscle function powered by sperm cells to deliver targeted drugs.

Biohybrid sensors: These sensors combine biological receptors (such as enzymes or antibodies) with synthetic materials for high-sensitivity detection of certain chemicals or biological signals, making them useful in environmental monitoring, disease diagnosis, or personalized medicine applications.

4.2 Xenobots

Imagine a world where little entities composed of living cells can autonomously form, move by swimming or rolling, and even reproduce. The xenobots represent a groundbreaking organism that is challenging the distinction between biology and technology.

These minuscule wonders are neither animals nor machines. These are autonomous robots, created using frog stem cells and carefully engineered by computers to carry out precise functions. Xenobots emerge through an innovative mechanism known as evolutionary design. Scientists apply algorithms to model numerous possible shapes and motions, ultimately choosing the most successful ones for actual implementation. It represents accelerated evolution, enabling scientists to investigate a wide range of designs and identify valuable functional features (Akolpoglu et al., 2020; Kučuk, Primožič, Knez, & Leitgeb, 2023). The fabrication of Xenobots

Computational Design: Utilizing evolutionary AI, scientists perform simulations on thousands of body shapes and swimming motions using evolutionary AI software. An algorithm then selects those designs that most effectively meet desired functional goals as "breeding stock."

Cell Harvesting: Stem cells derived from frog embryos are harvested and cultured in the lab, before being meticulously arranged into their desired shapes by microscopic needles as predicted by AI simulations. Eventually, Patterning occurs: Once all stem cells have been assembled into their desired patterns predicted by AI simulations.

Cell Fusion and Development: Over several days, stem cells fuse into functioning Xenobots and grow.

5. LITERATURE REVIEW

Recent developments in the area of nano- and biorobots have shown promising applications in numerous areas. In 2023, Yuan developed nanorobots that move by magnets. These nanorobots are designed to quickly remove reactive α-dicarbonyl species and prevent the formation of advanced glycation end products.

In Zhang's (2023) study, twin-engine self-adaptive micro/nanorobots were utilized for gastrointestinal inflammation therapy. The study showed that these robots were able to accumulate drugs more effectively at locations of inflammation. Guo (2023) conducted a comprehensive analysis of the development of intelligent moving micro/nanomotors and their use in biosensing and disease therapy. The study emphasized the need to resolve material and biocompatibility challenges. In Liu's (2023) study, the focus was on the development of responsive magnetic nanocomposites for shape-morphing microrobots. The study highlighted the potential of magnetic fields in enabling control without the need for tethers and emphasized the significance of future fabrication approaches. This research collectively emphasizes the promise of nano- and biorobots in addressing diverse biomedical concerns. (Li 2022, Refaai 2022) has focused on drug delivery and the neutralization of pathogenic bacteria and toxins. Sun 2022 has developed combinedly biological and artificial components that have been shown to enhance drug retention, precise operation of minimally invasive surgery, and medical sensing.

(Wang 2021) The possible applications of magnetic actuation-based micro/nanorobots include targeted delivery, minimally invasive surgery, cellular and intracellular monitoring, intelligent sensing, and detoxification. Despite these progressions, obstacles such as biosafety and clinical translation still need to be resolved. Research has investigated the production and possible uses of nanobots, namely in the field of biomedicine. The emphasis has been on their ability to be programmed and their potential for delivering drugs and treating diseases. The detailed literature review of existing work is shown in Table 2.

Table 2. Literature review of existing work

S.No.		Main findings
1	Guo *et al.*(2023)	**The article discusses the utilization of micro/nanomotors in biosensing, cancer treatment, gynecological illness treatment, and assisted fertilization.**
2	Peng *et al.*(2023)	An independent nanorobot with the ability to actively deliver drugs to mitochondria, inducing mitochondrial-mediated apoptosis and dysregulation to enhance the anticancer effect in vitro and inhibit cancer cell metastasis.
3	Li *et al.*(2023)	The biodegradability of microrobots is essential to prevent the accumulation of hazardous residue within the human body during therapeutic applications.
4	Gotovtsev (2023)	The review focused mostly on evaluating the system utilizing microbial cells. This technique hinges on harnessing microbes as the core element of a robot, responsible for tasks such as locomotion, conveyance of products, and, in some cases, the production of advantageous substances. The inclusion of living cells in these microrobots brings both advantages and disadvantages.
5	Popescu (2023)	Nanoparticles manufactured by environmentally friendly methods and combined with biopolymers offer a possible solution to the issue of chemical waste resulting from biosensor fabrication.
6	Bhandari *et.al* (2023)	Nanomaterials have the potential to serve as nanosensors, nanocides, nanofertilizers, nanobarcodes, and nano-remediators in contemporary agricultural methods.
7	Kučuk *et.al* (2023)	Biopolymeric nanoparticles have great potential as delivery vehicles for a wide range of medicinal medicines.
8	Li, *et.al* (2022)	Biohybrid micro- and nanorobots can do several medical activities, including targeted medication administration, single-cell manipulation, and cell microsurgery. They are equipped with onboard actuation, sensing, and control, and may execute numerous functions.
9	Jahromi, *et.al* (2021)	Various therapeutic approaches including living immune cells, immune cells modified at the surface, cell membranes of immunocytes, extracellular vesicles or exosomes produced from leukocytes, and artificial immune cells have been studied, and a small number of these have been commercially released.
10	Hu *et.al* (2021)	DNA nanorobots are very suitable as biomedical robots for the field of precision medicine.
11	Primožič, *et.al* (2021)	The utilization of nanoparticles and nanocomposites in food packaging enhances the mechanical durability and characteristics of the water and oxygen resistance of the packaging, while potentially offering additional advantages such as antibacterial qualities and light obstruction. There are uncertainties regarding the migration of nanoparticles from packaging to food, as migration assays and risk assessment procedures are not well-defined. The use of nanomaterials in the food packaging sector is limited due to the assumed toxicity, insufficient data from clinical trials, and the absence of risk assessment studies.
12	Shen *et.al* (2021)	DNA nanostructures can interact with various entities such as tiny molecules, nucleic acids, proteins, viruses, and cancer cells.

6. FABRICATION OF ROBOT AND BIOROBOTS

Fabrication methods for nano- and biorobots are constantly evolving as new materials and techniques are developed. Some of the most common methods include:

(i) Top-down approaches
Top-down approaches involve physically removing or sculpting material to create the desired nanorobotic structure. Some common top-down fabrication approaches include:

- Electron beam lithography (EBL) employs a concentrated stream of electrons to selectively expose a substrate coated with resist, thereby transferring the intended pattern onto the substrate.
- Focused Ion Beam (FIB) milling uses a concentrated ion beam to physically eliminate material from a substrate, enabling the fabrication of intricate and extremely accurate structures.
- Nanoimprint lithography (NIL) uses a mold with the desired nanostructure to imprint the pattern onto a substrate.
    (ii) Bottom-up approaches
Bottom-up approaches comprise assembling atoms or molecules into the anticipated nanorobotic structure. Some common bottom-up fabrication methods include:

- Self-assembly: Self-assembly relies on the natural interactions between molecules to drive the formation of the desired nanostructure.
- Directed assembly: Directed assembly involves using external forces, such as electric fields or magnetic fields, to guide the assembly of molecules into the desired nanostructure.
- Chemical synthesis: Chemical synthesis involves using chemical reactions to create the desired nanorobotic structure.
    (iii) Hybrid approaches
Hybrid approaches combine top-down and bottom-up methods or biofabrication techniques with top-down/bottom-up approaches to create complex nanorobotic structures with multifunctional capabilities. The choice of fabrication method depends on the specific design and functionality of a nanorobot being created. Top-down approaches tend to be more established and provide higher precision; bottom-up methods may allow for creating more complex or multipurpose nanobots; bio-fabrication techniques still have yet to mature fully but have great potential in terms of producing biocompatible and biodegradable devices. As nanorobotics manufacturing advances rapidly, new methods are constantly being devised. Over time, we can expect increasingly sophisticated nanorobots with multiple functions (Chelliah et al., 2021; Naghdi et al., 2023; Rodrigues et al., 2021).

6.1 Biofabrication

Biofabrication involves using living cells or biological materials to create the desired nanorobotic structure. Some common bio-fabrication methods include:

- Tissue engineering: Tissue engineering involves using scaffolds or cell culture techniques to create three-dimensional tissue structures.

      °   Genetic engineering: Genetic engineering involves modifying the genetic makeup of cells to give them new or enhanced functions.

      °   Synthetic biology: Synthetic biology involves designing and engineering new biological systems from scratch.

Fabrication processes involve precise manufacturing techniques at the nanoscale, using techniques such as nanolithography, self-assembly, or merging biological entities with artificial materials. Further research and exploration in nanofabrication technology are necessary to fully comprehend these fabrication methods used in creating nanorobots. The classification of nanomaterials according to their physical dimensions is shown in Figure 2.

6.2 Fabrication of Biorobots

The process of biorobot fabrication includes the integration of synthetic materials and biological entities to produce hybrid systems that possess improved functionality and attributes. The fabrication techniques and approaches described in the provided documents will be elaborated upon in this section. The authors discuss the utilization of enzymes as power sources for self-propelled nanoparticles at the nanoscale. Enzymes of this nature facilitate biocompatible reactions and can be integrated into nanoparticles exhibiting diverse geometries and compositions. This facilitates the development of active matter systems, which find applicability in biomedical contexts including drug delivery systems.

Moving to the microscale, the article mentions the use of single cells, such as bacteria or spermatozoa, as substitutes for enzymes. These self-propelling cells can transport cargo, serve as drug delivery systems, aid in in vitro fertilization practices, or remove biofilms. At the macro level, the integration of millions of cells forming tissues is discussed. Muscle cells, for example, can be combined to power biorobotic devices or actuators. The article mentions untethered biorobots that can crawl or swim due to the contractions of the tissue. The authors also mention ongoing developments in integrating different types of tissue to create more realistic biomimetic devices (Kim et al., 2022; Sun et al., 2020). Hu *et.*

Figure 2. The classification of nanomaterials according to their physical dimensions

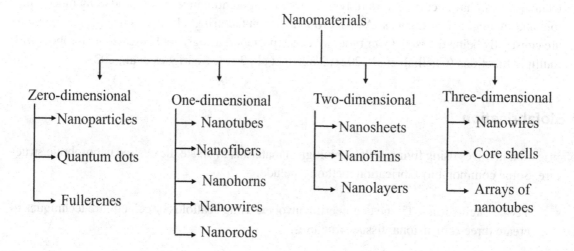

all highlights the use of advanced manufacturing technologies such as 3D bioprinting and microfluidic manufacturing for fabricating cell-laden hydrogel structures.

These technologies enable precise and customizable fabrication of biohybrid constructs. Furthermore, the article describes the development of engineered functional materials by combining living cells with hydrogels; such materials can then transport drugs, peptides, or artificial substances, making them suitable for transplantation, tissue engineering, or controllable biomedicine applications. Lin *et. all* describes biohybrid micro-robots powered by living cells. These authors outline its many advantages over rigid drives - including microscale self-assembly, high-energy efficiency, flexibility, self-repair capabilities, and multiple degrees of freedom (Koleoso et al., 2020; Zheng et al., 2021). The review article discusses challenges related to biohybrid micro-robot development, such as ethical considerations surrounding primary cell extraction from mammals or muscle tissues. Furthermore, the authors highlight the need for intelligent perception and control functions within biohybrid robots as well as highlighting current systems' limited lifespan and functionality. Overall, biorobot fabrication involves the combination of biological entities with artificial materials using various techniques and approaches. These may include using enzymes and cells at various scales - nanoparticles, self-propelling cells, and tissue constructs - while advanced manufacturing technologies like 3D bioprinting and microfluidic manufacturing play a key role in fabricating complex hybrid structures. Yet ethical considerations, intelligent control issues, lifespan issues, and functionality issues still need to be resolved before designing biorobots (Markande et al., 2021; Zhou et al., 2021).

7. CONCLUSION

In summary, the literature review regarding biohybrid robotics and nanofabrication technology advancements underscores the significant achievements and the future uses of these technologies across diverse domains, including but not limited to biomedical research, wearable electronics, human-machine interfaces, personal health monitoring systems, space exploration, and biomanufacturing. The literature on nanofabrication technology encompasses an examination of the evolution of hybrid intelligent nanorobotics, a paradigm shift in biomedical research concerning the design of intelligent systems. Drug delivery, disease diagnosis, imaging, surgery, tissue engineering, monitoring, and sensing, targeted therapy, and microsurgery are all potential applications of nanorobots in this field. The aforementioned applications exemplify the capacity of nanorobots to radically transform biomedical research and substantially enhance patient outcomes.

In contrast, biohybrid robotics integrates synthetic materials with biological entities to augment the functionality and characteristics of robotic systems. Biological entities and synthetic materials are integrated at various scales, including the nanoscale, microscale, and macroscale. The benefits of biohybrid robotics include enhanced functionality and biomimicry within the field of robotics. Biomimicry in robotics, enhanced performance, and improved functionality are all potential advantages of biohybrid robotics.

Soft electronics find diverse utility in human-machine interfaces, wearable electronics, and personal health-care monitoring systems. These technologies facilitate the creation of conformable and comfortable wearable devices, wearable sensors capable of acquiring physiological signals, and interfaces capable of seamless integration with the human body. Ultra-thin electronics, encompassing devices and components that are exceptionally thin, provide enhanced surface conformability for purposes such as personal health-care monitoring systems, wearable electronics, and human-machine interfaces.

In general, the literature review offers significant contributions by examining the development, utilization, and prospects of biohybrid robotics, soft electronics, ultra-thin electronics, hydrogels containing living cells, responsive biohybrid systems, biohybrid microrobots, and magneto/catalytic nanostructured BioBots. These technologies are capable of bringing about significant changes across multiple sectors, enhancing interactions between humans and machines, and potentially reshaping the course of human history on an international scale. Nevertheless, additional investigation and progress are required to rectify constraints, enhance functionality, and surmount obstacles linked to these technologies.

REFERENCES

Akolpoglu, M. B., Dogan, N. O., Bozuyuk, U., Ceylan, H., Kizilel, S., & Sitti, M. (2020). High-yield production of biohybrid microalgae for on-demand cargo delivery. *Advancement of Science*, 7(16), 2001256. doi:10.1002/advs.202001256 PMID:32832367

Bhandari, G., Dhasmana, A., Chaudhary, P., Gupta, S., Gangola, S., Gupta, A., Rustagi, S., Shende, S. S., Rajput, V. D., Minkina, T. M., Malik, S., & Sláma, P. (2023). A Perspective Review on Green Nanotechnology in Agro-Ecosystems: Opportunities for Sustainable Agricultural Practices & Environmental Remediation. *Agriculture*, 13(3), 668. doi:10.3390/agriculture13030668

Chelliah, R., Wei, S., Daliri, E. B. M., Rubab, M., Elahi, F., Yeon, S. J., Jo, K. H., Yan, P., Liu, S., & Oh, D. H. (2021). Development of Nanosensors Based Intelligent Packaging Systems: Food Quality and Medicine. *Nanomaterials (Basel, Switzerland)*, 11(6), 1515. doi:10.3390/nano11061515 PMID:34201071

Gotovtsev, P. M. (2023). Microbial Cells as a Microrobots: From Drug Delivery to Advanced Biosensors. *Biomimetics*, 8(1), 8. doi:10.3390/biomimetics8010109 PMID:36975339

Guo, Y., Jing, D., Liu, S., & Yuan, Q. (2023). Construction of intelligent moving micro/nanomotors and their applications in biosensing and disease treatment. *Theranostics*, 13(9), 2993–3020. doi:10.7150/thno.81845 PMID:37284438

Hu, Y. (2021). Self-Assembly of DNA Molecules: Towards DNA Nanorobots for Biomedical Applications. *Cyborg and Bionic Systems (Washington, D.C.)*, 2021, 2021. doi:10.34133/2021/9807520 PMID:36285141

Jahromi, L. P., Shahbazi, M., Maleki, A., Azadi, A., & Santos, H. A. (2021). Chemically Engineered Immune Cell-Derived Microrobots and Biomimetic Nanoparticles: Emerging Biodiagnostic and Therapeutic Tools. *Advancement of Science*, 8(8), 8. doi:10.1002/advs.202002499 PMID:33898169

Kim, D. S., Yang, X., Lee, J. H., Yoo, H. Y., Park, C., Kim, S. W., & Lee, J. (2022). Development of GO/Co/Chitosan-Based Nano-Biosensor for Real-Time Detection of D-Glucose. *Biosensors (Basel)*, 12(7), 464. doi:10.3390/bios12070464 PMID:35884266

Koleoso, M., Feng, X., & Xue, Y. (2020). *Materials Today Bio Micro/nanoscale magnetic robots for biomedical applications*. https://doi.org/ doi:10.1016/j.mtbio.2020.100085

Kučuk, N., Primožič, M., Knez, Ž., & Leitgeb, M. (2023). Sustainable Biodegradable Biopolymer-Based Nanoparticles for Healthcare Applications. *International Journal of Molecular Sciences*, 24. PMID:36834596

Li, J., Dekanovsky, L., Khezri, B., Wu, B., Zhou, H., & Sofer, Z. (2022). Biohybrid Micro- and Nano-robots for Intelligent Drug Delivery. *Cyborg and Bionic Systems (Washington, D.C.)*, 2022, 2022. doi:10.34133/2022/9824057 PMID:36285309

Li, J., & Yu, J. (2023). Biodegradable Microrobots and Their Biomedical Applications: A Review. *Nanomaterials (Basel, Switzerland)*, 13(10), 13. doi:10.3390/nano13101590 PMID:37242005

Markande, A., Mistry Kruti, U., & Shraddha, J. A. (2021) magnetic nanoparticles from bacteria. In Biobased Nanotechnology for Green Applications. SpringerNature Switzerland AG.

Naghdi, T., Ardalan, S., Asghari Adib, Z., Sharifi, A. R., & Golmohammadi, H. (2023). Moving toward Smart Biomedical Sensing. *Biosensors & Bioelectronics*, 223, 115009. doi:10.1016/j.bios.2022.115009 PMID:36565545

Peng, X., Tang, S., Tang, D., Zhou, D., Li, Y., Chen, Q., Wan, F., Lukas, H., Han, H., Zhang, X., Gao, W., & Wu, S. (2023). Autonomous metal-organic framework nanorobots for active mitochondria-targeted cancer therapy. *Science Advances*, 9(23), 9. doi:10.1126/sciadv.adh1736 PMID:37294758

Popescu, M., & Ungureanu, C. (2023). Biosensors in Food and Healthcare Industries: Bio-Coatings Based on Biogenic Nanoparticles and Biopolymers. *Coatings*, 13(3), 486. doi:10.3390/coatings13030486

Primožič, M., Knez, Ž., & Leitgeb, M. (2021). (Bio)Nanotechnology in Food Science—Food Packaging. *Nanomaterials (Basel, Switzerland)*, 11. PMID:33499415

Rodrigues, C., Souza, V. G. L., Coelhoso, I., & Fernando, A. L. (2021). Bio-Based Sensors for Smart Food Packaging—Current Applications and Future Trends. *Sensors (Basel)*, 21(6), 2148. doi:10.3390/s21062148 PMID:33803914

Shen, L., Wang, P., & Ke, Y. (2021). DNA Nanotechnology-Based Biosensors and Therapeutics. *Advanced Healthcare Materials*, 10. PMID:34085411

Sun, M., Liu, Q., Fan, X., Wang, Y., Chen, W., Tian, C., Sun, L., & Xie, H. (2020). Autonomous biohybrid urchin-like microperforator for intracellular payload delivery. *Small*, 16(23), 1906701. doi:10.1002/smll.201906701 PMID:32378351

Zheng, S., Wang, Y., Pan, S., Ma, E., Jin, S., Jiao, M., Wang, W., Li, J., Xu, K., & Wang, H. (2021). Biocompatible nanomotors as active diagnostic imaging agents for enhanced magnetic resonance imaging of tumor tissues in vivo. *Advanced Functional Materials*, 31(24), 2100936. doi:10.1002/adfm.202100936

Zhou, H., Mayorga-Martinez, C. C., Pané, S., Zhang, L., & Pumera, M. (2021). Magnetically Driven Micro and Nanorobots. *Chemical Reviews*, 121(8), 4999–5041. doi:10.1021/acs.chemrev.0c01234 PMID:33787235

Chapter 9
HRI in ITs Using ML Techniques

V. Saran

Karpagam Academy of Higher Education, India

R. Chennappan

https://orcid.org/0000-0002-6252-4614
Karpagam Academy of Higher Education, India

ABSTRACT

Machine learning (ML), deep learning, fuzzy logic, and traditional neural networks are just a few of the subsets that make up artificial intelligence (AI). These subgroups possesses unique qualities and skills that could improve the effectiveness of modern medical sciences. Human intervention in clinical diagnostics, medical imaging, and decision-making is facilitated by these clever solutions. The development of information technology, the concept of intelligent healthcare has become more and more popular. Intelligent healthcare is a revolutionary approach to healthcare that leverages state-of-the-art technology such as AI and the internet of things (IoT) to improve overall efficacy, convenience, and personalisation of the medical system.

I. INTRODUCTION

Artificial intelligence (AI) and medical technology have brought in a new age in healthcare in the twenty-first century. Diagnostics, patient care, and medical practices are changing as a result of the integration of intelligent systems and the unrelenting search of innovation. Known as Intelligent Healthcare, this paradigm shift offers opportunities for proactive and personalised health management in addition to increased efficiency and accuracy. Moerenhout, T., et al (2018). This introduction explores the innovative advancements that are reshaping healthcare and takes the reader on a voyage into the centre of this revolutionary convergence.

Modern healthcare has significantly advanced thanks to the collaboration of artificial intelligence and medical technology. AI has become a potent tool for data analysis, pattern recognition, and decision-making, whereas medical technology has historically fueled advances in diagnosis and treatment. A

DOI: 10.4018/979-8-3693-5767-5.ch009

revolutionary force that could completely reshape the healthcare system is unleashed by the union of these two domains.

Precision medicine, which considers individual variations in patients' genetics, environments, and lifestyles, is the primary idea behind this groundbreaking advancement. Healthcare professionals can now more precisely personalise therapies thanks to AI-driven technologies that enable the extraction of relevant insights from large datasets. This promises increased effectiveness as well as a move towards preventative treatment, which identifies and treats possible health problems before they materialise details from Dongari, S., et al (2023).

The field of diagnostics and medical imaging is one of the most innovative uses of AI in healthcare. Medical image analysis may now be done more accurately and efficiently than ever before because to machine learning algorithms that have been trained on large datasets. Artificial Intelligence is quickly turning into a vital tool for medical practitioners, helping them with everything from seeing minute irregularities in X-rays to assisting with cancer early detection through advanced imaging methods.

1.1 Realtime Usage of Intelligent Healthcare

Intelligent healthcare is no longer limited to conventional hospital settings. Healthcare professionals may remotely monitor patients thanks to AI-powered remote patient monitoring devices that track vital signs in real time. This improves the standard of care for long-term illnesses while simultaneously promoting the growth of telehealth, which allows for virtual consultations and diagnosis.

Barbosa, H. C., et al (2021), to find patterns and trends in patient data, artificial intelligence (AI) uses its predictive skills. AI is capable of forecasting possible health problems and suggesting preventive actions by evaluating past medical records and merging them with current data. By focusing on preventive measures to delay the start of diseases, this movement in healthcare towards proactive approaches encourages people to take responsibility for their health.

Intelligent healthcare encompasses all aspect of the healthcare system, not just programmes that interact with patients. Artificial intelligence (AI)-powered systems improve resource allocation, expedite appointment scheduling, and optimise administrative duties. In addition to lessening the workload for medical personnel, this efficiency improves resource efficiency and cost-effectiveness.

1.2 AI in Healthcare

Ethical issues grow more pressing when AI is incorporated more fully into healthcare. Critical issues that require close consideration include the ethical use of patient data, maintaining security and privacy, and resolving any biases in algorithms. It is essential for the long-term development of intelligent healthcare technology to strike a balance between innovation and ethical issues.

Intelligent healthcare holds great potential for change, but there are also impending difficulties. Important obstacles include the requirement for regulatory frameworks, data format standardisation, and system interoperability. Furthermore, it's critical to guarantee that these technologies are available to people from a variety of socioeconomic backgrounds in order to avoid making healthcare inequities worse. The research that is now being conducted to improve AI algorithms, increase datasets, and remove the obstacles in the way of developing a more inclusive and fair healthcare system bodes well for the future.

Figure 1 depicts AI employed in medical technology and intelligent healthcare. Here, we examine the potentially revolutionary nature of this synergy as we examine the innovative development of AI and medical technology for Intelligent Healthcare. Intelligent Healthcare is a promising new field that is expected to transcend science fiction and bring about a time when healthcare is not just personalised, but also proactive, available to all. From the microcosm of AI-powered diagnostics to the macrocosm of a redesigned healthcare ecosystem, the upcoming chapters of this story will elucidate the nuances of every aspect of this paradigm shift, Dzobo, K., et al (2020).

A new era of intelligent healthcare has begun with the merging of medical technology and artificial intelligence (AI), driven by the unwavering goal of improving healthcare outcomes. The medical field is undergoing a transformation thanks to the symbiotic link between cutting-edge technology and sophisticated algorithms, which promises previously unheard-of levels of precision, efficiency, and individualised care. This article explores the innovative advancements that have occurred at the nexus of AI and medical technology, and how they are transforming patient care, treatment modalities, diagnostics, and the larger healthcare ecosystem.

1.3 Technical Innovations

With technological advancements pushing the limits of medical knowledge, the healthcare sector is at a turning point in its history. An increasingly data-driven, networked, and intelligent healthcare system is replacing traditional methods of patient care. The proliferation of wearable technology, the exponential expansion of medical data, and the growing processing capacity of AI systems are all driving forces behind this change.

Healthcare data is growing at an exponential rate; this phenomenon is commonly known as "big data," and it has created opportunities for previously unheard-of insights and well-informed decisions.

Figure 1. Intelligent healthcare using artificial intelligence

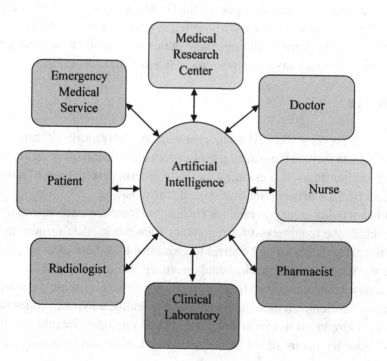

A thorough picture of both individual and public health can be obtained by successfully using the massive amounts of data generated by genomic sequencing, medical imaging, electronic health records (EHRs), and real-time patient monitoring. Healthcare workers are able to make data-driven decisions and anticipate future health hazards with the use of AI algorithms and big data analytics, which enable the extraction of significant patterns.

The emergence of precision medicine is one of the main characteristics of the union of AI and medical technology. Precision medicine develops treatment regimens that are specific to each patient, as opposed to using a one-size-fits-all strategy. In order to develop individualised treatment plans that maximise therapeutic success and minimise side effects, AI algorithms examine genetic, clinical, and lifestyle data. A paradigm change in illness prevention and management is brought about by this move towards personalised healthcare, which also improves patient outcomes.

With the introduction of AI-based diagnostic technologies, medical imaging has seen a revolutionary revolution. After being trained on enormous medical picture datasets, machine learning algorithms show amazing precision in identifying abnormalities and deciphering intricate diagnostic images. These AI-enhanced diagnostic technologies provide healthcare practitioners with faster and more accurate insights, enabling them to forecast cardiovascular risks and spot cancer early. This leads to increased diagnostic precision as well as better prognosis and early intervention possibilities.

Technological developments in healthcare have made it easier to create AI-powered remote patient monitoring systems that track health in real-time and continuously. Wearable technology with sensors can track vital signs, exercise levels, and other health parameters, according to Shaik, T., et al. (2023). With the use of AI algorithms, this data is analysed to find minute variations or abnormalities that enable early intervention and individualised treatment plan revisions. Enhancing patient engagement and increasing access to healthcare services, telehealth solutions with AI-driven diagnostics allow for remote consultations. Though there is much promise in the combination of AI and medical technology, there are drawbacks. Careful consideration is needed for correcting biases in AI models, handling sensitive patient data, and maintaining algorithmic transparency. Additionally, there are serious cybersecurity risks, which highlight the necessity of taking strong precautions to protect patient privacy and the integrity of healthcare systems. Maintaining the responsible implementation of intelligent healthcare technology requires striking a balance between innovation and ethical considerations.

The nexus between AI and medical technology marks a turning point in the development of healthcare. The healthcare environment is changing as a result of the convergence of big data analytics, AI-driven diagnostics, precision medicine, and remote patient monitoring. This convergence promises more precise diagnosis, individualised treatment plans, and better patient outcomes. Ensuring that the advantages of intelligent healthcare are accessible, equitable, and responsibly implemented requires us to give ethical issues top priority as we traverse this revolutionary journey. In order to provide a thorough examination of the rapidly developing field of intelligent healthcare, the ensuing chapters will focus on particular applications, case studies, and potential future developments of this dynamic interaction between medical technology and AI.

This paper consists of five parts. The section I discussed about the introduction of the work. The drawbacks and limitations of present systems are discussed in the section II. The section III explained about our new suggested system design work for medical technology and AI healthcare system. Section IV contains our newly constructed system's overall output result and discussion. Section V covers this paper's conclusion.

II. LITERATURE REVIEW

This literature review part discusses the shortcomings and restrictions of the current AI-based healthcare and medical technology solutions. Distinguished contributions from a broad spectrum of scholars in this field are included in the discussion of the literature review.

Wang .G & et al., (2022), enables people to interact with their avatars in a world facilitated by technologies like digital twins, blockchain, augmented reality, mixed reality, virtual reality, and high-speed internet. The metaverse offers the integration of physical and virtual realities, all enhanced by virtually limitless data. Recently, social media and entertainment platforms have evolved from the metaverse; however, if this concept is extended to the healthcare industry, patient outcomes and clinical practice may be significantly impacted. We as a team of academics from academia, industry, medicine, and regulation see special prospects for metaverse methods in the healthcare sector. The creation, testing, assessment, control, translation, and improvement of AI-based medicine, particularly imaging-guided diagnosis and treatment, can be aided by a metaverse of "medical technology and AI" (MeTAI).

Tian .S & et al., (2019) focuses on how information technology is developing, the idea of smart healthcare has progressively gained prominence. With the use of cutting-edge information technologies like big data, cloud computing, artificial intelligence, and the internet of things (IoT), smart healthcare transforms the conventional medical system from the inside out and offers more individualised, convenient, and effective care. In an attempt to present the notion of smart healthcare, we first enumerate the essential technologies that facilitate smart healthcare and outline the state of smart healthcare as of right now in a number of significant domains. Next, we outline the current issues with smart healthcare and make an attempt to provide fixes. Lastly, we consider the future and assess the potential for smart healthcare.

Ahmed.Z & et al, (2020), discusses the One of the most recent and significant advances in healthcare is precision medicine, which holds promise for enhancing the conventional symptom-driven practice of medicine by enabling early interventions through enhanced diagnostics and customising more effective and cost-effective treatments. The power of electronic health records must be harnessed to integrate disparate data sources, identify patient-specific patterns of disease progression, and enable real-time decision support in order to improve the networking and interoperability of clinical, laboratory, and public health systems as well as to effectively balance ethical and social concerns regarding the privacy and protection of healthcare data, useful analytical tools, technologies, databases, and methodologies are needed. By effectively classifying subjects to comprehend certain scenarios and enhance decision-making, the development of multifunctional machine learning systems for clinical data extraction, aggregation, management, and analysis can assist physicians. The goal of offering real-time, more personalised, and affordable population medicine at a reduced cost may be significantly improved by implementing artificial intelligence in healthcare, which is an appealing idea.

Manickam.P & et al, (2022), Addresses the A contemporary method founded in computer science, artificial intelligence (AI) creates algorithms and programmes to provide machines intelligence and efficiency for carrying out jobs that often call for highly trained human intellect. Artificial intelligence (AI) encompasses a number of subsets, each with special powers and functions that might enhance the performance of contemporary medical sciences. These subsets include machine learning (ML), deep learning (DL), conventional neural networks, fuzzy logic, and speech recognition. Medical imaging, clinical diagnosis, and decision-making are all made easier for humans by such intelligent systems. A software programme and network-linked biomedical gadget are combined to create the Internet of Medical Things (IoMT), a next-generation bio-analytical tool that advances human health in the same period. They go over the value of AI in enhancing IoMT and point-of-care (POC) device capabilities in cutting-edge healthcare domains like diabetes management, cancer diagnosis, and cardiac monitoring in this review.

Talukder.A & et al, (2021), told that, For sustainable societies and economies, providing high-quality healthcare to all communities is a key objective. Physicians are facing increased workloads and time constraints, despite working in highly developed healthcare facilities. Artificial Intelligence (AI), Big Data, Web technologies, and telemedicine can all be powerful tools for improving diagnosis accuracy and care delivery. Because of the COVID-19 pandemic, telemedicine via the internet is growing in popularity. We describe a unique smartphone-based care solution in this work that collects patient data using progressive web applications (PWA), combines it with a variety of medical knowledge sources, and uses artificial intelligence (AI) to facilitate patient stratification and differential diagnosis. The programme has been developed with special attention to cyber security and can recommend courses of action and therapy. A smart hospital may easily incorporate the smart care system, which is built on next-generation web technologies including PWA, WebBluetooth, Web Speech API, WebUSB, and WebRTC.

Banerjee.A & et al, (2020) deals with The biomedical and healthcare technologies have opened up exciting opportunities due to the recent revolutions in Internet of Things (IoT) and big data analytics. The subjects covered in this chapter's interesting examples include theoretical, methodological, empirical, and validated concepts. First, a brief explanation of how IoT and big data are used to analyse a large image database that is created daily from a variety of sources using big data along with machine learning and other artificial intelligence techniques to create structured information for use in remote diagnostics is given. Examples are provided. The use of artificial intelligence in robotic health care to further emergent trends in telemedicine has been explored. Moreover, information on wearable technology that is now on the market focuses on biomedical and healthcare applications that can gather data, analyse it using standard procedures, and use machine intelligence to forecast health-related problems.

Rani.S & et al (2023), discusses about nowadays, Observing and monitoring the numerous clinical parameters of patients in their daily lives through the use of various technologies is the industry's primary goal. As these apps enable more affordable healthcare services, distant patient observation apps are growing in popularity. Equal consideration must also be given to the data management procedure used with these apps. Healthcare apps that are enabled by cloud computing offer a range of ways to store patient records and provide the necessary data to meet the needs of all parties involved; nevertheless, these solutions are hampered by security risks, slower reaction times, and issues that impair the system's availability. This chapter suggests an intelligent distributed framework for IoT-based remote healthcare service deployment in order to address these issues. Distributed Database Management Systems (DDBMS) are employed in the suggested model to provide patients and healthcare professionals with quick and secure access to data. The system's many entities are connected through the Internet of Things. To guarantee the security of the patient's medical records, the blockchain idea is employed. In the suggested model, clinical records obtained via DDBMS and encrypted using blockchain would be intelligently analysed.

Tripathy.S,S & et al, (2023) deals with The healthcare industry has not been immune to the widespread usage of internet-enabled gadgets. Regardless of a person's medical concerns, their overall health is being observed. The introduction of these medical gadgets helps patients as well as doctors, hospitals, and insurance companies. It facilitates hassle-free, dependable, and quick healthcare. Individuals are able to monitor their own blood pressure, pulse rate, and other health parameters and take proactive steps. Hospitals are also utilising the Internet of Things (IoT) for a number of purposes, including electrocardiograms (ECGs), blood sugar and oxygen monitoring, and more. IoT in healthcare also lowers the cost of many diseases by quickly and thoroughly analysing data. Machine learning methods based on symptom analysis have emerged as a potential notion for disease prediction. In certain circumstances, real-time analysis can also be necessary. Fog computing becomes indispensable in such a latency-sensitive scenario. Fog reduces latency by eliminating the need to establish contact with the cloud on a per-occassion basis. Applications in the healthcare industry must respond quickly. Therefore, it is crucial that fog computing be implemented in this field. Our efforts are concentrated on enhancing the system's effectiveness in providing accurate heart disease diagnosis and recommendations. Using a machine-learning module, it assesses the system.

Ganji.K & Parimi.S (2022) deals with Undoubtedly, the COVID-19 pandemic transformed the world's lifestyle, particularly in the realm of health care. The future holds a significant transformation for the delivery of healthcare services. In particular, during pandemics, the goal is to examine the growing use of digital and IoT technologies in conjunction with health care systems. To categorise users' perceptions of using IoT-based smart health-care monitoring wearables according to their knowledge and experience, an efficient artificial neural network (ANN)-based predictive model is implemented.

Bohr, A., & Memarzadeh, K. (2020) said that, From entertainment to commerce to healthcare, big data and machine learning are influencing most facets of contemporary life. Search terms and symptoms are known to Google, Amazon and Netflix, respectively, and Netflix is aware of the films and TV shows that its users enjoy watching and when they want to buy certain products. In addition to being useful for behavioural targeting and understanding, all of this data may be utilised for extremely detailed personal profiling, which may also be useful for forecasting trends in healthcare. The potential for significant advancements in artificial intelligence (AI) to enhance healthcare in every facet, from diagnosis to treatment, is highly encouraging. AI technologies are often seen as enhancing and supporting human labour rather than taking the place of doctors and other healthcare professionals. When it comes to a range of duties, including clinical documentation, patient outreach, administrative workflow, and specialised support like image analysis, medical device automation, and patient monitoring, AI is prepared to assist healthcare professionals. Several significant uses of artificial intelligence (AI) in healthcare will be covered in this chapter, including those that are directly related to the field as well as those that are part of the healthcare value chain, such medication development and ambient assisted living.

Gilbert, S., & et al (2021) discussed about The ability to continuously enhance performance based on updates from automated learning from data is one of the main advantages of artificial intelligence (AI) and machine learning (ML) technologies in the healthcare industry. Nonetheless, the regulations that currently govern health care machine learning models were created for a bygone period of slowly evolving medical equipment, necessitating significant documentation updates, reshaping, and revalidation for each major model update produced by the ML algorithm. For models that are meant to be retrained and updated infrequently, this poses modest issues; but, for models that are meant to learn from data in real-time or almost real-time, it poses significant challenges. The announcement of action plans by regulators reflects a significant shift in their regulatory strategies. We analyse the laws and advancements in this field from this point of view. After reviewing the current state of affairs and recent advancements, we contend that these creative approaches to healthcare demand creative methods to regulation, and that patients will profit from these approaches.

Yaeger, K. A., & et al (2019) explores The way medical devices are regulated in the US has changed as technology has advanced. The FDA in the United States has further defined risk categories and intended uses for software applications and computer-based devices that are being incorporated into routine clinical practice. This is done to promote innovation in medical technology while also better ensuring patient safety. Yet, regulatory agencies will need to act quickly when new software technologies like artificial intelligence (AI) are created, improved, and applied in the healthcare industry. Within this analysis, we go over how the US FDA has regulated medical devices over time first with hardware and then with software and how they currently stand on artificial intelligence-enhanced devices.

Minopoulos, G. M., & et al (2022) discussed The medical techniques used today still face many obstacles and restrictions in their efforts to identify and treat diseases. It is projected that the use of new technology in the healthcare sector would make it possible to implement cutting-edge medical procedures for a smart healthcare system that is both successful and efficient. The rapid diagnosis of illnesses, diseases, infections, or abnormalities can be greatly aided by the IoT, WSN, Big Data Analytics (BDA), and Cloud Computing (CC). The discovery of new drugs and antibiotics may be accelerated by complex methods like AI, ML, and DL. In addition, medical professionals can use the integration of visualisation techniques like Virtual Reality (VR), Augmented Reality (AR), and Mixed Reality (MR) with Tactile Internet (TI) to give patients the most accurate diagnosis and treatment possible. This study suggests a new system architecture that integrates multiple next-generation technologies. In order to create a smart healthcare system that may be installed in hospitals or medical centres, the goal is to explain how a variety of cutting-edge technologies are integrated with support from innovative networks. A system like this will be able to offer medical personnel with fast, precise data so they can target patients and administer treatment with accuracy.

Amann, J., et al (2020), deals with Regarding artificial intelligence (AI) and its application to healthcare, one of the most hotly contested subjects is explainability. Its failure to explain itself continues to draw criticism, despite evidence that AI-driven systems can do better than humans in some analytical tasks. Explainable technology, however, raises a number of social, legal, ethical, and medical issues that need careful consideration rather than being limited to a technological one.

Nasar, M & et al, (2021), deals with a new paradigm for healthcare systems is desperately needed, as seen by the sharp rise in the number of people with chronic illnesses, including the aged and disabled. With decreased reliance on conventional physical healthcare facilities like hospitals, assisted living facilities, and long-term care facilities, the developed model will be more individualised. Major advancements in contemporary technology, particularly in the areas of artificial intelligence (AI) and machine learning (ML), have led to a growing interest in and necessity for the smart healthcare system.

III. SYSTEM DESIGN

The proposed study describes an experiment utilising deep learning techniques for the medical technology and healthcare industry, with Ant Colony Optimisation (ACO) handling the optimisation process and employing artificial intelligence (AI) in the form of convolution neural networks (CNNs) for detection and classification. With the help of this study, an artificial intelligence-based intelligent healthcare system with lifetime aid for patients will be built.

3.1 AI for Intelligent Healthcare

The latest technical advances in AI have stunned people all around the world, and medical experts are trying to embrace it. With the help of AI's new indications, doctors can discover new uses for current drugs and leverage biological information to develop new ones. By evaluating drugs on cells, AI also

Figure 2. CNN employs patient data to determine appropriate treatment

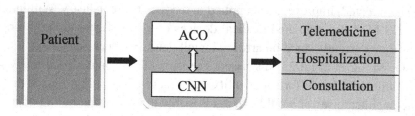

plays a critical role in identifying which chemicals need more investigation and work. We used the CNN method with ACO from the AI technique in this proposed work, which is shown in the Figure 2.

3.2 Ant Colony Optimization (ACO)

A metaheuristic called Ant Colony Optimisation (ACO) was developed after observing how some ant species leave and follow pheromone trails. Making use of tailored (false) pheromone data Based on the ant's search history and any potentially accessible heuristic data provided. Artificial ants in ACO are stochastic techniques for generating solutions that propose potential fixes for the issue instance in question.

To locate the area in the input text data that is affected, the ACO method is employed. Where different iterations of the ACO algorithm are present and it clearly explained in below figure 3. In each iterations, a number of ants construct comprehensive answers based on heuristic input and the knowledge collected by prior populations of ants. A component of a solution leaves a pheromone trail that is a representation of these collected experiences. This initial stage involves setting up all settings and pheromone variables. A group of ants builds a solution to the problem at hand after initiating it by using pheromone values and extra information. In this optional stage, the ants refine the created solution. At this step, pheromone variables are modified depending on observations from search behaviour in ants.

Now that we are aware of how the ants acted in the example above, we can construct an algorithm. For simplicity's sake, one food source, one ant colony, and two potential travel routes have been examined. The complete scenario may be simulated using weighted graphs: the paths are the edges, the ant colony and food source are the vertices (or nodes), and the weights associated with the pathways are the pheromone levels.

Take into account the graph as $G = (V, E)$, where V, E are the graph's vertices and edges, respectively. Assuming that we take it into mind, the vertices are (Vs - Source vertex, an ant colony, and Vd - Destination vertex, a food source). The lengths of the two edges, E1 and E2, are L1 and L2, respectively. Now, it may be hypothesised that, depending on their strength, Vertices E1 and E2 have related pheromone values of R1 and R2, respectively. As a result, the probability that each ant will first select a path (between E1 and E2) is as follows:

$$Pi = \frac{Ri}{R1 + R2}; i = 1, 2$$

Apparently, if R1>R2 and vice versa, there is a larger possibility of choosing E1. As you return along that path, say along, the pheromone value for that route now changes, Ei.

Figure 3. ACO algorithm pseudocode

> **Ant Colony Optimization:**
>
> Initialise the pheromone experiments and relevant parameters;
>
> Do not terminate, but instead
>
> Create an ant colony;
>
> Determine the fitness ratings for each ant;
>
> Using selection techniques, determine the optimal option;
>
> Pheromone trial updates;
>
> Stop while
>
> Process over

1. Considering the length of the path

$$Ri \leftarrow Ri + \frac{K}{Li}$$

In the aforementioned update, "K" and "i=1, 2" serve as model parameters.

2. Dependent on the pheromone's rate of evaporation

$$Ri \leftarrow (1 - v) * Ri$$

"V," a parameter with a range of [0, 1], controls the evaporation of pheromones. Similarly, "i" is equal to 1 plus 2 plus 3.

The optimized values are fed to the classification based Artificial Neural Network techniques.

3.3 Convolutional Neural Network (CNN)

Convolutional layers, max-pooling, and sparse connectivity are the three basic facets of CNN architecture. The local connection of neurons on neighbouring layers takes advantage of the spatial dependency of the image's pixels.

The convolution kernel, from which the term "convolution neural network" derives, is the most crucial component of the CNN. An n-column two-dimensional matrix makes up the convolution kernel and matching weights for each point. Convolution kernels resemble neurons, and their size is referred to as the receptive field of a neuron. The receptive field of the CNN is filled with the addition of the weight values of the convolution kernel and the pixel values at the pertinent locations in the image. The system's k convolution kernels do this. Once all of the pixels in the image have been counted, to the following point in the image in accordance with the step size. The original image's feature map is now the output pixel matrix.

$$Ow = [\frac{i_w - n + 2_p}{s}] + 1 \tag{1}$$

$$Oh = [\frac{i_h - n + 2_p}{s}] + 1 \tag{2}$$

The most common pooling techniques are maximal combining, averaging, etc. Here, we select the highest pooling approach and the associated calculation. The area of the image the highest value per pixel in the range of 2 is chosen using a filter of size 2 2, and it is then kept as the distinctive feature of this area. A feature map is created once the filter repeats this process for the subsequent range. The convolution network's performance has increased with end-to-end training, not just in terms of overall data classification but also in terms of the local task's progress toward producing structured output. As a result, it has seen extensive use in the fields of data detection and categorization. CNNs will quickly run into problems when using a typical multi-layer perceptron, meaning all layers are completely connected, because the data dimensions are too big. There is no available space information on CNN. The final efficiency may be quite low if we use the pooling layer. CNNs suffer a significant of information in the pooling layer, which lowers the spatial resolution, because only the most active neurons can be communicated between each layer during the transfer of neurons.

The above Figure 4 shows the pseudocode for convolutional Neural Network. CNNs won't be able to discern between variations in postures and other features as a result. Overtraining from every viewpoint is one technique to overcome this issue, but it typically takes more time and computer power. When dealing with extremely complicated field-of-view data that has a lot of overlap, mutual masking, and diverse backgrounds, traditional CNN cannot be identified properly. We will go over the rationale behind removing the fully connected layers, the parameters for the skip and pooling layers in the suggested approach, and the use of two convolutional layers as a skip layer. More specifically, the tail of CNN is frequently added with several fully-connected layers.

IV. RESULT AND DISCUSSION

The suggested model's recognition performance is calculated using the accuracy, sensitivity, specificity, and execution time parameters. To assess how well the process of existing classifiers like ANN, BPNN, and CNN. The proposed system of Ant Colony Optimization (ACO) with Convolution Neural Network (CNN) achieves the highest accuracy.

Figure 4. Pseudocode for CNN

> **CNN Algorithm Pseudocode:**
>
> **Input:**
>
> d: dataset,
>
> 1: real labels in the dataset
>
> W: Word-to-Vec matrix
>
> **Output:**
>
> Test dataset score for the CNN-trained model
>
> Assume that f is the 3D matrix of features.
>
> **For** i in dataset **do**
>
> Allow f to represent the sample i's feature set matrix.
>
> **For** j in i **do**
>
> Vj - vectorize(j, w)
>
> **Append** vj to f
>
> **Append** fi to f
>
> ftrain, ftest, Itrain, Itest Create a train and test subsets from the feature set and labels
>
> M-CNN (ftrain, Itrain)
>
> Score-analysis (i, Itest, M)
>
> **Return** score

4.1 Performance Evaluation Matrix

The effectiveness of the trained machine learning models may be evaluated by using performance evaluation metrics. In order to determine how well the machine learning model will work, it will be able to do this by using an unknown dataset. A confusion matrix is utilised to illustrate and provide a summary of the efficacy of a categorization strategy. This matrix of perplexity is seen in Figure 5.

True Positive (TP): Effective detection of the bounding box on the ground.

False Positive (FP): Incorrectly detecting an object that is present or incorrectly detecting one that is not.

False Negative (FN): A bounding box for ground truth that is hidden.

True Negative (TN): Because there are numerous bounding boxes in every image that shouldn't be recognised, a true negative (TN) outcome is irrelevant while discussing object detection.

Accuracy: By deviating from 100% in the error rate, the accuracy formula represents accuracy. Before we can assess accuracy, we must first assess error rate. Next, we divide the observed value by the actual value to get a percentage that represents the mistake rate.

Figure 5. Confusion matrix

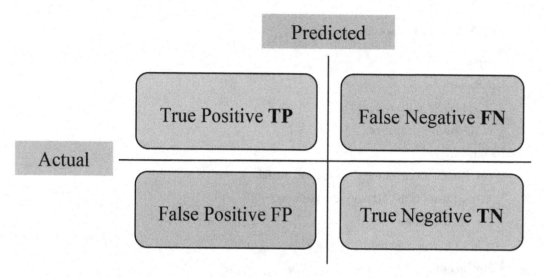

$$Accuracy = \frac{TP + TN}{TP + TN + FN + FP}$$

The output graph of accuracy from table 1 for existing and proposed system is clearly shown in the Figure 6. From this graph, our proposed ACO+CNN system outperforms the better accuracy of 96.23% comparing with the other current systems.

Sensitivity: Sensitivity is defined as the sum of positives divided by the number of accurate positive forecasts.

$$Sensitivity = \frac{TP}{TP + FN}$$

The results for sensitivity with the proposed and existing systems are shown in the aforementioned Figure 7 and are taken from Table 2. It shows the ACO+CNN gives higher sensitivity result of 94.52%, it is comparatively better than the current approaches.

Specificity: Calculating specificity involves dividing the entire number of accurate negative predictions by the total number of negatives. The specificity is calculated using the formula shown below.

Table 1. Accuracy

Algorithm	Accuracy (%)
ANN	84.64
BPNN	86.47
CNN	88.28
ACO+CNN	96.23

Figure 6. Accuracy graph

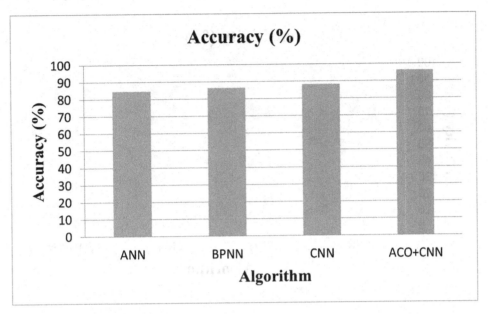

Table 2. Sensitivity

Algorithm	Sensitivity (%)
ANN	83.72
BPNN	86.25
CNN	87.61
ACO+CNN	94.52

$$\text{Specificity} = \frac{TN}{\left(TN + FP\right)}$$

The results for specificity with the proposed and existing systems are shown in the aforementioned Figure 8 and are taken from Table 3. It shows the ACO+CNN gives higher specificity result of 92.46%, it is comparatively better than the current approaches.

Time of Execution: The total time duration is how long the something lasts, from the beginning to the end.

The Table 4 explains the execution time of proposed system comparison with existing systems; the indicated results are plotted in the Figure 9. The proposed ACO+CNN take 12.75 milliseconds time duration. The results show that suggested system consumes minimum time duration comparing with the other existing systems.

The overall performances of a novel proposed hybrid ACOCNN system was well and good for the diseases diagnosis in medical technology using the Artificial Intelligence in intelligent healthcare sector.

Figure 7. Sensitivity graph

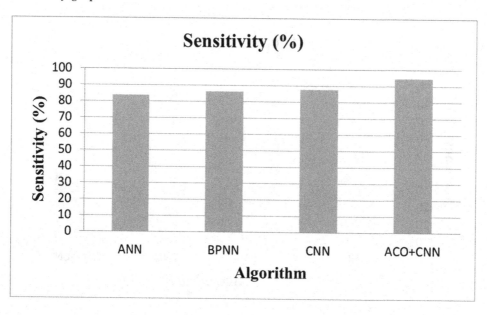

Table 3. Specificity

Algorithm	Specificity (%)
ANN	84.34
BPNN	82.69
CNN	89.18
ACO+CNN	92.46

Figure 8. Specificity graph

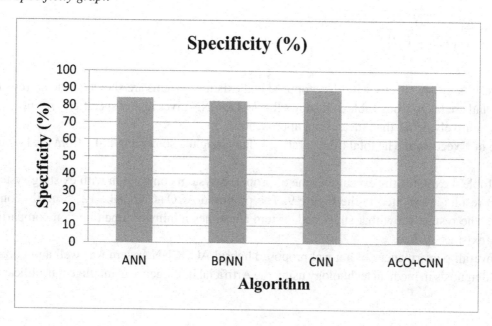

Table 4. Execution time

Algorithm	Execution Time (ms)
ANN	24.53
BPNN	18.58
CNN	20.43
ACO+CNN	12.75

Figure 9. Execution time graph

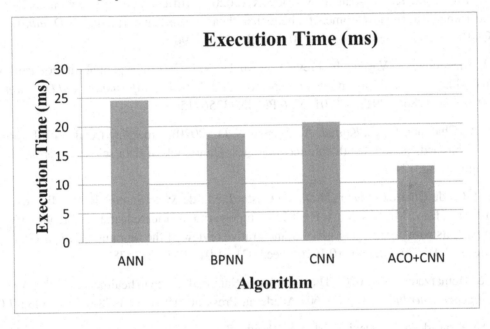

V. CONCLUSION

All things considered, smart healthcare has a bright future. People can take better care of their health on their own by using smart healthcare. When needed, quick, appropriate medical care can be obtained; the type of care provided will be more customised. Smart healthcare could result in reduced costs for healthcare institutions, lighter workloads for employees, unified material and information management, and improved patient outcomes. Research institutions may perform studies more quickly, more cheaply, and more effectively overall with the use of smart healthcare. Smart healthcare can help with macro decision-making by promoting preventative care uptake, lowering social medical costs, and improving the current condition of medical resource disparity. AI-based findings verify earlier hypotheses and evaluate risk in the diagnosis process. Many physicians use machine learning algorithms for prediction since they generate accurate data. As with any innovation, there are challenges along the way. In the case of artificial intelligence, these include heterogeneity, connectivity, and intricate data management issues. The answers to these problems have been discussed in this investigation. AI doesn't appear to be multitasking, and it hasn't advanced to the point where a physician can be fully replaced by it. Despite this disadvantage, further developments in the medical field have shown AI's supremacy. In addition to being a cutting-edge method, AI subsets like ACO with CNN are also frequently employed in the healthcare field. In comparison to the current methods, it attains a high accuracy of 96.23% while requiring only 12.75 milliseconds to diagnose a condition. The review's conclusions inspire young scientists to investigate and develop combinational strategies that integrate IoMT, AI, and nano-enabled sensing for efficient biosensing a crucial component of tailored illness control and treatment.

REFERENCES

Ahmed, Z., Mohamed, K., Zeeshan, S., & Dong, X. (2020). Artificial intelligence with multi-functional machine learning platform development for better healthcare and precision medicine. *Database (Oxford)*, *2020*, baaa010. doi:10.1093/database/baaa010 PMID:32185396

Amann, J., Blasimme, A., Vayena, E., Frey, D., & Madai, V. I. (2020). Explainability for artificial intelligence in healthcare: A multidisciplinary perspective. *BMC Medical Informatics and Decision Making*, *20*(1), 1–9. doi:10.1186/s12911-020-01332-6 PMID:33256715

Banerjee, A., Chakraborty, C., Kumar, A., & Biswas, D. (2020). Emerging trends in IoT and big data analytics for biomedical and health care technologies. Handbook of data science approaches for biomedical engineering, 121-152.

Barbosa, H. C., de Queiroz Oliveira, J. A., da Costa, J. M., de Melo Santos, R. P., Miranda, L. G., de Carvalho Torres, H., ... Martins, M. A. P. (2021). Empowerment-oriented strategies to identify behavior change in patients with chronic diseases: An integrative review of the literature. *Patient Education and Counseling*, *104*(4), 689–702. doi:10.1016/j.pec.2021.01.011 PMID:33478854

Bohr, A., & Memarzadeh, K. (2020). The rise of artificial intelligence in healthcare applications. In *Artificial Intelligence in healthcare* (pp. 25–60). Academic Press. doi:10.1016/B978-0-12-818438-7.00002-2

Dongari, S., Nisarudeen, M., Devi, J., Irfan, S., Parida, P. K., & Bajpai, A. (2023). Advancing Healthcare through Artificial Intelligence: Innovations at the Intersection of AI and Medicine. *Tuijin Jishu/Journal of Propulsion Technology, 44*(2).

Dzobo, K., Adotey, S., Thomford, N. E., & Dzobo, W. (2020). Integrating artificial and human intelligence: A partnership for responsible innovation in biomedical engineering and medicine. *OMICS: A Journal of Integrative Biology, 24*(5), 247–263. doi:10.1089/omi.2019.0038 PMID:31313972

Ganji, K., & Parimi, S. (2022). ANN model for users' perception on IOT based smart healthcare monitoring devices and its impact with the effect of COVID 19. *Journal of Science and Technology Policy Management, 13*(1), 6–21. doi:10.1108/JSTPM-09-2020-0128

Gilbert, S., Fenech, M., Hirsch, M., Upadhyay, S., Biasiucci, A., & Starlinger, J. (2021). Algorithm change protocols in the regulation of adaptive machine learning–based medical devices. *Journal of Medical Internet Research, 23*(10), e30545. doi:10.2196/30545 PMID:34697010

Manickam, P., Mariappan, S. A., Murugesan, S. M., Hansda, S., Kaushik, A., Shinde, R., & Thippe-rudraswamy, S. P. (2022). Artificial intelligence (AI) and internet of medical things (IoMT) assisted biomedical systems for intelligent healthcare. *Biosensors (Basel), 12*(8), 562. doi:10.3390/bios12080562 PMID:35892459

Minopoulos, G. M., Memos, V. A., Stergiou, C. L., Stergiou, K. D., Plageras, A. P., Koidou, M. P., & Psannis, K. E. (2022). Exploitation of Emerging Technologies and Advanced Networks for a Smart Healthcare System. *Applied Sciences (Basel, Switzerland), 12*(12), 5859. doi:10.3390/app12125859

Moerenhout, T., Devisch, I., & Cornelis, G. C. (2018). E-health beyond technology: Analyzing the paradigm shift that lies beneath. *Medicine, Health Care, and Philosophy, 21*(1), 31–41. doi:10.1007/s11019-017-9780-3 PMID:28551772

Nasr, M., Islam, M. M., Shehata, S., Karray, F., & Quintana, Y. (2021). Smart healthcare in the age of AI: Recent advances, challenges, and future prospects. *IEEE Access : Practical Innovations, Open Solutions, 9*, 145248–145270. doi:10.1109/ACCESS.2021.3118960

Rani, S., Chauhan, M., Kataria, A., & Khang, A. (2023). IoT equipped intelligent distributed framework for smart healthcare systems. In *Towards the Integration of IoT, Cloud and Big Data: Services, Applications and Standards* (pp. 97–114). Springer Nature Singapore. doi:10.1007/978-981-99-6034-7_6

Shaik, T., Tao, X., Higgins, N., Li, L., Gururajan, R., Zhou, X., & Acharya, U. R. (2023). Remote patient monitoring using artificial intelligence: Current state, applications, and challenges. *Wiley Interdisciplinary Reviews. Data Mining and Knowledge Discovery, 13*(2), e1485. doi:10.1002/widm.1485

Talukder, A., & Haas, R. (2021, June). AIoT: AI meets IoT and web in smart healthcare. In *Companion Publication of the 13th ACM Web Science Conference 2021* (pp. 92-98). Academic Press.

Tian, S., Yang, W., Le Grange, J. M., Wang, P., Huang, W., & Ye, Z. (2019). Smart healthcare: Making medical care more intelligent. *Global Health Journal (Amsterdam, Netherlands), 3*(3), 62–65. doi:10.1016/j.glohj.2019.07.001

Tripathy, S. S., Imoize, A. L., Rath, M., Tripathy, N., Bebortta, S., Lee, C. C., & Pani, S. K. (2023). A novel edge-computing-based framework for an intelligent smart healthcare system in smart cities. *Sustainability (Basel), 15*(1), 735. doi:10.3390/su15010735

Wang, G., Badal, A., Jia, X., Maltz, J. S., Mueller, K., Myers, K. J., & Zeng, R. (2022). Development of metaverse for intelligent healthcare. *Nature Machine Intelligence, 4*(11), 922–929. doi:10.1038/s42256-022-00549-6 PMID:36935774

Yaeger, K. A., Martini, M., Yaniv, G., Oermann, E. K., & Costa, A. B. (2019). United States regulatory approval of medical devices and software applications enhanced by artificial intelligence. *Health Policy and Technology, 8*(2), 192–197. doi:10.1016/j.hlpt.2019.05.006

Chapter 10

Human–Robot Safety Guarantees Using Confidence–Aware–Game–Theoretic Human Models With EMG Signal

R. Gokulakrishnan

Karpagam Academy of Higher Education, India

C. Balakumar

(iD) https://orcid.org/0009-0002-4655-2786

Karpagam Academy of Higher Education, India

ABSTRACT

This research provides a method for gesture recognition that integrates two separate recognizers. These two recognizers use the CAR equation to ascertain the hands sign. The robot's two main parts are its sending and receiving ends. Within the process of developing the same, three domains were specifically combined: biomedicine, which involved registering biosignals using analog channels composed of instrumental amplifiers; software development, involving microcontrollers, core processing (DSP), and the resulting control of the robot hand; PC software for tracking the registered biosignals; and mechatronics, involving the design and mechanical construction of the robot hand. The hand can control how much pressure is given to things because of the force sensor (FSR) in each finger. While developing a hand and wrist prototype that can rotate in response to EMG signal pulses, this was discovered.

I. INTRODUCTION

We investigate safety considerations for people in highly dynamic interactions with robots, such as autonomous vehicles merging approaching. Safety monitors are becoming a desired additional layer of safety, even though there are numerous ways that planning strategies incorporate safety constraints. With the use of these strategies, the planner may not only operate the robot but also anticipate when

DOI: 10.4018/979-8-3693-5767-5.ch010

a collision is likely to ability to detect approaching collisions is one of their key features. Thus, we attempt to determine ways to reduce the conservativeness of safety monitors without compromising their capacity to carry out their primary responsibility of guaranteeing safety explained by Lees, Michael J., and Monique C. Johnstone (2021). We suggest two methods to settle this dispute: Initially, we employ providing with limit protect by excluding to be extremely. The functionality of the model in Figure 1 can be viewed online.

We are interested in guaranteeing vehicles merging different methods different are now considered an extra safety precaution that is wanted. With these methods, the planner can still operate the robot while keeping an eye out for potential collisions and taking over in case of one. A key component of predicting approaching usually, the worst-case scenario is used to determine this approach that is widely used to analyze interactions that protects from human directives is called reverse reachability analysis. Hao, Jianli, et al (2020) said about Safety monitors, in their overzealous interference, wind up preventing progress while simultaneously ensuring safety. Hence, we search for methods to make safety monitors less cautious so that they can continue to successfully fulfill their primary responsibility of guaranteeing safety. The challenge here is that any human behavior model that is substituted for the zero-sum game assumption runs the risk of being inaccurate and causing a loss of safety. We suggest using two distinct approaches to mediate this conflict:

- Maintain the limit collection we protect against by removing those that the human behavior model determines to be highly unlikely;
- Examine the model's efficacy on the internet. The robot automobile (white) and a nearby human-driven car (orange) merge into a roundabout. To the left Humans adapt to the robot despite its extraordinary caution and protection against the complete backwards reaching tube (BRT).

By assuming the effects of the robot on the human, our Bayesian BRT minimizes the set of hazardous states. The robot quickly returns to the full BRT when it detects a deviation in human behavior from the model. Suits the human, and uses this data to adjust the limitation in the worst case, return when the model is completely erroneous.

Figure 1. Basic of human robot interaction

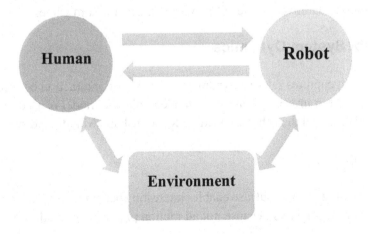

1.1 HRI Can Be Divided Roughly Into Four Areas of Application

1. These include handling items production obtaining and delivering, parts, prescriptions, hospitals, warehouses. These referred to as telerobots, can environment own locations. Telerobots are computer programs that enable them to carry out a restricted range of tasks automatically.
2. Remote of spacecraft, aircraft, land vehicles, and underwater vehicles for unreachable areas. When carry out manipulation mobility activities sync with Teleoperators are people who continuously control movements from a distance by using human intervention. If a computer is routinely reprogrammed by a human supervisor to perform certain duties within the greater objective, the computer is referred to as a telerobot.
3. Automated automobiles human occupants, such as automated commercial airplanes and railroads.
4. Human-robot social interaction, such as when robots are used to amuse, instruct, comfort, and assist the elderly, the disabled, and kids with autism said by Onnasch, Linda, and Eileen Roesler (2021).

1.2 Human Supervisory Control of Robots for Routine Industrial Tasks

There are undoubtedly fully automated vehicle and other product production lines that readers are aware of. Insofar as supervisory control functions such as planning, training, monitoring automated control, problem-solving, and experience-gathering require human operators, these machines are referred to as telerobots. The Boston-based company Rethink Robotics this meant work to humans the visible eyes of instead permitting, this enables its present focused on. Baxter can also be programmed to move its hands manually in order to teach it tasks involving manipulation or humans.

The questions of which recognize inaccurate remain unsolved. Despite the fact that these "human-in-isolation" models are popular and models would step to prevent the planner from merging in front of the human humans and robots, consider robot as well as human influence by Kumar, Shitij (2020). To further spectrum against which can defend itself, we propose extending these kinds of models to safety monitoring as well. Because human models are not perfect, relying only on them could result in the elimination of constraints that humans ultimately carry out. Nevertheless, this would mean less conservatism. Our methodology assesses adjusts constraint using determine whether prediction accuracy is decreasing and, if so, to step up conservatism. Expanding on earlier work in safe planning, our method for automated adaptation cognitive abilities of players in a general-sum game.

1.3 Human-Robot System Dynamics

A dynamical system that captures relative dynamics in a pairwise interaction between the robot cars is considered in our simulation investigations. An extended unicycle model is used to simulate a fidelity bicycle model is used to model the robot car said by Khoramshahi, Mahdi, and Aude Billard (2019).

1.3.1 Robot Planner

This leads to some modeling mistake but also enables interesting behaviors from the robot and emphasizes how important it is to verify the real because not all motion planners are made equal. The present uses method dangerous regardless of how complicated the upstream robot planner is. Stated differently, when the human-driven vehicle gets close to the BRT boundary, the safety controller takes over the planner controls by Sanchez-Ibanez, Jose Ricardo (2021).

1.3.2 Simulated Humans

We model three different human types: unmodeled non-Stackelberg human driving with constant controls, a (noisy) suboptimal Stackelberg human, and a rational Stackelberg human. Metrics we assess the robot's mobility plan's safety performance, conservatism, and total reward when depending on different safety monitors explained by Vianello, Alvise, et al (2019).

1.3.3 Reward Improvement Percent

The robots performed trajectory's % reward increase when it uses our Bayesian BRT instead of a baseline BRT technique (the appropriate baseline is chosen based on our case study). In terms of math, given any baseline safety technique discussed by Putra, Maha, and Nurevi Damayanti (2020), This study describes a method for assessing danger awareness. More precisely, we are shown to be able to model how this concept affects people's decisions using a binary variable known as the danger awareness coefficient. This work also proposes a method by which the robot constantly learns the value of this coefficient by making on-the-spot observations. It should be noted that a probabilistically safe strategy is one in which the likelihood of a human-robot collision is less than a certain value.

The five portions of this essay are summarized below: Session 2 illustrates the shortcomings of the current system in terms of artificial intelligence, machine learning, and human-robot interaction. In the third session, the suggested Human Robot Interaction analysis method was presented. The anticipated system's output is shown in Session 4. A summary of the recommended methodology for human-robot interaction wraps up Session 5.

II. LITERATURE REVIEW

In this paper, we present an overview summarizing in order to identify the scientific and technological advances that have enabled HRI to develop into a separate field. Rather than just restating and reformulating previous research, we argue that HRI with important structure the assessment from the standpoint of the designer, the topic entails considerable interdisciplinary mixing from multiple scientific and engineering fields.

An author Połap, Dawid (2022) focus components mold after a study of important elements in the development of HRI as a field. The application domains that underpin a large portion of contemporary HRI are then described. Many of these issues have significant societal ramifications and are quite difficult. We classify application domains into the two broad categories of remote and proximate interactions that were previously discussed, and we highlight significant, thought-provoking, or influential work that falls into each of these two categories. We then proceed to describe barrier challenges and common solution concepts that cut across application areas and types of interactions. The review is then summarized and we quickly point out relevant work from other domains involving the interaction of humans and technology.

2.1 Safety for Robots Operating Around Humans

According to Zacharaki, Angeliki, et al (2020) the robot uses algorithms future might, and prepares its movements despite being cautious actions by the robot, especially when interaction is close at hand. In previous studies, "human-in-isolation" and empirical models were used to limit the forward accessible set of human controls and, eventually, them. Since naturally adapt its settings, more is added on the other hand, reverse based on follow regulations. This method, complete frequently suffers from misclassifying safe states as dangerous due to the full control power. Lessening this conservatism, subsequent work has attempted to use data-driven human trajectory forecasts to constrain the range however, this approach depends too much on the data-driven forecasts and is not able to spot hence, deteriorates does so endangering.

2.2 Structured Human Decision-Making Models

The researcher Shrestha, Yash Ra et al (2019) field of navigation has witnessed a great deal of study and implementation of "human-in-isolation wherein agent thought to behave to that describe individuals make decisions about what to do by considering the behavior of others. Research published recently value generic provide modeling in study, leveraging these game theoretic human models. Ancient Greek and Chinese texts mention human-like that employed as workers and attendants, illustrating human obsession with creating a mechanical. In the past century, robots have captured the attention of both imagination and research and development funds more than any other type of automaton. Some contemporary models are even beginning to resemble the romanticized fictions from ages before showcased in these publications, which lay the groundwork for neurocognitive perspectives on human-robot interactions.

When the author Eckstein, Maria K., and Anne GE Collins (2020) humans and an industrial robot arm perform identical tasks connected the action observation (AON) exhibit different reactions. This was discovered in an early study on the adaptability of this network. Supporting these results was human and robotic individuals the AON is actually action independent produced according to (fMRI) investigations published by researchers. These results expanded on earlier discoveries about duplicated them increased more has been explained as a result of the unfamiliarity of robotic motion, along with other first surprising findings. It is unknown to what extent human observers imbue inanimate objects with emotions and goals, even if witnessing robotic movements stimulates brain regions associated with action. Conditions in which brain reactions linked with empathy could arise while comparing humans and robots experiencing simulated misery, or when trying to understand the intentions of, are still being explorer participants in an MRI only metalizing-related actions they someone else was in charge.

2.3 Task Dynamic Analysis

A key toolset used by Chen, Ruifeng, et al (2020) relatively separate field study in the division labor machines popular hierarchy of "man" and "machine" talents is outdated and hasn't been updated in a long time. Try creating a robot that can gently assist the elderly and crippled with getting in and out of bed or using the restroom if ergonomics is merely a fancy word for outdated human factors. Many human caretakers today fill this task, often at the expense of back pain. Planning and simulating tasks in terms of force, energy, time, space, and money has never been more challenging, even with the use of virtual reality visualization technologies. Furthermore, experience shows that the most effective method for determining a robot's physical form is task context. An additional query is whether it makes sense to construct humanoid-looking general-purpose robots. Analyzing HRI tasks to identify the ideal physical form presents another difficulty.

2.4 Teaching a Robot and Avoiding Unintended Consequences

The author O'Brolcháin, Fiachra (2019) robot hand can be controlled by a human using geometric commands, but it requires symbolic language to tell it when and how to move as well as what to avoid. The rapid progress in speech interpretation promises the ease of operating robots. Unexpected results are, nevertheless, highly likely. Before signaling that the robot is ready to move forward, human supervisors could see what the robot does in response to spoken instructions via a real-time virtual reality simulation. This technique, which updates its models of the regulated process continuously, would essentially be a continuation of predictor displays through the use of model extrapolation.

2.5 Interfacing Mutual "Mental" Models to Avoid Working at Cross-Purposes

There has long been a theory that humans and robots are just internal (mental) replicas of one another. Modern computer vision techniques enable data to be stored as conduct and monitor human behaviour explained by Lai, Hsueh-Yi (2023). Human-robot interaction (HRI) challenges involve deriving human mental models according to AI experience visual pattern recognition, language translation.

2.6 Role of Robots in Education

The researcher Alam, Ashraf (2022) tells about adults with cognitive disabilities or young children who are non-readers cannot even choose to study this way. When students interact with a real teacher or cleaner, learning is nearly always improved. Since Paper experiments, in which children were taught to educate a mechanical "turtle," the robot has been taken into consideration in conversations concerning the future of education. In addition to adding amusement, it can serve as a talking or teaching avatar, depict react replies by offering advice or criticism. Robots that can learn from humans and other common workshop topic. One significant difficulty is figuring out all ages and abilities learn from robots.

2.7 Lifestyle, Fears, and Human Values

The Constant, Aymery, et al (2020) explain about his science fiction/horror performance first caused entertainment, it appeared ridiculous. He being to introduce robot these days, we watch new horror movies where robots invade people's personal space and take over jobs. Due both its advantages and disadvantages for society, robots have gained a lot of attention in the media. Naturally, there are trade-offs that merit careful consideration. These include the following: robots enhancing human security versus turning into spies, killers, and tiny UAVs; robots improving human security versus taking jobs away from humans; and robots acting as helpful assistants that increase it. I think the average less realities human factors scientists are thus, human have a duty to take part making about matters awareness campaigns.

III. SYSTEM DESIGN

Consider an interaction between a human and a robot when both are heading to two different ongoing destinations. We will use driving following formulation of problem for a general interaction to illustrate the use of the proposed method. In Figure 2, the HRI is shown.

3.1 Experimental Setup

A roundabout, an intersection, and a highway. We modelled the driving environment and sensitive participants drove in person. After a predetermined period of time, each interaction ended. After watching an animated video, online participants choose their course of action from a list of options. Points were given to who, both physically and digitally, avoided collisions, stayed on the road, and moved up in their lane. We presented the subject's current score at each stage of the trial.

3.2 Danger Signalling System

Even if the majority of studies in the literature suggests otherwise, we assume that people can be influenced by the actions of robots. More specifically, we believe it is ludicrous to presume that humans behave irrationally in human-robot scenarios and deliberately disregard the robots, which greatly diminishes the efficacy of robots.

The principal aim of utilizing danger signalling is to enhance the robot's capacity to gauge the human's level of danger awareness and to ensure that the robot continues to operate efficiently towards the goal state, unaffected by the potentially hazardous behaviors of the human. We suppose that the robot has a suitable pre-collision that signals or indicators to warn the human of the impending collision in order to resolve this problem. The binary variable dR indicates the on/off status of the warning signals.

Figure 2. Human robot interaction structure

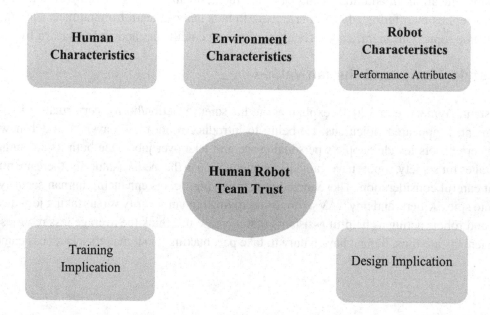

Where dR = 0 if the signal is off and dR = 1 if it is on.

3.3 Human Action Prediction

Our presumption is that the robot is familiar with QH g ($\bullet$) and QH s ($\bullet$), as was previously mentioned. Assuming this, the robot is able to forecast the human's behavior as actions dependent on subjective Operating As an Assuming that the pedestrian wishes to go from one sidewalk to the other, the goal objective function QH g ($\bullet$) is easily formulated. For pedestrians1, the safety objective function QH s ($\bullet$) can be learned based on their behavioral pattern.

The robot uses a game-theoretic predictive model of human behavior to plan utilizing model-predictive control. The robot determines an open-loop control trajectory u* R at each time step by solving:

maxuR EuH [RR(x0, uR, uH) | P(uH | x0, uR)]

The conditional expectation in this case replaces game-theoretic human paths with fixed rationality. In figure 3, the algorithm planning scheme is explained. Re-planning in a receding-horizon method, original the next time Remember that, consistent with earlier always engage in the interaction as a fixed follower.

3.4 Dual Control Effect

Therefore, the human's future uncertainty, as determined by their belief states, may be influenced by optimal policy uR,* t:= πR 0 (^xt, ^bt). Therefore, the optimal policy uR,* t, as stated explicitly in Definition 1, possesses the dual control characteristic. The concept of optimality, which the policy delivers,

Figure 3. Algorithm planning scheme

1. Observe the human's state $x_H[t]$ and the state of the robot $x_R[t]$.

2. Compute the mixture distribution $P(u_H|x_H[t],x_R[t];\beta)$ for every $u_H \in U_H$ and β via.

3. Compute the probability distribution of human's states $P(x_H[k])$ for $k \in \{t+1,t+T_R\}$ via.

4. Compute the probability of collision $P_{Coll}[k]$ for $k \in \{t+1,t+T_R\}$ via.

5. Determine the action of the robot $u*_R[t]$ and the on/off status of the danger signal d_R via.

6. Observe the human's action $u_H[t]$.

7. Update the belief about the danger awareness coefficient via, i.e.,

allows the robot to optimize its predicted performance target while simultaneously actively learning the uncertainty of the human. To emphasize, the optimal strategy is to automatically communicate with human agents in order degree of improves efficiency.

By choosing its own actions, the self-driving automobile was able to solve the Stackelberg game. The robot was awarded for preventing crashes and slowing down.

We chose:

$r_R(s, U_R, U_H) = - S_H - 10 \cdot 1\{\text{collision in } s\}$

$r_H(s, U_R, U_H) = S_H - 10 \cdot 1\{\text{on road in } s\} - 100 \cdot 1\{\text{collision in } s\}$

The robot combines and attempts to encourage others to give way in order to slow down their progress in their lane.

3.5 Human Decides

The hazard awareness coefficient β and the objective QH g (●) and QH s (●) serve as the foundation for human decision-making. We consider the robot to be conversant with QH g (●) and QH s (●). This supposition is reasonable given that the robot can either directly acquire these functionalities from the system designers or acquire them through prior interactions with humans. However, in practice, any supposition about the value of the coefficient β will often be wrong; humans can perceive safety in a different way or pay less attention than normal when they are among robots because they think the robot should remain a safe distance from them therefore, reliable human in future, the robot must have quick reasoning over the value of β.

3.6 Agent Dynamics

Our dynamical system model of each agent has the notations xR ∈ Rn for the robot state and xH ∈ Rm for the human state.

xi = f (xi, ui) i ∈ [R, H]

Furthermore, we permit $\xi(\tau ; xi, ui(\bullet), t)$ to represent the agent state at time τ. Control ui(●) is implemented over the time horizon [t, τ], starting at time t at the state xi. The thought a task objective, desired for which It is imperative that, in the course of doing its duties, the robot never breaches any safety laws. C denotes a group of states that the robot should stay out of in order to preserve safety. These conditions could indicate actual collisions with people. The objective of this work is to the accessible here, xr and xh stand for respectively T(), the track reward, pays the agent for the distance both ahead of the human player and down the track.

Using centerline points evenly distributed along the course, the nearest point is returned as the distance between the person and the robot at that precise moment. The safety cost is repaid based on the level of risk, projected speeds, current player conditions, and anticipated future opponent behavior. Furthermore, the safety cost for the separation between agents and the distance to the racecourse borders is simulated using a sigmoid curve, where agents who are closer together pay a higher cost than those that are farther apart.

Additionally, this final step is to take and scale the difference between the opponent's expected position, xo, and the agent's position. First, we will suggest a probabilistic Boltzmann model to forecast human behavior in the sections that follow. Next, a mechanism for updating the belief on the danger awareness coefficient's value will be suggested. Lastly, how the robot can forecast the likelihood of a collision in the future will be covered.

IV. RESULT AND DISCUSSION

The proof that complexity exists the degree to which a traffic signal's transient adjustments to traffic circumstances are intricate can significantly affect how the signal flows within a lane. It is possible to move the traffic signal as well as the two road segments' locations. Furthermore, a single city has the power to alter the traffic light control system. In the urban environment, there are several traffic patterns the road. Compared to a typical transportation system, the spatial linkages between these two sectors are more intricate.

4.1 Initial Human

Trials and Encounters to establish baseline methodologies for policy benchmarking and initial assessments for curriculum training, fifty distinct human opponent runs with full state information were recorded. Through the study of human runs from two player opponents, closely related races showed indications of similar risk levels and differences in danger. The normal course of races was to level off at a constant danger threshold. Risk was always changing since it was frequent for novice players to face opponents who had aggressive driving and control skills. This shows that risk can be used as a useful metric to encourage a range of behaviors and may also be useful in enhancing the accuracy of future agent.

4.2 Initial Training

So far, a rudimentary policy that was educated through simulation has been put into place for both and human participants. Constant initiatives include developing curricula, improving incentives, and optimizing safety and risk awareness. The neural network's initial training yielded error rates of 12% for calculated risk, 43% for adaptability, and 24% for future state. In order to acquire more accurate adaption parameters and collect more human trials to produce a more plausible future state in figure 4, these are the topics of ongoing study.

The primary innovations we offer in this study are efficient, probabilistically-safe motion planning while maintaining tractable methods for joint planning and prediction. Our real-time robust motion planning is done using the reachability-based FaSTrack framework. Robots employ a basic model that disregards potential interaction effects in order to forecast human movements and ensure real-time feasibility. Since the presumed model will be a simplification of actual human motion, we employ confidence-aware predictions that are more cautious more humans diverge from the model. Lastly, groups of robots consecutively plan using a priority ordering that has been previously set.

We will compare and contrast our innovative predictor with throughout the study. That being said, all prediction techniques eventually yield a collection can employ detection $t + \tau$, which represents the probable human situations at a future time, is defined as follows: $\mathcal{K} t (\tau), \forall \tau \in [0, N]$.

Hardware demonstration by Figure 5 preserving against purposeful humans, external disruptions, and internal dynamics. Each quad copter's projected trajectory is displayed, and a box containing the tracking error bound is displayed around it. Before each human is displayed in pink the probability distribution

Figure 4. Fine tuning for risk awareness and safety parameters

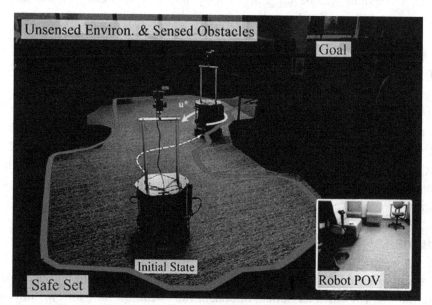

Figure 5. Hardware demonstration of real-time multi-agent

over their future mobility. Robots plan dangerous motions with confidence described (top) our method (bottom), robot plans are safer even with a stated model since we only trust eliminate utterly implausible order for the robot to escape the anticipated human state when utilizing the worst-case prediction, it must completely exit the surroundings.

The robot's objective $u_R(t)$, $t \in [0, \cdot T]$, so attains its goal g_R by $\cdot T$ while avoiding collisions with the person and any known static impediments. Recessing horizon planning will be the method used to

Figure 6. Bridges robust control and intent-driven

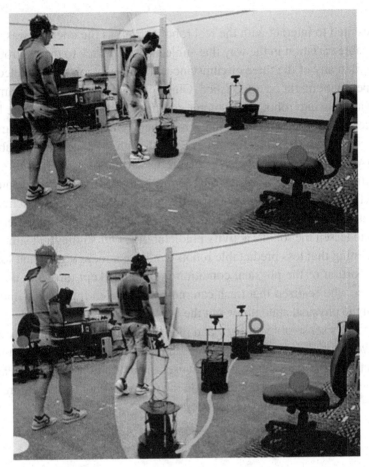

overcome this problem in this work. Unfortunately, the robot cannot plan collision-free routes because the human's future states are unknown in advance. Instead, it must anticipate human movements.

Our goal in this work is to find a method that combines intent-driven predictors in Figure 6 with robust control to create a predictor that is safer to reduce conservatism while still being more resilient to given models and priors. Our main notion is to trust inform the extremely, and so calculate a constrained forward reachable set. We will not, however, base our predictions on the precise likelihood of every action in our model, in contrast to intent-driven predictors. Instead, we split the collection of human behaviors into two distinct sets of likely and unlikely acts using similarly whole forward reachable we then use this approach to forecast human motion we treat plausible prediction problem that arises from using this limited control set is easily formulated and solved using current robust control techniques and tools. In order to ensure safety for continuous-time, nonlinear dynamical systems, we apply Hamilton Jacobi (HJ) reachability analysis.

V. CONCLUSION

The subjects were excited to interact with the real robot as well as the one that was displayed on video. There was no noticeable variation in the way that subjects waved back to the real robot and the one that was shown on camera in any of the three circumstances. If the easy job is completed, the book-moving paradigm of the experiment might be established. Our proposal was to integrate game theoretic models with confidence-awareness into robot safety monitors. Limiting the set of possible human controls according to the degree to which autonomously dangerous the whole investigations demonstrate on derivative actual and our control constraint architecture. Even though they were first unable to decide which pile of books to move, the majority of participants in the virtual group (90%) moved a pile of books after getting instruction. Progress toward ongoing human-robot communication can be made with our research. For powerful robots, we first demonstrated that a shared basis can function effectively in the short term, but eventually, people grow acclimated to these strong behaviors. Afterward, we proposed three modifications to lessen the robot's activity predictability. Our studies and simulations corroborate these changes, suggesting that less predictable robots can have a more significant long-term influence. However, a sizable portion of the physical condition individuals kept throwing books in the garbage. Even when only those who realized that trash can move were taken into consideration, a much higher number of people in the physical state threw out the books. This suggests that the physical presence of oneself increased people's sense of legitimacy and motivated them to fulfill a request that was unusual. As demonstrated by centered fast updating of the human's forward accessible tube allows for safer human-robot interactions. Our paradigm's potential uses in autonomous navigation and assisted including human-robot interaction should be fascinating in the future. We will also explore large-scale human population work and closed-loop human-robot interactions.

REFERENCES

Alam, A. (2022). Social robots in education for long-term human-robot interaction: Socially supportive behaviour of robotic tutor for creating robo-tangible learning environment in a guided discovery learning interaction. EC*S Transactions, 107(1)*, 12389–12403. doi:10.1149/10701.12389ecst

Chen, R., Xu, C., Dong, Z., Liu, Y., & Du, X. (2020). DeepCQ: Deep multi-task conditional quantification network for estimation of left ventricle parameters. *Computer Methods and Programs in Biomedicine, 184*, 105288. doi:10.1016/j.cmpb.2019.105288 PMID:31901611

Constant, A., Conserve, D. F., Gallopel-Morvan, K., & Raude, J. (2020). Socio-cognitive factors associated with lifestyle changes in response to the COVID-19 epidemic in the general population: Results from a cross-sectional study in France. *Frontiers in Psychology, 11*, 579460. doi:10.3389/fpsyg.2020.579460 PMID:33132989

Eckstein, M. K., & Collins, A. G. (2020). Computational evidence for hierarchically structured reinforcement learning in humans. *Proceedings of the National Academy of Sciences of the United States of America, 117*(47), 29381–29389. doi:10.1073/pnas.1912330117 PMID:33229518

Hao, J., Li, M., Chen, W., Yu, L., & Ye, L. (2022). Experimental research on space distribution of reverse flow U-tubes in steam generator primary side. *Nuclear Engineering and Design, 388*, 111650. doi:10.1016/j.nucengdes.2022.111650

Khoramshahi, M., & Billard, A. (2019). A dynamical system approach to task-adaptation in physical human–robot interaction. *Autonomous Robots, 43*(4), 927–946. doi:10.1007/s10514-018-9764-z

Kumar, S., Savur, C., & Sahin, F. (2020). Survey of human–robot collaboration in industrial settings: Awareness, intelligence, and compliance. *IEEE Transactions on Systems, Man, and Cybernetics. Systems, 51*(1), 280–297. doi:10.1109/TSMC.2020.3041231

Lai, H. Y. (2023). Breakdowns in team resilience during aircraft landing due to mental model disconnects as identified through machine learning. *Reliability Engineering & System Safety, 237*, 109356. doi:10.1016/j.ress.2023.109356

Lees, M. J., & Johnstone, M. C. (2021). Implementing safety features of Industry 4.0 without compromising safety culture. *IFAC-PapersOnLine, 54*(13), 680–685. doi:10.1016/j.ifacol.2021.10.530

O'Brolcháin, F. (2019). Robots and people with dementia: Unintended consequences and moral hazard. *Nursing Ethics, 26*(4), 962–972. doi:10.1177/0969733017742960 PMID:29262739

Onnasch, L., & Roesler, E. (2021). A taxonomy to structure and analyze human–robot interaction. *International Journal of Social Robotics, 13*(4), 833–849. doi:10.1007/s12369-020-00666-5

Połap, D., Włodarczyk-Sielicka, M., & Wawrzyniak, N. (2022). Automatic ship classification for a riverside monitoring system using a cascade of artificial intelligence techniques including penalties and rewards. *ISA Transactions, 121*, 232–239. doi:10.1016/j.isatra.2021.04.003 PMID:33888294

Putra, M., & Damayanti, N. (2020). The Effect of Reward and Punishment to Performance of Driver Grabcar in Depok. *International Journal of Research and Review, 7*(1), 312–319.

Sanchez-Ibanez, J. R., Perez-del-Pulgar, C. J., & García-Cerezo, A. (2021). Path planning for autonomous mobile robots: A review. *Sensors (Basel), 21*(23), 7898. doi:10.3390/s21237898 PMID:34883899

Shrestha, Y. R., Ben-Menahem, S. M., & Von Krogh, G. (2019). Organizational decision-making structures in the age of artificial intelligence. *California Management Review, 61*(4), 66–83. doi:10.1177/0008125619862257

Vianello, A., Jensen, R. L., Liu, L., & Vollertsen, J. (2019). Simulating human exposure to indoor airborne microplastics using a Breathing Thermal Manikin. *Scientific Reports, 9*(1), 8670. doi:10.1038/s41598-019-45054-w PMID:31209244

Zacharaki, A., Kostavelis, I., Gasteratos, A., & Dokas, I. (2020). Safety bounds in human robot interaction: A survey. *Safety Science, 127*, 104667. doi:10.1016/j.ssci.2020.104667

Chapter 11
Image Steganography–Embedding Secret Data in Images Using Convolutional Neural Networks

N. Nissi Angel

Department of ECE, Velagapudi Ramakrishna Siddhartha Engineering College, Vijayawada, India

Gunnam Suryanarayana

Department of ECE, Velagapudi Ramakrishna Siddhartha Engineering College, Vijayawada, India

Siva Ramakrishna Pillutla

School of Electronics Engineering, VIT-AP University, Amaravati, India

Kathik Chandran

Jyothi Engineering College, India

ABSTRACT

Image steganography methods use manual features for hiding payload data in cover images. These manual features allow less payload capacity and also cause image distortion. In this chapter, the authors detail a CNN-based network image steganography. The major contributions are twofold. First, they presented a CNN-based encoder-decoder architecture for hiding image. Secondly, they introduce a loss function, which checks joint end-to-end encoder-decoder networks. They evaluate this architecture on publicly available datasets CIFAR10. The results indicate an increase in payload capacity with high peak signal-to-noise ratio and structural similarity index values.

DOI: 10.4018/979-8-3693-5767-5.ch011

1. INTRODUCTION

Massive amounts of data have been flowing over the internet in the modern world as its accessibility has improved, making the internet more appealing to individuals. In addition to the aforementioned critics, the emergence of wireless communication technology has resulted in the devices connecting to the internet. Individuals fear as a result of data risks and securities. Because of the sensitive nature of private information, this flexibility raises a slew of issues in terms of data privacy and security. Consider the following example: personal data in a medical report reflect the entire state of the patient as well as the necessary diagnosis. In this context, intruders may generate incorrect data during transmission, resulting in incorrect patient diagnosis.

To overcome these challenges, one project that has sparked concern is data protection. Data hiding or embedding is the process of embedding any digital data within another digital data without causing any distortion for identification or copyright. As a result, images, texts, audios, and videos are used as a medium for secret communication.

Watermarking and steganography are two major perspectives that would be considered when studying data shielding or hiding in general. Watermarking's main applications/objectives are ownership assertion, content authentication, fingerprinting, and the message inserted is related to the cover (Kumar & Kumar, n.d.). Whereas the steganography process conceals unrelated/irrelevant data. The goals of steganalysis involve identifying potentially concealed content within data packets, determining the presence of encoded payloads, and subsequently extracting these payloads. Consequently, the primary hurdles in achieving effective steganography can be outlined as follows: Numerous researchers have dedicated their efforts to developing steganography algorithms aimed at impeding the detection and extraction of concealed information through steganalysis. They also strive to make it challenging for the Human Visual System (HVS) to discern any subtle alterations that may occur in the original data, whether it be in the form of audio, images, or videos, after the concealment process. Furthermore, there has been a noticeable uptick in the utilization of multi-level steganography techniques in recent times (Alanzy et al., 2023). However, steganography is preferred over watermarking due to its inherent properties such as hiding capacity or payload, imperceptibility and robustness. Furthermore, cryptography can be used in this context, but it has its own drawbacks. The study of secure communication methods, such as encryption, that only the message's sender and intended recipient can access, is known as cryptography. In the realm of cryptography, two key texts come into play: plaintext and ciphertext. Plaintext refers to data that has not undergone encryption, remaining in its original, unencrypted form. Ciphertext, on the other hand, represents the encrypted counterpart of plaintext, created through the application of encryption techniques (Simmons, 2019).

Steganography is the process/practice of concealing a secret message behind other data, so the only data types used in this procedure are secret and covered data. In this case, the primary goal is to hide a secret object it may be text/image in a cover object without modifying its characteristics. The two major factors affecting change are the cover object and the amount of data to be concealed. There are many traditional methods to implement steganography. LSB (Least Significant Bit method) is a simple method of hiding images. This method allows the secret image to hide in the least significant bits of the cover image (Chan & Cheng, 2004; Kavitha et al., 2012). Though it is easy to hide, it's pretty simple to spot the secret image. PVD (pixel value differencing) is a method which divides the image into blocks of two consecutive pixels, and then calculates the difference value for each block. PVD (Liu et al., 2020; Sahu & Swain, 2016) has great data hiding capacity but it degrades the image quality. Another method is DTC which is a frequency domain technique in which the image is transformed into frequency domain from spatial domain. This method is more complicated than the spatial domain (Patel & Dave, 2012). The DFT-based technique is similar to the DCT-based technique (Hemachandran, 2016), but it is less resistant to severe geometric distortions because it employs the Fourier transform instead of the cosine.

The primary benefit of the Discrete Fourier Transform (DFT) lies in its substantial improvement in computational precision without requiring additional computation time. It offers significantly enhanced efficiency and swiftness.

For image steganography, discrete wavelet transforms can be employed. A higher quality photograph uses quite a lot of disc space. DWT is used to shrink an image's size without sacrificing quality, increasing resolution (Chen & Lin, 2006; Narasimmalou & Allen Joseph, 2012). Spread spectrum image steganography (SSIS) disguises a message within an image as Gaussian noise (Marvel et al., 1999). The degradation of the image is invisible to the human eye at low noise power levels, but becomes visible at higher levels as "snow" or speckles. There are many other methods such as Image steganography based on canny edge detection, dilation operator and hybrid coding (Gaurav & Ghanekar, 2018) and a genetic algorithm based steganography using discrete cosine transformation (Khamrui & Mandal, 2013). Instead of plain regions, the best steganography involves hiding data in high frequency filled and noisy portions of an image while causing less perturbation. We present an efficient, automated steganography method in this paper that causes the least amount of distortion to the cover image while concealing a secret image. To accomplish this, a convolution neural network is created that merges data and extracts the best attributes from both cover and secret images (Rahim & Nadeem, 2018; Subramanian et al., 2021). The main advantage of our method is that it is not specific and can be applied to a wide range of images.

2. METHODOLOGY

Figure 1 depicts the overall workflow of the proposed method, which is comprised of three modules: preparation, hiding, and extraction networks. The preparation network extracts secret image features, and the hiding network combines cover image and secret image features. The extraction network, on the other hand, extracts the secret image hidden in the stego image, which is the output of the hiding network.

2.1 Preparation Network

The preparation network extract features from the cover and secret images rather than processing them in their raw form. High resolution images have many noisy regions and also redundant data, we can extract the important features from those regions. The input size should be in x*x*y format, which represents the width, height, depth dimensions. The width and height both are represented by x because they should be in same size.

The size of the secret image can be any size but should be equal to or less than cover image. The preparation network extracts the useful features. This network consists of convolutional layers, each composed of filters. As we move up the layers, the filter sizes grow larger. Smaller filters primarily capture basic image features like edges, while larger filters, when employed, allow for the extraction of a broader range of features.

Figure 1. Network architecture of image steganography using convolution layers

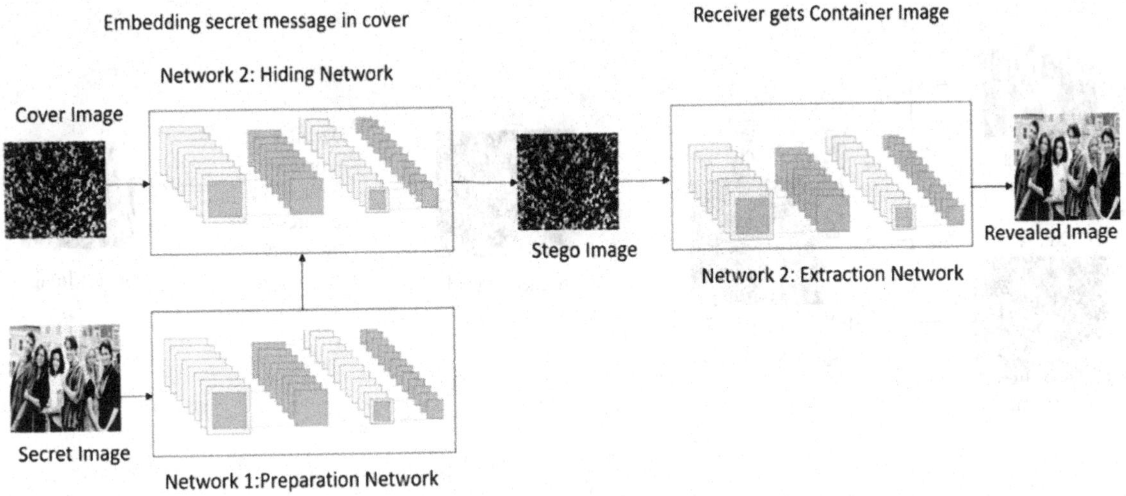

2.2 Hiding Network

The hiding network is also functions as the encoder component of the architecture. This network mainly extracts the important features of the cover image. This hiding network also combines the features of the cover image with those of the secret image. The features extracted from cover image are noisy regions. Based on the size of the filter used the features are extracted. In the hiding network, first the filter is increased and then it is decreased. Later ReLu activation is introduced to attain maximum linearity. Finally, a 3-unit filter is employed to transform the feature vector image into a stego image.

2.3 Extraction Network

The work of extraction network is to find the hidden image and to extract it from the cover image. This CNN network gives the best results. The filters and filter size that are used in the network are fine tuned.

Within the extraction network, the encoder phase comprises five convolutional layers where the filter sizes progressively increase. This is followed by a decoding phase consisting of five convolutional layers, with filter sizes decreasing. Additionally, three filters are applied to eliminate the perfectly concealed image from the cover image. The following equation reduces error and trains the system accordingly, (c and s represent the cover and secret images, respectively).

$$L(c, c0, s, s0) = \|c - c0\| + \beta\|s - s0\| \tag{1}$$

In Figure 2, the dimensions of the secret image are denoted as N*N, while the dimensions of the cover image are specified as M*M. The hidden image can be of any size that is equal to or smaller than the header image(cover). After the extraction of features, the secret image is hidden in the noisy regions of cover image. The result obtained after this encoding process is referred to as the embedded or hybrid image. Subsequently, this hybrid image is then subjected to this network that is designed to eliminate the hidden secret image. The output is measured in a metric PSNR (Setiadi, 2021) and payload capacity.

Figure 2. Encoding and decoding of secret image

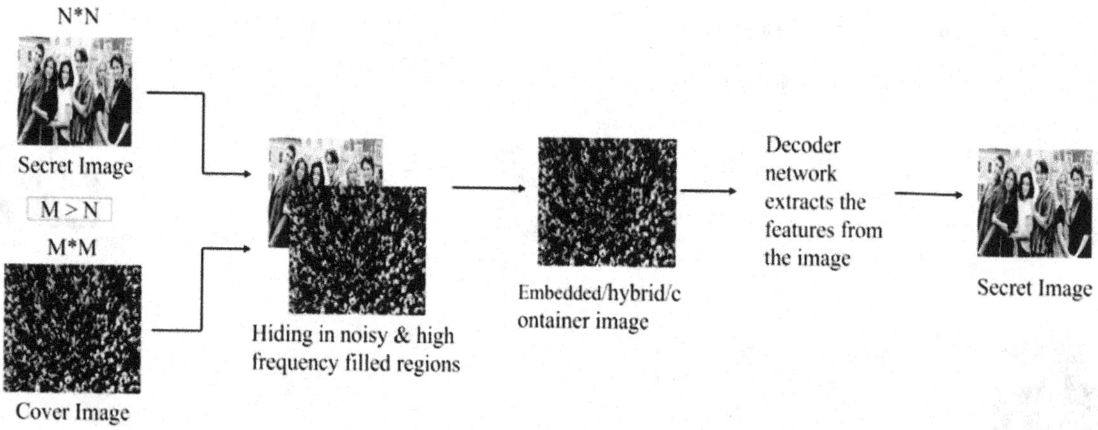

3. RESULTS

In this section, we outline the criteria that guided our design choices. We also present the qualitative and statistical outcomes of our approach on a range of publicly available datasets, which include ImageNet.

The considered dataset has been divided into three distinct sets: the training set, the testing set, and the validation set. All adjustments and alterations were exclusively applied using the validation set, and the outcomes of the simulation are reported based on the test set. From the corresponding dataset, we choose two RGB images at random; one serves as the cover image, while the other is converted to grayscale for the payload. 6000 images from the ImageNet database were chosen at random to serve as the training set, and 2000 additional images were chosen at random to act as the test set. The network learning rate is held at 0.001 and $\alpha = 0.75$, lr is adapted utilizing Adam optimization. s. From the results we can observe that our method is able to hide the image with less humanly detectable perturbations with better results. Table 1 shows the comparison of the proposed method's PSNR and payload capacity with those of other traditional methods.

Table 1. Comparison of our method's PSNR and payload capacity with those of other traditional methods

Technique	Capacity	PSNR
DCT	20.71	28.30
LSB	33	30.99
LSB and MSB modification	22.52	32
PVD	5.56	31.99
k-LSB	21.33	32.45
Proposed method	24	38.83

Figure 3. Sample results of our output in ImageNet

4. CONCLUSION

According to the research presented above, image steganography is a widely used technology for securing any type of data, including text, digital, audio, and video. Steganography is accomplished using a variety of techniques. Many techniques have been developed in various domains. The method we proposed outperforms all other traditional techniques in terms of payload capacity and PSNR. Hence the algorithm we proposed produces high imperceptibility, data embedding, and good data reconstruction without distortion and provides better security and quality.

REFERENCES

Alanzy, M., Alomrani, R., Alqarni, B., & Almutairi, S. (2023). Image Steganography Using LSB and Hybrid Encryption Algorithms. *Applied Sciences (Basel, Switzerland)*, *13*(21), 11771. doi:10.3390/app132111771

Chan, C.-K., & Cheng, L.-M. (2004). Hiding data in images by simple LSB substitution. *Pattern Recognition*, *37*(3), 469–474. doi:10.1016/j.patcog.2003.08.007

Chen, P.-Y., & Lin, H.-J. (2006). A DWT based approach for image steganography. *International Journal of Applied Science and Engineering*, *4*(3), 275–290.

Gaurav, K., & Ghanekar, U. (2018). Image steganography based on Canny edge detection, dilation operator and hybrid coding. *Journal of Information Security and Applications, 41*, 41–51. doi:10.1016/j.jisa.2018.05.001

Hemachandran, K. (2016). Study of Image Steganography using LSB, DFT and DWT. *International Journal of Computers and Technology, 11*, 2618–2627.

Kavitha, K.K., Koshti, A., & Dunghav, P. (2012). Steganography using least significant bit algorithm. *International Journal of Engineering Research and Applications*.

Khamrui, A., & Mandal, J. K. (2013). A genetic algorithm-based steganography using discrete cosine transformation (GASDCT). *Procedia Technology, 10*, 105–111. doi:10.1016/j.protcy.2013.12.342

Kumar & Kumar. (n.d.). *Techniques of Digital Watermarking*. Academic Press.

Liu, H. H., Su, P. C., & Hsu, M. H. (2020). An improved steganography method based on least-significant-bit substitution and pixel-value differencing. *KSII Transactions on Internet and Information Systems, 14*(11), 4537–4556.

Marvel, L. M., Boncelet, C. G., & Retter, C. T. (1999). Spread spectrum image steganography. *IEEE Transactions on Image Processing, 8*(8), 1075–1083. doi:10.1109/83.777088 PMID:18267522

Narasimmalou, T., & Allen Joseph, R. (2012). Discrete wavelet transform based steganography for transmitting images. In *IEEE-International Conference on Advances In Engineering, Science And Management (ICAESM2012)* (pp. 370-375). IEEE.

Patel, H., & Dave, P. (2012). Steganography technique based on DCT coefficients. *International Journal of Engineering Research and Applications, 2*(1), 713–717.

Rahim, R., & Nadeem, S. (2018). End-to-end trained CNN encoder-decoder networks for image steganography. *Proceedings of the European Conference on Computer Vision (ECCV) Workshops*.

Sahu, A. K., & Swain, G. (2016). A review on LSB substitution and PVD based image steganography techniques. *Indonesian Journal of Electrical Engineering and Computer Science, 2*(3), 712–719. doi:10.11591/ijeecs.v2.i3.pp712-719

Setiadi, D. R. I. M. (2021). PSNR vs SSIM: Imperceptibility quality assessment for image steganography. *Multimedia Tools and Applications, 80*(6), 8423–8444. doi:10.1007/s11042-020-10035-z

Simmons, G. (2019). *Secure communications and asymmetric cryptosystems*. Routledge. doi:10.4324/9780429305634

Subramanian, N., Cheheb, I., Elharrouss, O., Al-Maadeed, S., & Bouridane, A. (2021). End-to-end image steganography using deep convolutional autoencoders. *IEEE Access : Practical Innovations, Open Solutions, 9*, 135585–135593. doi:10.1109/ACCESS.2021.3113953

Chapter 12
Impacts of Nano–Materials and Nano Fluids on the Robot Industry and Environments

Nalla Bhanu Teja

Department of Mechanical Engineering, Aditya College of Engineering, Surampalem, India

V. Kannagi

Department of Electronics and Communication Engineering, R.M.K. College of Engineering and Technology, Puduvoyal, India

A. Chandrashekhar

Department of Mechanical Engineering, Faculty of Science and Technology, ICFAI Foundation for Higher Education, Hyderabad, India

T. Senthilnathan

Department of Applied Physics, Sri Venkateswara College of Engineering, Sriperumbudur, India

Tarun Kanti Pal

Department of Mechanical Engineering, College of Engineering and Management, Kolaghat, India

Sampath Boopathi

iD https://orcid.org/0000-0002-2065-6539

Department of Mechanical Engineering, Muthayammal Engineering College, Namakkal, India

ABSTRACT

The integration of nanotechnology into robotics has revolutionized the design, manufacturing, and performance of robotic systems. Nano-materials, with their unique properties at the nanoscale, enhance strength, flexibility, and functionality, revolutionizing the construction and operation of robots. Nano fluids, with their superior heat transfer properties, address overheating issues, improving performance, extended operational lifespans, and increased adaptability in diverse environmental conditions. The chapter also explores the environmental impact of robotics, highlighting the integration of nano-materials and nano fluids in eco-friendly solutions. The chapter delves into the challenges and future directions of the synergy between nanotechnology and robotics, discussing potential breakthroughs, ethical considerations, and the need for ongoing research. It provides a comprehensive analysis of the impacts of nano-materials and nano fluids on the robot industry and their environments.

DOI: 10.4018/979-8-3693-5767-5.ch012

INTRODUCTION

The integration of nanotechnology and robotics has led to significant innovation across various industries. The use of nanomaterials and nanofluids in robotics can enhance mechanical performance and promote environmental sustainability. This chapter examines the profound impacts of these materials on the robot industry, highlighting their impact on technological advancements and ecological considerations (Jakkula & Sethuramalingam, 2023).

The integration of nanotechnology and robotics is revolutionizing the field, enhancing the mechanical, electrical, and thermal capabilities of robots. Nanomaterials, with dimensions at the nanoscale, offer unique properties that can significantly improve the performance of these systems. By integrating these materials into the design and manufacturing processes, engineers and researchers are pushing the boundaries of what was once thought possible. This convergence leads to more efficient and versatile robots, with nanocomposites offering superior strength, durability, and lightweight properties. This results in robots with enhanced agility, increased load-bearing capacities, and prolonged operational lifespans. The integration of nanomaterials enhances robot performance in various industries (He et al., 2018).

Nano-fluids, colloidal suspensions of nanoparticles in conventional fluids, have emerged as a game-changer in thermal management for robotic systems. The exceptional thermal conductivity and heat transfer capabilities of nano-fluids enable more efficient cooling of robotic components, preventing overheating and optimizing energy consumption. As robots become more sophisticated and undertake tasks with higher energy demands, the implementation of nano-fluids becomes crucial in ensuring the reliability and longevity of these systems. This chapter will delve into specific case studies and applications where nano-fluids have proven instrumental in revolutionizing the thermal dynamics of robots (Hassani et al., 2020).

While the integration of nanomaterials and nano-fluids brings about remarkable advancements in robotic technology, it is imperative to assess their environmental impact. As robots become ubiquitous across industries, understanding the ecological footprint of these nanotechnology-infused systems is crucial. This chapter will explore the sustainability aspects of using nano-materials, considering factors such as recyclability, resource consumption, and end-of-life disposal. By addressing environmental concerns, the robotics industry can steer towards eco-friendly practices and contribute to a more sustainable future (Malik et al., 2023).

This chapter explores the impact of nano-materials and nano-fluids on the robot industry and its environment. It delves into their applications in various robotic domains, including manufacturing, healthcare, and agriculture. The chapter also examines the challenges and future prospects of integrating nanotechnology into robotics. Nanotechnology is a pioneering force in the robotics industry, with the potential to redefine its capabilities. The chapter focuses on the environmental impact of this symbiotic relationship, as nanomaterials are widely used in robot design, manufacturing, and operation. The chapter aims to provide an in-depth exploration of how nanotechnology is shaping the robot industry while raising questions about its environmental footprint (Ramesh et al., 2020).

Nanotechnology, operating at the scale of one billionth of a meter, has catalyzed a paradigm shift in the development of robotic systems. The integration of nanomaterials into the fabric of robotics introduces a new era of possibilities, ranging from improved structural integrity to enhanced functionality. Robots are no longer bound by the constraints of traditional materials; instead, they are endowed with unprecedented strength, durability, and versatility. This chapter aims to dissect the manifold ways in which nanotechnology is propelling the robot industry forward, with a particular emphasis on the transformative potential that nanomaterials bring to the design and performance of robotic systems (Sethuramalingam et al., 2023).

As the robot industry expands its horizons, driven by technological advancements, the integration of nanomaterials becomes a central theme. Nanotechnology is not merely a tool for incremental improvement; it is a catalyst for disruptive innovation. From autonomous manufacturing robots to sophisticated medical nanobots, the applications span across diverse sectors, each bringing its unique set of challenges and opportunities. This chapter will navigate through the varied landscapes of the robot industry, highlighting important nanotechnological contributions that are reshaping the way robots operate, communicate, and interact with their environment (Kiran & Prabhu, 2020).

While the marriage of nanotechnology and robotics ushers in a wave of transformative possibilities, it is crucial to pause and assess the environmental implications of this revolution. The production, utilization, and disposal of nanomaterials raise questions about their long-term impact on ecosystems. This chapter will delve into the life cycle assessment of nanotechnology in the robot industry, considering factors such as resource extraction, energy consumption, and waste management. By critically evaluating the environmental footprint, we aim to provide insights into how the robot industry can navigate toward sustainable practices without compromising technological advancements (Singh et al., 2021).

The inherent tension between technological innovation and environmental stewardship becomes apparent in the context of nanotechnology in the robot industry. Striking a balance between pushing the boundaries of what is achievable and adopting eco-conscious practices is a challenge that demands careful consideration. This chapter will explore case studies, best practices, and emerging frameworks that showcase how the robot industry can embrace nanotechnology responsibly, mitigating adverse environmental effects while fostering groundbreaking advancements (Sonika, 2023).

The chapters will explore the application of nanotechnology in the robot industry, focusing on technological advancements and environmental implications. They will explore nanomaterial-enhanced components and eco-friendly manufacturing processes, providing a comprehensive understanding of the relationship between nanotechnology, the robot industry, and environmental impact. The aim is to provide a nuanced perspective for future developments.

Objectives of the Chapter

- The study explores the integration of nanomaterials in robot design and manufacturing, evaluating their impact on structural integrity, durability, and performance, and highlighting case studies and advancements in nanotechnology.
- The study aims to analyze the environmental impact of nanomaterials in the robot industry, focusing on resource extraction, energy consumption, and waste management, to determine the sustainability of these innovations.
- Explore nanotechnology's applications in manufacturing, healthcare, agriculture, and beyond, highlighting its contributions to advancements and examining the challenges and opportunities of their integration into diverse robotic applications.
- The study explores the environmental challenges of integrating nanotechnology into the robot industry, focusing on resource consumption, recyclability, and responsible waste management, offering insights for sustainable practices in nanorobotics.
- The study aims to develop guidelines and recommendations for responsible and sustainable nanorobotics development, guiding the robot industry to balance technological innovation with envi-

ronmental stewardship, ensuring nanotechnology advancements positively impact both industries and the environment.

NANOTECHNOLOGY IN ROBOTICS

This chapter explores the fundamentals of nanotechnology and its integration into robotics. It delves into the core principles and concepts of nanotechnology, aiming to provide readers with a comprehensive understanding of how nanoscale phenomena are harnessed to enhance the capabilities of robotic systems (Sharmin, 2020; Sheikh et al., 2020).

- *Nanotechnology Primer:* Nanotechnology involves the manipulation and utilization of materials and devices at the nanoscale, typically ranging from 1 to 100 nanometers. At this scale, the properties of materials undergo significant changes, enabling scientists and engineers to exploit novel phenomena not observed at larger scales. This section will provide a primer on nanotechnology, introducing concepts such as quantum effects, surface area dominance, and unique mechanical, thermal, and optical properties that become pronounced at the nanoscale (Boopathi, Umareddy, et al., 2023; Boopathi & Davim, 2023b).
- *Building Blocks of Nanomaterials:* The fundamental building blocks of nanotechnology are nanomaterials, which serve as the cornerstone for advancements in robotics. Nanomaterials, such as nanoparticles, nanotubes, and nanocomposites, exhibit exceptional mechanical, electrical, and thermal properties. This chapter will explore the diverse nature of nanomaterials and elucidate their role in augmenting the structural and functional aspects of robotic systems. Readers will gain insights into the synthesis, manipulation, and application of nanomaterials in the context of robotics (Fowziya et al., 2023; Vijayakumar et al., 2024).
- *Tools and Techniques in Nanofabrication:* Nanofabrication techniques play a pivotal role in shaping the landscape of nanotechnology. This section will introduce readers to state-of-the-art tools and techniques employed in the creation of nanoscale structures. From top-down approaches like photolithography to bottom-up methods such as self-assembly, we will examine how these techniques contribute to the precision and reproducibility required in nanorobotics. Understanding these techniques is essential for grasping the practical aspects of implementing nanotechnology in robotic systems (Boopathi & Davim, 2023b, 2023a).
- *Quantum Mechanics at the Nanoscale:* Quantum mechanics governs the behavior of matter at the nanoscale and is instrumental in shaping the properties of nanomaterials. This chapter will provide an overview of main quantum mechanical principles relevant to nanotechnology, including quantum tunneling, confinement effects, and quantum dots. Understanding these principles is crucial for appreciating the unique characteristics and potential applications of nanomaterials in robotics (Paul et al., 2024).
- *Integration of Nanotechnology in Robotics:* Building on the fundamentals established, this section will explore how nanotechnology is seamlessly integrated into robotics. From nanoscale sensors and actuators to the development of nanobots for medical applications, we will examine the diverse applications and contributions of nanotechnology to the field of robotics. By elucidating these connections, readers will gain a holistic understanding of how nanotechnology shapes the future of robotic systems (Mohanty et al., 2023).

In subsequent chapters, we will delve into specific applications of nanotechnology in robotics, showcasing real-world examples and emerging trends that exemplify the transformative power of the fusion between these two cutting-edge fields.

Applications of Nanomaterials in Robotics: Harnessing Unique Properties

This chapter explores the diverse applications of nanomaterials in robotics, highlighting their unique properties that enhance the capabilities of robotic systems. It highlights the impact of nanomaterials on the design, performance, and functionality of robots, highlighting the wide range of applications they offer (Joe et al., 2023).

- *The Versatility of Nanomaterials:* Nanomaterials encompass a wide array of substances, including nanoparticles, nanocomposites, and nanotubes, each with its unique set of properties. This section will provide an overview of the versatility of nanomaterials, emphasizing their adaptability and applicability across various robotic applications. From enhancing structural strength to enabling precise sensing, nanomaterials serve as the fundamental building blocks that drive innovation in modern robotics.
- *Improved Structural Integrity and Durability:* One of the primary applications of nanomaterials in robotics lies in improving the structural integrity and durability of robotic components. Nanocomposites, for instance, exhibit remarkable strength and lightweight characteristics. This chapter will delve into how the integration of nanomaterials in the fabrication of robotic exoskeletons, frames, and limbs enhances mechanical properties, making robots more resilient and capable of withstanding challenging environments (Boopathi, Thillaivanan, et al., 2022).
- *Sensing and Actuation at the Nanoscale:* Nanomaterials play a crucial role in the development of sensors and actuators that operate at the nanoscale. Quantum dots and nanosensors, for instance, enable robots to perceive their surroundings with unprecedented precision. This section will explore how nanomaterials contribute to advancements in sensing technologies, allowing robots to detect and respond to stimuli with enhanced sensitivity and efficiency (S. Karthik et al., 2023).
- *Nanomaterials in Biomedical Robotics:* In the realm of biomedical robotics, nanomaterials hold immense promise for transformative applications. Nanoparticles can be engineered for targeted drug delivery, and nanobots can navigate through the human body to perform intricate medical procedures. This chapter will delve into specific examples of how nanomaterials are revolutionizing biomedical robotics, highlighting their potential in diagnostics, therapy, and minimally invasive surgeries (Sengeni et al., 2023).
- *Energy Efficiency and Environmental Sustainability:* Nanomaterials contribute to the pursuit of energy-efficient and environmentally sustainable robotic systems. With enhanced thermal conductivity and lightweight properties, nanomaterials facilitate better heat dissipation, leading to energy savings in robotic operations. Moreover, the recyclability of certain nanomaterials aligns with the growing emphasis on sustainable practices in robotics. This section will explore how nanomaterials contribute to eco-friendly advancements in the robot industry (Boopathi, 2022a, 2022b).

It explores the transformative impact of nanomaterials in robotics through case studies and real-world examples, aiming to provide readers with a nuanced understanding of their role in shaping the future of robotic systems.

IMPACT OF NANO-MATERIALS ON ROBOTIC SYSTEMS

Figure 1 depicts the significant impact of nano-materials on various aspects of robotic systems.

Figure 1. Impact of nano-materials on robotic systems in various aspects

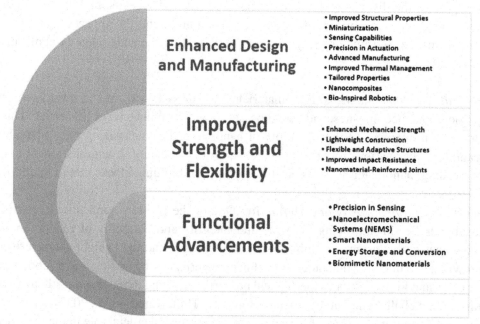

Enhanced Design and Manufacturing

Improved Structural Properties: Nano-materials, such as carbon nanotubes and graphene, offer exceptional strength-to-weight ratios and structural integrity. Integration of these materials in robotic frames and components results in lighter yet more robust structures, enhancing overall durability (Chakraborty et al., 2021; Mouchou et al., 2021).

   Miniaturization and Increased Efficiency: Nanomaterials enable the development of smaller and more efficient robotic components, allowing for the miniaturization of robots. This facilitates the creation of compact, agile robots capable of navigating confined spaces and executing precise tasks with higher efficiency.

   Enhanced Sensing Capabilities: Quantum dots and nanosensors made from nanomaterials provide robots with improved sensing capabilities at the nanoscale. The heightened sensitivity of these sensors allows robots to perceive and respond to their environment more accurately, enhancing their adaptability.

   Precision in Actuation: Nanomaterials contribute to the development of high-precision actuators, enabling robots to perform intricate movements and tasks. This precision is particularly crucial in applications such as surgical robotics, where nanomaterial-based actuators can enhance the accuracy of procedures.

Advanced Manufacturing Processes: Nanomaterials are integrated into manufacturing processes, such as 3D printing and nanoscale assembly techniques, allowing for the creation of intricate robotic components. These advanced manufacturing methods enable the production of complex structures that were previously unattainable, pushing the boundaries of design possibilities (Revathi et al., 2024; Senthil et al., 2023).

Improved Thermal Management: Nano-materials exhibit superior thermal conductivity, addressing challenges related to heat dissipation in robotic systems. Efficient thermal management ensures that robots can operate for extended periods without overheating, contributing to enhanced reliability and longevity (Boopathi, Jeyakumar, et al., 2022).

Customization and Tailored Properties: The ability to engineer nanomaterials with specific properties allows for the customization of robotic components based on application requirements. Tailored properties, such as electrical conductivity or magnetism, enable the creation of specialized robotic systems for diverse tasks.

Nanocomposites for Multifunctionality: Nanocomposites, combining nanomaterials with traditional materials, result in multifunctional robotic components. This integration allows for the simultaneous improvement of mechanical, thermal, and electrical properties, contributing to the overall efficiency of robotic systems (Gowri et al., 2023; K. Karthik et al., 2023; Murali et al., 2023).

Facilitation of Bio-Inspired Robotics: Nanomaterials play a crucial role in the development of bio-inspired robotic systems by mimicking biological structures at the nanoscale. This bio-mimicry enhances the performance of robots in tasks inspired by nature, such as agile locomotion or adaptive responses to environmental stimuli (Koshariya, Khatoon, et al., 2023; Maheswari et al., 2023).

Streamlined Production Processes: The integration of nanomaterials often simplifies and streamlines the production processes for robotic components. Reduced material complexity and improved manufacturing efficiency contribute to cost-effectiveness in the production of advanced robotic systems.

Nanomaterials significantly impact robotic systems in design and manufacturing, offering enhanced structural properties, improved efficiency, advanced sensing capabilities, and innovative manufacturing processes. These advancements contribute to the development of more capable, efficient, and versatile robotic systems across various applications.

Improved Strength and Flexibility

- **Enhanced Mechanical Strength:** Nano-materials, such as carbon nanotubes and nanocomposites, contribute to significant improvements in the mechanical strength of robotic components. The incorporation of these materials in structural elements enhances the overall robustness of robots, allowing them to withstand higher loads and impacts (Oluwasanu et al., 2019).

- **Lightweight Construction:** Nano-materials offer high strength-to-weight ratios, enabling the creation of lightweight yet durable robotic structures. This lightweight construction enhances the mobility and agility of robots, making them more versatile in various applications, including exploration and search-and-rescue missions (Boopathi, 2023a).

- **Flexible and Adaptive Structures:** Nanomaterials, particularly those with flexible properties, allow for the development of robotic components that can bend, stretch, and adapt to varying conditions. This flexibility is advantageous in applications where robots need to navigate complex environments or perform tasks that require adaptive movements.

- **Improved Impact Resistance:** Nanomaterials enhance the impact resistance of robotic systems, reducing the risk of damage in scenarios where robots may encounter collisions or harsh conditions. This impact resistance is particularly crucial in applications like industrial automation and field robotics, where robots operate in dynamic and unpredictable environments.
- **Nanomaterial-Reinforced Joints and Connectors:** Nano-materials are employed to reinforce joints and connectors in robotic systems, ensuring greater durability and longevity. This reinforcement enhances the structural integrity of robotic limbs and joints, reducing wear and tear during repetitive movements.

Functional Advancements

- **Precision in Sensing:** Nanomaterial-based sensors enable robots to achieve high precision in sensing various stimuli, such as temperature, pressure, and chemical changes. This precision enhances the ability of robots to gather accurate data from their environment, contributing to improved decision-making and adaptability (Kumar et al., 2023).
- **Nanoelectromechanical Systems (NEMS):** NEMS, built using nanomaterials, enable the integration of electromechanical functionalities at the nanoscale. This advancement allows for the creation of ultra-sensitive sensors, actuators, and transducers, enhancing the overall performance of robotic systems.
- **Smart Nanomaterials for Adaptive Behavior:** Smart nanomaterials with responsive properties, such as shape memory alloys and piezoelectric materials, enable robots to exhibit adaptive behaviors. These materials can change their physical properties in response to external stimuli, allowing for dynamic adjustments in the robot's form and function (Maguluri et al., 2023; Maheswari et al., 2023).
- **Improved Energy Storage and Conversion:** Nanomaterials contribute to the development of advanced energy storage and conversion systems for robotic applications. Nanostructured materials in batteries and energy storage devices enhance the energy density, leading to longer operational periods for robots between charging cycles (Satav et al., 2023; Syamala et al., 2023).
- **Biomimetic Nanomaterials for Soft Robotics:** Nanomaterials with biomimetic properties are utilized in soft robotics, enabling robots to mimic the flexibility and dexterity of natural organisms. Soft robotic structures built with nanomaterials find applications in fields like medical robotics and human-machine interaction.

The integration of nanomaterials into robotic systems enhances their strength, flexibility, and functional advancements, enhancing the robustness and adaptability of robots and paving the way for the development of more sophisticated and capable robotic systems across various applications.

NANO FLUIDS IN ROBOTIC THERMAL MANAGEMENT

Importance of Thermal Management in Robotics

Effective thermal management is a critical aspect of ensuring the optimal performance, reliability, and longevity of robotic systems. As robots become more sophisticated and engage in tasks with higher energy demands, managing heat dissipation becomes imperative. This section will underscore the importance of thermal management in robotics, exploring the impact of excessive heat on components and the overarching significance of maintaining optimal operating temperatures for enhanced efficiency and functionality (Du et al., 2021; Ramalingam & Rasool Mohideen, 2021).

Characteristics of Nano Fluids

Nano fluids, colloidal suspensions of nanoparticles in conventional fluids, present a cutting-edge solution to the thermal management challenges faced by robotic systems. This segment will delve into the unique characteristics of nano fluids that set them apart in heat transfer applications. Attributes such as enhanced thermal conductivity, stability, and the ability to disperse uniformly will be discussed, highlighting how these properties make nano fluids well-suited for cooling and thermal regulation in robotics (Sevinchan et al., 2018).

Applications in Cooling Systems

Nano fluids have found widespread applications in robotic cooling systems, contributing to superior heat dissipation and overall thermal control. This section will explore specific use cases of nano fluids in cooling robotic components, ranging from microprocessors and motors to entire robotic frames. Real-world examples and case studies will be presented to showcase the efficacy of nano fluids in optimizing thermal performance, thus ensuring the reliability and sustained operation of robotic systems (Harris et al., 2022).

Overcoming Thermal Challenges

Robotic systems often face formidable thermal challenges, from heat concentration in compact spaces to the need for continuous operation under varying environmental conditions. Nano fluids offer innovative solutions to these challenges. This part of the chapter will elucidate how nano fluids overcome thermal obstacles in robotics. Topics such as the role of nano fluids in preventing overheating, reducing thermal resistance, and facilitating efficient heat dissipation will be explored, providing insights into how these advanced fluids address important thermal management concerns.

USE OF NANO-FLUID IN ROBOT THERMAL MANAGEMENTS

Nano fluids, colloidal suspensions of nanoparticles in a base fluid, are known for their unique thermal properties. They are utilized in robotic systems for thermal management, addressing challenges in heat dissipation, cooling, and temperature control (Harris et al., 2022; Sevinchan et al., 2018). Figure 2 depicts the various roles of nano fluids in the thermal management of robots.

Enhanced Heat Transfer: Nano fluids exhibit significantly higher thermal conductivity compared to traditional fluids. By incorporating nano fluids in the cooling systems of robotic components, heat transfer is greatly enhanced. This ensures more efficient dissipation of heat generated during the operation of motors, processors, and other heat-emitting components.

Figure 2. Roles of nano fluids in the robot thermal managements

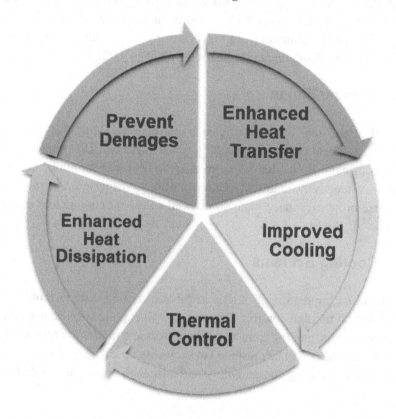

Improved Cooling of Electronics: Robotics often involve the use of electronic components that generate heat during operation. Nano fluids are employed in cooling systems to extract and carry away heat from electronic circuits more effectively than conventional coolants. This prevents overheating and ensures optimal performance and longevity of electronic components (Myilsamy & Sampath, 2021; Sampath & Myilsamy, 2021).

Thermal Control in Motors and Actuators: Motors and actuators in robotic systems can experience temperature fluctuations during operation. Nano fluids help maintain consistent temperatures by efficiently absorbing and dissipating heat. This is particularly crucial in precision applications where thermal stability is essential for accurate movements.

Miniaturization of Cooling Systems: The high thermal conductivity of nano fluids allows for the miniaturization of cooling systems. In small-scale robotic applications or devices with limited space, nano fluids enable the design of compact and efficient cooling solutions, ensuring that size constraints do not compromise thermal management.

Heat Dissipation in 3D Printing: In the field of robotics, especially in the manufacturing of robotic components using 3D printing, controlling heat is critical. Nano fluids can be utilized in the cooling systems of 3D printers to manage the heat generated during the printing process. This contributes to the production of high-quality, precisely manufactured robotic parts (Boopathi, Khare, et al., 2023; Boopathi & Kumar, 2024; Mohanty et al., 2023).

Thermal Management in Exoskeletons: Exoskeletons, worn by humans to augment strength and endurance, can generate heat due to the motors and actuators involved. Nano fluids play a role in the thermal management of these exoskeletons, ensuring that the temperature remains within a comfortable and safe range for the user.

Efficient Cooling in Harsh Environments: In robotic applications deployed in harsh environments, such as space exploration or industrial settings with high temperatures, nano fluids provide efficient cooling. Their superior thermal properties enable robots to operate reliably in extreme conditions without succumbing to heat-related issues.

Medical robotics, particularly in surgeries and diagnostic procedures, benefit from precise thermal management using nano fluids. These fluids regulate the temperature of components, ensuring optimal operation without discomfort or tissue damage. They also prevent hot spots, which can cause uneven temperature distribution and overheating. By integrating nano fluids with advanced materials, these fluids enhance the thermal performance of the components, leading to more heat-resistant and thermally efficient robotic systems.

Nano fluids play a significant role in robotic thermal management, enhancing heat transfer, cooling electronic components, and enabling miniaturization of cooling systems. They optimize thermal performance in robotic systems across various applications. This section explores the application of nano fluids in robotics, their impact on thermal dynamics, and their role in overcoming thermal challenges. It discusses the importance of thermal management, unique characteristics of nano fluids, their use in cooling systems, and their transformative role in enhancing thermal efficiency.

ENVIRONMENTAL CONSIDERATIONS IN ROBOTICS

Ecological Impact of Robotics

Resource Consumption: Robotics often involves the use of materials and energy-intensive manufacturing processes, contributing to resource depletion. Exploration of the ecological impact includes assessing the extraction and utilization of raw materials in robot production (Cui et al., 2021).

Electronic Waste (E-Waste): The disposal of obsolete or malfunctioning robotic components contributes to the growing problem of electronic waste. Examining the life cycle of robots involves considering strategies for minimizing e-waste and promoting responsible recycling (Harikaran et al., 2023; Selvakumar, Adithe, et al., 2023; Selvakumar, Shankar, et al., 2023; Sengeni et al., 2023).

Energy Consumption: The energy demands of robotic systems, particularly in industries like manufacturing and transportation, have implications for carbon footprints. Evaluating the environmental impact includes analyzing the energy sources powering robots and exploring avenues for energy efficiency (Naveeenkumar et al., 2024).

Sustainable Technology in Robotics

Renewable Energy Integration: Adopting sustainable practices involves exploring the integration of renewable energy sources, such as solar or wind power, to reduce reliance on conventional energy grids.

Life Cycle Assessment: Conducting a life cycle assessment of robotic systems helps identify and mitigate environmental impacts at every stage, from design and manufacturing to operation and disposal.

Efficient Materials Usage: Sustainable technology in robotics emphasizes the use of eco-friendly and recyclable materials to reduce the environmental footprint of manufacturing processes and end-of-life disposal.

Role of Nanotechnology in Eco-Friendly Robotics

Figure 3 depicts the significant role of nanotechnology in the development of eco-friendly robotics.

Nanomaterials for Light weighting: Nanotechnology contributes to eco-friendly robotics by providing lightweight nanomaterials that enhance structural integrity without excessive resource consumption (Gellers, 2020).

Energy-Efficient Nanodevices: The development of energy-efficient nanodevices, facilitated by nanotechnology, plays a crucial role in minimizing the energy consumption of robotic systems.

Recyclable Nanomaterials: Nanomaterials engineered for recyclability contribute to sustainable practices in robotics, ensuring that materials used in robots can be efficiently repurposed at the end of their life cycle.

Nanofluids for Thermal Efficiency: The use of nanofluids in robotics, enabled by nanotechnology, enhances thermal efficiency, reducing the need for excessive cooling mechanisms and contributing to energy savings.

Figure 3. Role of nanotechnology in eco-friendly robotics

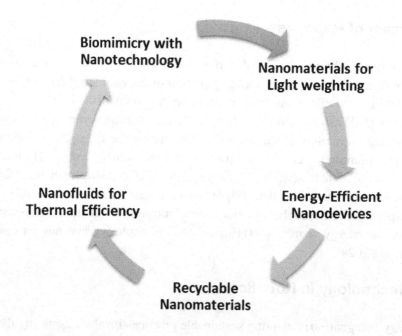

Biomimicry with Nanotechnology: Nanotechnology facilitates biomimicry in robotic design, allowing robots to emulate natural structures and processes, leading to more energy-efficient and environmentally friendly robotic systems.

Nanotechnology improves eco-friendly robotics by providing lightweight, energy-efficient nanomaterials that maintain structural integrity without excessive resource consumption. It also facilitates the development of energy-efficient nanodevices and contributes to sustainable robotic practices by ensuring efficient material repurposing.

OPERATIONAL EFFICIENCY AND ADAPTABILITY

This chapter emphasizes the importance of operational efficiency and adaptability in robotics, focusing on energy-efficient manufacturing processes, improved operational lifespan, and adaptability to diverse environmental conditions. These aspects are crucial for the sustainability of robotic systems and their applicability across various industries (Bragança et al., 2019; Marinoudi et al., 2019).

Energy-Efficient Manufacturing Processes

Nanotechnology in Materials and Manufacturing: The integration of nanotechnology into robotic manufacturing processes has revolutionized the energy efficiency of creating robotic components. Nanomaterials, such as carbon nanotubes and graphene, exhibit extraordinary strength and conductivity, enabling the production of lightweight yet robust components with minimal material usage. Nanotechnology has also facilitated precision engineering at the nanoscale, allowing for the creation of intricate robotic parts with reduced waste.

Additive Manufacturing (3D Printing): Additive manufacturing, commonly known as 3D printing, plays a pivotal role in energy-efficient robotic manufacturing. This process allows for the layer-by-layer construction of complex structures, minimizing material wastage and energy consumption. Additionally, 3D printing enables the customization of robotic components, tailoring designs to specific applications and optimizing their overall energy efficiency.

Sustainable Materials and Recycling Practices: The pursuit of energy efficiency extends to the choice of materials used in robotic manufacturing. Sustainable materials, such as bioplastics and recycled metals, contribute to reducing the environmental impact of the production process. Moreover, implementing recycling practices for robotic components at the end of their operational lifespan aligns with a circular economy, minimizing waste and conserving resources.

Integration of Renewable Energy Sources: Energy-efficient manufacturing extends beyond material usage to the power sources driving production facilities. Integration of renewable energy sources, such as solar or wind power, not only reduces the carbon footprint of robotic manufacturing but also aligns with sustainable practices. Implementing energy-efficient technologies, such as regenerative braking systems in manufacturing equipment, further contributes to the overall efficiency of the production process.

Improved Operational Lifespan

Robust Materials and Design: Enhancing the operational lifespan of robotic systems involves the use of robust materials and thoughtful design. Nanomaterials, with their exceptional strength and durability, contribute to the creation of components that can withstand the rigors of continuous operation. Additionally, adopting modular design principles facilitates the replacement of specific components, prolonging the overall lifespan of the robotic system (Javaid et al., 2021).

Predictive Maintenance and Monitoring: Implementing predictive maintenance strategies is crucial for optimizing operational lifespan. Advanced sensors and monitoring systems, often utilizing nano-technology, enable real-time data collection on the health and performance of robotic components. This data-driven approach allows for proactive maintenance, addressing potential issues before they escalate and ensuring the longevity of the robotic system (Boopathi, 2021, 2023b).

Upgradability and Adaptation; Designing robotic systems with upgradability in mind is another strategy to extend operational lifespan. This involves creating platforms that can accommodate hardware and software upgrades, allowing robots to evolve with technological advancements. The ability to adapt to new requirements and integrate the latest technologies enhances the relevance and longevity of robotic systems in dynamic operational environments.

Adaptability to Diverse Environmental Conditions

Sensory Systems and Environmental Perception: Achieving adaptability to diverse environmental conditions requires sophisticated sensory systems that enable robots to perceive and respond to their surroundings. Advanced sensors, often leveraging nanotechnology, provide robots with enhanced environmental perception capabilities. These sensors can detect changes in temperature, humidity, terrain, and other variables, allowing robots to adapt their behavior in real-time (Atilano et al., 2019).

Machine Learning and AI Algorithms: Machine learning and artificial intelligence (AI) algorithms empower robots to adapt and learn from their experiences in different environments. These technologies enable robots to analyze data, make informed decisions, and optimize their performance based on the specific conditions they encounter. Adaptive learning algorithms enhance the versatility of robotic systems across various applications (Maheswari et al., 2023; Ramudu et al., 2023; Syamala et al., 2023).

Soft Robotics for Flexible Environments: Soft robotics, a field that integrates flexible and deformable materials, enables robots to navigate and operate in complex and dynamic environments. Soft robotic components, often inspired by natural organisms, enhance adaptability by allowing robots to squeeze through tight spaces, conform to uneven surfaces, and interact safely with humans. This adaptability is particularly advantageous in applications like search-and-rescue missions or healthcare settings.

Robotic Swarms and Collaborative Adaptation: The concept of robotic swarms involves multiple robots working collaboratively to achieve a common goal. In diverse environmental conditions, robotic swarms can adapt and coordinate their actions based on the collective intelligence of the group. This collaborative approach enhances adaptability, allowing the swarm to accomplish tasks more efficiently and effectively than individual robots operating in isolation.

The robotics field is undergoing significant transformation due to its emphasis on operational efficiency and adaptability, encompassing energy-efficient manufacturing, sustainable materials, predictive maintenance, and adaptability to diverse environments, thereby maximizing the potential of robotic systems across various industries.

CHALLENGES IN THE INTEGRATION OF NANOTECHNOLOGY AND ROBOTICS

The integration of nanotechnology and robotics holds great potential for transformative advancements in various industries. However, it also presents challenges like ethical considerations, safety concerns, and regulatory hurdles. Researchers and engineers must navigate these issues to ensure responsible development and deployment of these cutting-edge technologies (Jakkula & Sethuramalingam, 2023; Ness et al., 2023; Sevinchan et al., 2018).

Ethical Considerations

Privacy Concerns in Nanorobotics: The miniaturization afforded by nanotechnology raises concerns about privacy, especially in the context of nanorobotics used for surveillance or medical applications. Ethical dilemmas arise when considering the potential intrusion into personal spaces and the collection of sensitive information through nanoscale devices.

Autonomous Decision-Making: As robots equipped with nanotechnology become more autonomous, ethical questions arise about their decision-making capabilities. Determining the ethical framework for robots making decisions in dynamic and unpredictable environments poses challenges, especially when human lives or critical tasks are at stake.

Dual-Use Dilemma: The dual-use nature of nanotechnology and robotics, where the same technology can have both beneficial and harmful applications, raises ethical concerns. Striking a balance between promoting innovation and preventing misuse requires thoughtful consideration and robust ethical guidelines.

Equity and Access: Ensuring equitable access to nanotechnology-enhanced robotics is an ethical challenge, as disparities in technology adoption can exacerbate existing societal inequalities. Addressing issues related to accessibility, affordability, and the potential exacerbation of socio-economic divides is crucial in the ethical integration of nanotechnology and robotics.

Safety Concerns

Nanomaterial Toxicity: The safety of nanomaterials used in robotic systems is a significant concern, as the potential toxicity of certain nanoparticles raises health and environmental risks. Understanding the long-term effects of exposure to nanomaterials and implementing safety measures in their production and use are essential for responsible integration.

Human-Robot Interaction Safety: As nanotechnology enhances the capabilities of robots, ensuring the safety of human-robot interactions becomes more complex. Safety measures must be implemented to prevent accidents or harm during close collaborations between humans and nanotechnology-enhanced robotic systems (Puranik et al., 2024).

Cybersecurity Risks: With increased connectivity and the incorporation of nanoscale sensors and communication devices, cybersecurity risks become a critical safety concern. Safeguarding robotic systems from cyber threats, such as unauthorized access or manipulation of nanotechnology-driven functionalities, requires robust cybersecurity measures (Maguluri et al., 2023; Rahamathunnisa et al., 2023; Srinivas et al., 2023).

Unintended Consequences of Autonomy: As robots gain autonomy through advanced nanotechnology, the potential for unintended consequences and malfunctions increases. Establishing fail-safe mechanisms and comprehensive testing procedures is crucial to mitigate safety risks associated with the autonomous behavior of nanotechnology-driven robotic systems.

Regulatory Challenges

Lack of Standardization: The fast-paced nature of technological advancements in nanotechnology and robotics poses challenges for regulatory bodies to establish standardized guidelines. The lack of universally accepted standards hinders regulatory efforts, making it challenging to ensure consistency and safety across different applications and industries.

Ethical and Legal Frameworks: Developing ethical and legal frameworks that keep pace with the evolving capabilities of nanotechnology-driven robotics is a regulatory challenge. Establishing guidelines for responsible research, development, and deployment of these technologies is essential to prevent misuse and ethical violations (Boopathi & Khang, 2023).

Cross-Disciplinary Regulations: Nanotechnology and robotics span multiple disciplines, and existing regulatory frameworks may not adequately address the convergence of these fields. Regulatory bodies face the challenge of fostering collaboration across disciplines to create comprehensive regulations that account for the unique aspects of nanotechnology-enhanced robotics.

International Coordination: The global nature of technological innovation requires international coordination in regulatory efforts. Coordinating regulatory frameworks across borders is challenging, given differing cultural, ethical, and legal perspectives, and establishing effective international collaboration is essential for responsible integration.

The integration of nanotechnology and robotics holds immense potential for enhancing industries, healthcare, and daily life. However, ethical concerns, safety concerns, and regulatory hurdles must be addressed for responsible technology development, necessitating a collaborative approach between researchers, policymakers, and industry stakeholders.

FUTURE DIRECTIONS AND BREAKTHROUGHS

This chapter delves into the ongoing research and development efforts, potential breakthroughs, and ethical and social implications of nanotechnology-enhanced robotics, highlighting its potential to revolutionize industries and redefine human-machine interactions (Bragança et al., 2019; Cui et al., 2021; Malik et al., 2023; Ramesh et al., 2020).

Ongoing Research and Development

Nanorobotics in Medicine: Ongoing research focuses on the development of nanorobots for targeted drug delivery, cancer treatment, and minimally invasive surgeries. Nano-scale robots could navigate the human body, delivering drugs precisely to targeted cells or assisting in intricate medical procedures with unprecedented precision.

Swarm Robotics: Research in swarm robotics, utilizing nanoscale components, explores the collective intelligence of robotic swarms for tasks such as environmental monitoring, disaster response, and exploration. Swarm robotics holds potential breakthroughs in collaborative problem-solving, adaptability, and scalability.

Soft Robotics with Nanomaterials: Ongoing efforts investigate the integration of nanomaterials in soft robotics, enabling robots with flexible and deformable structures. This research aims to create robots that can navigate complex environments, interact safely with humans, and perform delicate tasks in fields like healthcare and manufacturing.

Energy Harvesting at the Nanoscale: Research focuses on nanotechnology-enabled energy harvesting mechanisms for robotic systems. Nanoscale devices could harness ambient energy sources, such as vibrations or thermal gradients, to power robotic components, reducing dependence on traditional energy sources.

Potential Breakthroughs

Nanoscale Energy Storage: Breakthroughs in nanotechnology may lead to advancements in energy storage, allowing for high-capacity and lightweight nanobatteries. This breakthrough could significantly extend the operational lifespan and efficiency of nanotechnology-enhanced robotic systems.

Nano sensors for Enhanced Perception: Advancements in nanosensors could revolutionize the perception capabilities of robots, enabling them to detect and respond to stimuli with unparalleled sensitivity. This breakthrough could enhance robotic applications in areas like environmental monitoring, healthcare, and industrial automation.

Quantum Computing in Robotics: Integration of quantum computing with nanotechnology could unlock unprecedented computational power for robotic systems. Quantum-enhanced robotics may lead to breakthroughs in optimization, machine learning, and complex decision-making processes.

Self-Healing Nanomaterials: Breakthroughs in self-healing nanomaterials could enhance the durability and longevity of robotic components. This advancement may lead to robotic systems capable of autonomously repairing minor damages, reducing maintenance requirements.

Ethical and Social Implications

Privacy Concerns with Nanorobotics: As nanorobots advance in medical applications, concerns about privacy arise regarding the collection and transmission of sensitive health data. Striking a balance between the potential benefits of nanorobotics in healthcare and ensuring patient privacy will be a critical ethical consideration.

Autonomy and Decision-Making: The increasing autonomy of robots, fueled by nanotechnology, raises ethical questions about the responsibility and accountability for robotic decisions. Establishing ethical guidelines for autonomous systems to ensure alignment with human values and legal frameworks becomes paramount.

Socio-Economic Impact: The widespread adoption of nanotechnology-enhanced robotics may lead to shifts in the job market and socio-economic structures. Ethical considerations include addressing potential job displacement, ensuring access to new technologies, and mitigating inequalities in technology adoption.

Ethical Use of AI in Nanorobotics: Integrating artificial intelligence (AI) with nanorobotics requires ethical considerations to prevent unintended consequences and misuse. Developing frameworks for responsible AI use in nanotechnology-enhanced robotics involves addressing biases, ensuring transparency, and preventing malicious applications (Boopathi & Khang, 2023; Koshariya, Kalaiyarasi, et al., 2023; Zekrifa et al., 2023).

The future of nanotechnology-enhanced robotics is promising, with advancements in medical nanobotics and energy storage. However, ethical and social implications must be considered, and stakeholders must collaborate across disciplines for responsible development, deployment, and societal integration.

CONCLUSION

The integration of nanotechnology into robotics has revolutionized the design, manufacturing, and performance of robotic systems. The use of nano-materials at the nanoscale has improved the fundamental attributes of robots, enhancing strength, flexibility, and functionality. This paradigm shift has not only revolutionized the construction of robots but also elevated their operational capabilities. The unique properties of nano-materials have opened new horizons for the robotic industry, enabling robots to navigate complex environments, perform intricate tasks, and adapt dynamically to changing conditions. The marriage of nano-materials with robotic frameworks has redefined possibilities and allowed for unprecedented innovations.

The integration of nano fluids into robotic systems has improved performance and extended operational lifespans, ensuring sustained functionality in diverse environmental conditions. This breakthrough has set the stage for energy-efficient and adaptable solutions, contributing to the eco-friendliness of advanced systems. Nano-materials and nano fluids offer eco-friendly solutions, mitigating resource consumption, minimizing electronic waste, and promoting energy efficiency in robotic systems. This chapter emphasizes the role of nanotechnology in addressing the environmental impact of robotics and promoting sustainability.

The integration of nanotechnology and robotics presents challenges such as ethical considerations, safety concerns, and the need for robust regulatory frameworks. However, potential breakthroughs like nanoscale energy storage and self-healing nanomaterials are promising. It's essential to focus on ongoing research and collaboration between researchers, engineers, and policymakers to navigate the complexities of nanotechnology-enhanced robotics. Ethical considerations must guide the trajectory, ensuring the integration benefits humanity positively and inclusively.

This chapter explores the significant impact of nano-materials and nano fluids on the robot industry and their environments, highlighting the revolutionary changes brought by nanotechnology. It highlights the strengths, flexibility, and adaptability of nanotechnology in robotics. As we approach a future characterized by nanotechnology and robotics convergence, it urges responsibility, innovation, and commitment to harnessing this transformative synergy's full potential.

ABBREVIATIONS

3D- Three Dimensional
    AI - Artificial intelligence
    E-Waste - Electronic Waste
    NEMS - Nanoelectromechanical Systems

REFERENCES

Atilano, L., Martinho, A., Silva, M., & Baptista, A. (2019). Lean Design-for-X: Case study of a new design framework applied to an adaptive robot gripper development process. *Procedia CIRP, 84,* 667–672. doi:10.1016/j.procir.2019.04.190

Boopathi, S. (2021). *Pollution monitoring and notification: Water pollution monitoring and notification using intelligent RC boat*. Academic Press.

Boopathi, S. (2022a). An investigation on gas emission concentration and relative emission rate of the near-dry wire-cut electrical discharge machining process. *Environmental Science and Pollution Research International, 29*(57), 86237–86246. doi:10.1007/s11356-021-17658-1 PMID:34837614

Boopathi, S. (2022b). Cryogenically treated and untreated stainless steel grade 317 in sustainable wire electrical discharge machining process: A comparative study. *Springer :Environmental Science and Pollution Research*, 1–10.

Boopathi, S. (2023a). An Investigation on Friction Stir Processing of Aluminum Alloy-Boron Carbide Surface Composite. In *Springer:Advances in Processing of Lightweight Metal Alloys and Composites* (pp. 249–257). Springer. doi:10.1007/978-981-19-7146-4_14

Boopathi, S. (2023b). Internet of Things-Integrated Remote Patient Monitoring System: Healthcare Application. In *Dynamics of Swarm Intelligence Health Analysis for the Next Generation* (pp. 137–161). IGI Global. doi:10.4018/978-1-6684-6894-4.ch008

Boopathi, S., & Davim, J. P. (2023a). Applications of Nanoparticles in Various Manufacturing Processes. In *Sustainable Utilization of Nanoparticles and Nanofluids in Engineering Applications* (pp. 1–31). IGI Global. doi:10.4018/978-1-6684-9135-5.ch001

Boopathi, S., & Davim, J. P. (2023b). *Sustainable Utilization of Nanoparticles and Nanofluids in Engineering Applications*. IGI Global. doi:10.4018/978-1-6684-9135-5

Boopathi, S., Jeyakumar, M., Singh, G. R., King, F. L., Pandian, M., Subbiah, R., & Haribalaji, V. (2022). An experimental study on friction stir processing of aluminium alloy (AA-2024) and boron nitride (BNp) surface composite. *Materials Today: Proceedings, 59*(1), 1094–1099. doi:10.1016/j.matpr.2022.02.435

Boopathi, S., & Khang, A. (2023). AI-Integrated Technology for a Secure and Ethical Healthcare Ecosystem. In *AI and IoT-Based Technologies for Precision Medicine* (pp. 36–59). IGI Global. doi:10.4018/979-8-3693-0876-9.ch003

Boopathi, S., Khare, R., KG, J. C., Muni, T. V., & Khare, S. (2023). Additive Manufacturing Developments in the Medical Engineering Field. In Development, Properties, and Industrial Applications of 3D Printed Polymer Composites (pp. 86–106). IGI Global.

Boopathi, S., & Kumar, P. (2024). Advanced bioprinting processes using additive manufacturing technologies: Revolutionizing tissue engineering. *3D Printing Technologies: Digital Manufacturing, Artificial Intelligence, Industry 4.0*, 95.

Boopathi, S., Thillaivanan, A., Mohammed, A. A., Shanmugam, P., & VR, P. (2022). Experimental investigation on Abrasive Water Jet Machining of Neem Wood Plastic Composite. *IOP: Functional Composites and Structures, 4*, 025001.

Boopathi, S., Umareddy, M., & Elangovan, M. (2023). Applications of Nano-Cutting Fluids in Advanced Machining Processes. In *Sustainable Utilization of Nanoparticles and Nanofluids in Engineering Applications* (pp. 211–234). IGI Global. doi:10.4018/978-1-6684-9135-5.ch009

Bragança, S., Costa, E., Castellucci, I., & Arezes, P. M. (2019). A brief overview of the use of collaborative robots in industry 4.0: Human role and safety. *Occupational and Environmental Safety and Health*, 641–650.

Chakraborty, A., Ravi, S. P., Shamiya, Y., Cui, C., & Paul, A. (2021). Harnessing the physicochemical properties of DNA as a multifunctional biomaterial for biomedical and other applications. *Chemical Society Reviews*, *50*(13), 7779–7819. doi:10.1039/D0CS01387K PMID:34036968

Cui, Y., Qin, Z., Wu, H., Li, M., & Hu, Y. (2021). Flexible thermal interface based on self-assembled boron arsenide for high-performance thermal management. *Nature Communications*, *12*(1), 1284. doi:10.1038/s41467-021-21531-7 PMID:33627644

Du, C., Ren, Y., Qu, Z., Gao, L., Zhai, Y., Han, S.-T., & Zhou, Y. (2021). Synaptic transistors and neuromorphic systems based on carbon nano-materials. *Nanoscale*, *13*(16), 7498–7522. doi:10.1039/D1NR00148E PMID:33928966

Fowziya, S., Sivaranjani, S., Devi, N. L., Boopathi, S., Thakur, S., & Sailaja, J. M. (2023). Influences of nano-green lubricants in the friction-stir process of TiAlN coated alloys. *Materials Today: Proceedings*. Advance online publication. doi:10.1016/j.matpr.2023.06.446

Gellers, J. C. (2020). *Rights for robots: Artificial intelligence, animal and environmental law* (1st ed.). Routledge. doi:10.4324/9780429288159

Gowri, N. V., Dwivedi, J. N., Krishnaveni, K., Boopathi, S., Palaniappan, M., & Medikondu, N. R. (2023). Experimental investigation and multi-objective optimization of eco-friendly near-dry electrical discharge machining of shape memory alloy using Cu/SiC/Gr composite electrode. *Environmental Science and Pollution Research International*, *30*(49), 1–19. doi:10.1007/s11356-023-26983-6 PMID:37126160

Harikaran, M., Boopathi, S., Gokulakannan, S., & Poonguzhali, M. (2023). Study on the Source of E-Waste Management and Disposal Methods. In *Sustainable Approaches and Strategies for E-Waste Management and Utilization* (pp. 39–60). IGI Global. doi:10.4018/978-1-6684-7573-7.ch003

Harris, M., Wu, H., Zhang, W., & Angelopoulou, A. (2022). Overview of recent trends in microchannels for heat transfer and thermal management applications. *Chemical Engineering and Processing*, *181*, 109155. doi:10.1016/j.cep.2022.109155

Hassani, S. S., Daraee, M., & Sobat, Z. (2020). Advanced development in upstream of petroleum industry using nanotechnology. *Chinese Journal of Chemical Engineering*, *28*(6), 1483–1491. doi:10.1016/j.cjche.2020.02.030

He, L., Xu, J., Dekai, Z., Qinghai, Y., & Longqiu, L. (2018). Potential application of functional micro-nano structures in petroleum. *Petroleum Exploration and Development*, *45*(4), 745–753. doi:10.1016/S1876-3804(18)30077-6

Jakkula, R. V. S. K., & Sethuramalingam, P. (2023). Analysis of coatings based on carbon-based nanomaterials for paint industries-A review. *Australian Journal of Mechanical Engineering*, *21*(3), 1008–1036. doi:10.1080/14484846.2021.1938953

Javaid, M., Haleem, A., Singh, R. P., & Suman, R. (2021). Substantial capabilities of robotics in enhancing industry 4.0 implementation. *Cognitive Robotics*, *1*, 58–75. doi:10.1016/j.cogr.2021.06.001

Joe, S., Bliah, O., Magdassi, S., & Beccai, L. (2023). Jointless Bioinspired Soft Robotics by Harnessing Micro and Macroporosity. *Advancement of Science*, *10*(23), 2302080. doi:10.1002/advs.202302080 PMID:37323121

Karthik, K., Teferi, A. B., Sathish, R., Gandhi, A. M., Padhi, S., Boopathi, S., & Sasikala, G. (2023). Analysis of delamination and its effect on polymer matrix composites. *Materials Today: Proceedings*. Advance online publication. doi:10.1016/j.matpr.2023.07.199

Karthik, S., Hemalatha, R., Aruna, R., Deivakani, M., Reddy, R. V. K., & Boopathi, S. (2023). Study on Healthcare Security System-Integrated Internet of Things (IoT). In Perspectives and Considerations on the Evolution of Smart Systems (pp. 342–362). IGI Global.

Kiran, J. S., & Prabhu, S. (2020). Robot nano spray painting-A review. *IOP Conference Series. Materials Science and Engineering*, *912*(3), 032044. doi:10.1088/1757-899X/912/3/032044

Koshariya, A. K., Kalaiyarasi, D., Jovith, A. A., Sivakami, T., Hasan, D. S., & Boopathi, S. (2023). AI-Enabled IoT and WSN-Integrated Smart Agriculture System. In *Artificial Intelligence Tools and Technologies for Smart Farming and Agriculture Practices* (pp. 200–218). IGI Global. doi:10.4018/978-1-6684-8516-3.ch011

Koshariya, A. K., Khatoon, S., Marathe, A. M., Suba, G. M., Baral, D., & Boopathi, S. (2023). Agricultural Waste Management Systems Using Artificial Intelligence Techniques. In *AI-Enabled Social Robotics in Human Care Services* (pp. 236–258). IGI Global. doi:10.4018/978-1-6684-8171-4.ch009

Kumar, M. R., Reddy, V. P., Meheta, A., Dhiyani, V., Al-Saady, F. A., & Jain, A. (2023). Investigating the Effects of Process Parameters on the Size and Properties of Nano Materials. *E3S Web of Conferences*, *430*, 01125.

Maguluri, L. P., Arularasan, A., & Boopathi, S. (2023). Assessing Security Concerns for AI-Based Drones in Smart Cities. In Effective AI, Blockchain, and E-Governance Applications for Knowledge Discovery and Management (pp. 27–47). IGI Global. doi:10.4018/978-1-6684-9151-5.ch002

Maheswari, B. U., Imambi, S. S., Hasan, D., Meenakshi, S., Pratheep, V., & Boopathi, S. (2023). Internet of things and machine learning-integrated smart robotics. In Global Perspectives on Robotics and Autonomous Systems: Development and Applications (pp. 240–258). IGI Global. doi:10.4018/978-1-6684-7791-5.ch010

Malik, S., Muhammad, K., & Waheed, Y. (2023). Nanotechnology: A revolution in modern industry. *Molecules (Basel, Switzerland)*, *28*(2), 661. doi:10.3390/molecules28020661 PMID:36677717

Marinoudi, V., Sørensen, C. G., Pearson, S., & Bochtis, D. (2019). Robotics and labour in agriculture. A context consideration. *Biosystems Engineering*, *184*, 111–121. doi:10.1016/j.biosystemseng.2019.06.013

Mohanty, A., Jothi, B., Jeyasudha, J., Ranjit, P., Isaac, J. S., & Boopathi, S. (2023). Additive Manufacturing Using Robotic Programming. In *AI-Enabled Social Robotics in Human Care Services* (pp. 259–282). IGI Global. doi:10.4018/978-1-6684-8171-4.ch010

Mouchou, R., Laseinde, T., Jen, T.-C., & Ukoba, K. (2021). Developments in the Application of Nano Materials for Photovoltaic Solar Cell Design, Based on Industry 4.0 Integration Scheme. *Advances in Artificial Intelligence, Software and Systems Engineering: Proceedings of the AHFE 2021 Virtual Conferences on Human Factors in Software and Systems Engineering, Artificial Intelligence and Social Computing, and Energy,* July 25-29, 2021, USA, 510–521.

Murali, B., Padhi, S., Patil, C. K., Kumar, P. S., Santhanakrishnan, M., & Boopathi, S. (2023). Investigation on hardness and tensile strength of friction stir processing of Al6061/TiN surface composite. *Materials Today: Proceedings.*

Myilsamy, S., & Sampath, B. (2021). Experimental comparison of near-dry and cryogenically cooled near-dry machining in wire-cut electrical discharge machining processes. *Surface Topography : Metrology and Properties, 9*(3), 035015. doi:10.1088/2051-672X/ac15e0

Naveeenkumar, N., Rallapalli, S., Sasikala, K., Priya, P. V., Husain, J., & Boopathi, S. (2024). Enhancing Consumer Behavior and Experience Through AI-Driven Insights Optimization. In *AI Impacts in Digital Consumer Behavior* (pp. 1–35). IGI Global. doi:10.4018/979-8-3693-1918-5.ch001

Ness, S., Shepherd, N. J., & Xuan, T. R. (2023). Synergy Between AI and Robotics: A Comprehensive Integration. *Asian Journal of Research in Computer Science, 16*(4), 80–94. doi:10.9734/ajrcos/2023/v16i4372

Oluwasanu, A. A., Oluwaseun, F., Teslim, J. A., Isaiah, T. T., Olalekan, I. A., & Chris, O. A. (2019). Scientific applications and prospects of nanomaterials: A multidisciplinary review. *African Journal of Biotechnology, 18*(30), 946–961. doi:10.5897/AJB2019.16812

Paul, A., Thilagham, K., KG, J.-, Reddy, P. R., Sathyamurthy, R., & Boopathi, S. (2024). Multi-criteria Optimization on Friction Stir Welding of Aluminum Composite (AA5052-H32/B4C) using Titanium Nitride Coated Tool. Engineering Research Express.

Puranik, T. A., Shaik, N., Vankudoth, R., Kolhe, M. R., Yadav, N., & Boopathi, S. (2024). Study on Harmonizing Human-Robot (Drone) Collaboration: Navigating Seamless Interactions in Collaborative Environments. In Cybersecurity Issues and Challenges in the Drone Industry (pp. 1–26). IGI Global.

Rahamathunnisa, U., Subhashini, P., Aancy, H. M., Meenakshi, S., Boopathi, S., & ... (2023). Solutions for Software Requirement Risks Using Artificial Intelligence Techniques. In *Handbook of Research on Data Science and Cybersecurity Innovations in Industry 4.0 Technologies* (pp. 45–64). IGI Global.

Ramalingam, S., & Rasool Mohideen, S. (2021). Composite materials for advanced flexible link robotic manipulators: An investigation. *International Journal of Ambient Energy, 42*(14), 1670–1675. doi:10.1080/01430750.2019.1613263

Ramesh, R. D., Santhosh, A., & Syamala, S. R. N. A. (2020). Implementation of Nanotechnology in the Aerospace and Aviation Industry. In *Smart Nanotechnology with Applications* (pp. 51–69). CRC Press. doi:10.1201/9781003097532-4

Ramudu, K., Mohan, V. M., Jyothirmai, D., Prasad, D., Agrawal, R., & Boopathi, S. (2023). Machine Learning and Artificial Intelligence in Disease Prediction: Applications, Challenges, Limitations, Case Studies, and Future Directions. In Contemporary Applications of Data Fusion for Advanced Healthcare Informatics (pp. 297–318). IGI Global.

Revathi, S., Babu, M., Rajkumar, N., Meti, V. K. V., Kandavalli, S. R., & Boopathi, S. (2024). Unleashing the Future Potential of 4D Printing: Exploring Applications in Wearable Technology, Robotics, Energy, Transportation, and Fashion. In Human-Centered Approaches in Industry 5.0: Human-Machine Interaction, Virtual Reality Training, and Customer Sentiment Analysis (pp. 131–153). IGI Global.

Sampath, B., & Myilsamy, S. (2021). Experimental investigation of a cryogenically cooled oxygen-mist near-dry wire-cut electrical discharge machining process. *Stroj. Vestn. Jixie Gongcheng Xuebao*, 67(6), 322–330.

Satav, S. D., Lamani, D., Harsha, K., Kumar, N., Manikandan, S., & Sampath, B. (2023). Energy and Battery Management in the Era of Cloud Computing: Sustainable Wireless Systems and Networks. In Sustainable Science and Intelligent Technologies for Societal Development (pp. 141–166). IGI Global.

Selvakumar, S., Adithe, S., Isaac, J. S., Pradhan, R., Venkatesh, V., & Sampath, B. (2023). A Study of the Printed Circuit Board (PCB) E-Waste Recycling Process. In Sustainable Approaches and Strategies for E-Waste Management and Utilization (pp. 159–184). IGI Global.

Selvakumar, S., Shankar, R., Ranjit, P., Bhattacharya, S., Gupta, A. S. G., & Boopathi, S. (2023). E-Waste Recovery and Utilization Processes for Mobile Phone Waste. In *Handbook of Research on Safe Disposal Methods of Municipal Solid Wastes for a Sustainable Environment* (pp. 222–240). IGI Global. doi:10.4018/978-1-6684-8117-2.ch016

Sengeni, D., Padmapriya, G., Imambi, S. S., Suganthi, D., Suri, A., & Boopathi, S. (2023). Biomedical waste handling method using artificial intelligence techniques. In *Handbook of Research on Safe Disposal Methods of Municipal Solid Wastes for a Sustainable Environment* (pp. 306–323). IGI Global. doi:10.4018/978-1-6684-8117-2.ch022

Senthil, T., Puviyarasan, M., Babu, S. R., Surakasi, R., Sampath, B., & ... (2023). Industrial Robot-Integrated Fused Deposition Modelling for the 3D Printing Process. In *Development, Properties, and Industrial Applications of 3D Printed Polymer Composites* (pp. 188–210). IGI Global.

Sevinchan, E., Dincer, I., & Lang, H. (2018). A review on thermal management methods for robots. *Applied Thermal Engineering*, 140, 799–813. doi:10.1016/j.applthermaleng.2018.04.132

Sharmin, I. (2020). *Preparation and evaluation of carbon nano tube based nanofluid in milling alloy steel*. Academic Press.

Sheikh, J. A., Waheed, M. F., Khalid, A. M., & Qureshi, I. A. (2020). Use of 3D printing and nano materials in fashion: From revolution to evolution. *Advances in Design for Inclusion: Proceedings of the AHFE 2019 International Conference on Design for Inclusion and the AHFE 2019 International Conference on Human Factors for Apparel and Textile Engineering*, July 24-28, 2019, Washington DC, USA 10, 422–429.

Singh, K., Sharma, S., Shriwastava, S., Singla, P., Gupta, M., & Tripathi, C. (2021). Significance of nano-materials, designs consideration and fabrication techniques on performances of strain sensors-A review. *Materials Science in Semiconductor Processing, 123*, 105581. doi:10.1016/j.mssp.2020.105581

Srinivas, B., Maguluri, L. P., Naidu, K. V., Reddy, L. C. S., Deivakani, M., & Boopathi, S. (2023). Architecture and Framework for Interfacing Cloud-Enabled Robots. In *Handbook of Research on Data Science and Cybersecurity Innovations in Industry 4.0 Technologies* (pp. 542–560). IGI Global. doi:10.4018/978-1-6684-8145-5.ch027

Syamala, M., Komala, C., Pramila, P., Dash, S., Meenakshi, S., & Boopathi, S. (2023). Machine Learning-Integrated IoT-Based Smart Home Energy Management System. In *Handbook of Research on Deep Learning Techniques for Cloud-Based Industrial IoT* (pp. 219–235). IGI Global. doi:10.4018/978-1-6684-8098-4.ch013

Vijayakumar, G. N. S., Domakonda, V. K., Farooq, S., Kumar, B. S., Pradeep, N., & Boopathi, S. (2024). Sustainable Developments in Nano-Fluid Synthesis for Various Industrial Applications. In Adoption and Use of Technology Tools and Services by Economically Disadvantaged Communities: Implications for Growth and Sustainability (pp. 48–81). IGI Global.

Zekrifa, D. M. S., Kulkarni, M., Bhagyalakshmi, A., Devireddy, N., Gupta, S., & Boopathi, S. (2023). Integrating Machine Learning and AI for Improved Hydrological Modeling and Water Resource Management. In *Artificial Intelligence Applications in Water Treatment and Water Resource Management* (pp. 46–70). IGI Global. doi:10.4018/978-1-6684-6791-6.ch003

Chapter 13
Predictive Modelling for Employee Retention:
A Three–Tier Machine Learning Approach With oneAPI

Ahamed Thaiyub
KPR Institute of Engineering and Technology, India

Akshay Bhuvaneswari Ramakrishnan
ⓘ https://orcid.org/0009-0000-1578-0984
SASTRA University, India

Shriram Kris Vasudevan
Intel Corporation, India

T. S. Murugesh
Government College of Engineering, Srirangam, India

Sini Raj Pulari
Bahrain Polytechnic, Bahrain

ABSTRACT

Organizations face enormous issues when it comes to employee turnover, which is why they need to develop accurate predictive models for retention. The purpose of this chapter is to present a three-tiered machine learning approach for predicting employee turnover that makes use of resume parsing, performance analysis, and advanced algorithms. In addition, the authors make use of Intel oneAPI, which is a unified programming model that is increasingly becoming the industry standard, in order to improve the scalability and performance of the solution. The system that is offered delivers full HR (human resource) analytics, which enables firms to make educated decisions regarding recruiting and retention tactics. The results of the experimental evaluation show that the solution is effective in providing an accurate forecast of attrition, which paves the way for proactive retention measures. The approach enhances system performance by utilizing oneAPI, which in turn ensures that it is scalable over a variety of different hardware architectures.

DOI: 10.4018/979-8-3693-5767-5.ch013

1. INTRODUCTION

The loss of employees results in significant financial expenses and impedes the smooth running of business operations, making it a significant worry for businesses all over the world. It is of the utmost importance for businesses that are interested in retaining key personnel and lowering the costs associated with turnover to have the ability to both predict and manage employee turnover (Jarrahi, 2018; Vardarlier & Zafer, 2020). This study provides a three-tier machine learning (ML) method to handle this difficulty. The solution makes use of advanced algorithms and predictive analytics to accurately forecast staff turnover. The three-tiered solution includes resume parsing, a review of employee performance, and a prediction of future attrition. In the initial level of the selection process, procedures for "resume parsing" are utilized in order to extract pertinent information from candidate resumes. This information includes skills, experience, and education. The second level does an analysis of past employee performance data, taking into consideration a variety of indicators and characteristics such as attendance records. The third layer makes use of advanced machine learning techniques to make predictions about future attrition based on the data that has been acquired. These algorithms include Gradient Boosting, Logit, KNN, Sequential, and XGBoost. The integration of these three tiers enables firms to make educated decisions regarding staff recruitment and retention. A full HR analytics platform is provided by the system, which helps in proactively detecting attrition concerns and putting targeted retention initiatives into action. This strategy ultimately assists companies in cutting the costs associated with employee turnover, keeping valuable people as staff, and preserving a stable and productive workforce. In this article, we go further into the specifics of the three-tiered machine learning solution, examining the methodology, data collecting, and pre-processing processes along the way. We discuss the outcomes of the experiment as well as the analysis conducted to demonstrate how accurate the proposed solution is in forecasting employee turnover. In addition, we address the benefits and implications of deploying such a solution in the real-world organizational settings, which will provide HR professionals and decision-makers with information that can help them improve their employee retention efforts.

2. LITERATURE REVIEW

Fallucchi et al. (2020) have recognized the primary contributors to the turnover rate of employees and proposed a real classification based on the statistical examination of the data collected. The Gaussian Nave Bayes classifier was seen as the algorithm that generated the best results for the specific dataset used. It suggested the best recall rate, a metric measuring a classifier's ability to find all of the positive events, and it attained an overall false negative rate that was equivalent to 4.5% of the total measurements. These are the two metrics which assessed a classifier's ability to find all of the positive instances. Researchers in Raza et al. (2022) have attempted to forecast employee turnover by employing ETC (Extra Trees Classifier), SVM (Support Vector Machine), LR (Logistic Regression), and DTC (Decision Tree Classifier), which are the four sophisticated machine learning algorithms. Using dataset 10folds, the applied machine learning methods have achieved an accuracy score of 88% in SVM, 74% in the LR technique, 84% in the DTC method and 93% with the proposed method. In the work Najafi-Zangeneh et al. (2021), a thorough comparison of the outcomes with the existing approaches was carried out. The comparison demonstrates that the proposed feature selection improves the predictor's overall performance. This study proposed a three-stage, pre-processing, processing, and post-processing framework for creating an accurate employee attrition prediction model and for corroborating the validity of the model's parameters. The binary and continuous feature sets were selected using the max-out feature selection approach. This method was applied to the set of feature sets. The objective of the researchers in Pratt et al. (2021) was to provide a comparison of several machine learning approaches in order to make an estimate of employees who are likely to leave their firm. In this paper, a total of six different ML algorithms were utilized. According to the findings of the work, the Random Forest method displayed the best ability to estimate the attrition rate of employees. The most accurate prediction was made with a score of 85.12, which was considered to be a good accuracy.

3. METHODOLOGY

The methodology that we propose contains three stages: the first one is "resume parsing" the second one "employee performance analysis," and the third is "future attrition prediction." We make use of a total of five distinct machine learning models. This section offers a comprehensive explanation of each tier, including the methods involved in data collection and pre-processing as well as the models that are implemented is as shown in Figure 1.

3.1 The Three-Tier Architecture

In this study, a three-tier machine learning approach is provided as depicted in Fig 2, which includes resume parsing, employee performance analysis, and attrition prediction in the future. In the overall architecture, each tier is essential to the correct prediction of staff attrition and the facilitation of well-informed decisions for retention tactics.

Tier 1: Parsing Resumes

The first tier is on employing sophisticated parsing techniques to obtain pertinent data from candidate resumes. The textual content of resumes is analyzed using Natural Language Processing (NLP) algorithms, allowing the extraction of crucial details that includes abilities, experience, education, and certifications. These extracted qualities offer a thorough overview of candidate qualifications and potential, serving as useful inputs for the following tiers.

Tier 2: Employee Performance Analysis

In the second stage, historical employee performance data are analyzed to find the trends and warning signs of attrition risks. Numerous criteria are taken into account, including attendance records, productivity levels, performance evaluations, and more pertinent elements unique to the firm. The performance data is analyzed and modelled using statistical and machine learning approaches, which reveals correlations and trends that might help forecast attrition.

Figure 1. Methodology flow

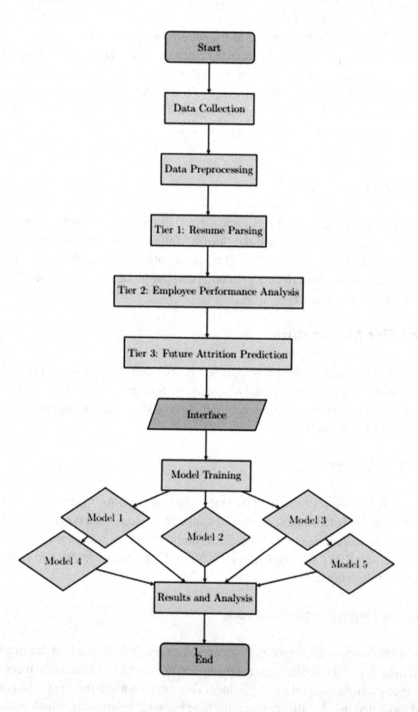

Tier 3: Prediction of Future Attrition

Based on the data obtained from the previous layers, the third and final tier aims to forecast future attrition. Predictive models are created using cutting-edge machine learning techniques including Gradient

Boosting, Logit (logistic regression), KNN (K-Nearest Neighbor), Sequential, and XGBoost. These models take into account the insights obtained from employee performance analysis as well as the features acquired from resume parsing. The solution forecasts the probability that a worker would leave the company in the future using historical data and sophisticated algorithms.

Figure 2. The three-tier architecture

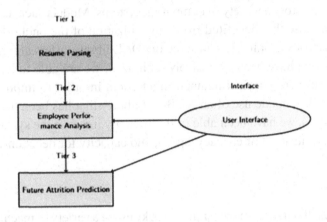

3.2 Data Collection

During the phase of data gathering, we extract essential information from GitHub repositories in order to construct a complete dataset for the prediction of employee turnover. GitHub is an outstanding platform that provides access to a wealth of useful information, such as code repositories, commit histories, issue tracking, and activities involving collaborative efforts.

3.3 Data Pre-Processing

Data pre-processing is a key part of our method because it changes the collected data and gets it ready for analysis and model building. It requires cleaning the data to deal with missing numbers, outliers, and duplicates, making sure that the data is correct. Techniques for selecting the most important traits are used to reduce the number of dimensions. To improve the performance of a model, feature engineering includes changing and making new features, like encoding categorical variables and scaling numerical features. Before information is taken from the resumes, a number of steps are taken to make sure the data is in a format that can be used for analysis. First, the necessary tools are brought in. These include NLTK (Natural Language Toolkit), spaCy, pyresparser, and pandas. Then, the Resume Parser class from pyresparser is used to read and pull out the data from the resume. The important information from the resumes is extracted from the parsed data, which is then saved to a file called "resume.txt" using the open() function. This step makes it possible to do more work and research. Also, skills are taken from the resume info that has been parsed. A CSV file called "skills.csv" stores a list of skills that has already been set up. The list of skills is read from the CSV file by the code. The application data that has been parsed is then looked over, and any skills that are found are added to a skill_list. These steps make sure that the data from the resumes is well organized and ready to be analyzed and modelled.

3.4 Optimization With oneAPI

In order to optimize and increase the performance of our staff retention system, we have taken benefit of the capabilities offered by oneAPI (Intel, n.d.). Intel's oneAPI is a unified programming model which allows efficient execution across a variety of computing systems. Models such as logit, KNN, and gradient boosting are among those that benefited from our utilization of the oneDAL library, which is part of the oneAPI Data Analytics Library. Because the oneDAL library offers algorithms and methods for data analytics activities that have been extensively optimized, we are able to handle and analyze massive amounts of data in a more efficient manner. In addition, in order to improve the functionality of the XGBoost model, we have made use of an XGBoost library that has been tuned for the oneAPI. By utilizing these technologies, we have been able to significantly improve the system that we use to retain our employees in terms of its level of efficacy, speed, and capacity for decision-making.

3.5 Models

In the Models component of our three-tier design, we make use of a variety of machine learning techniques to forecast employee attrition. We use a total of five distinct models, namely Gradient Boosting, Logit, K-Nearest Neighbors, Sequential, and XGBoost. Each of these models is distinguished by a distinct set of qualities and advantages.

3.5.1 Gradient Boosting

The concept of boosting is rethought by gradient boosting, which recasts it as a numerical optimization problem with the goal of reducing the loss value of the model through the addition of base learners through the use of gradient descent (Bentéjac et al., 2021; Johnson et al., 2017). The process of obtaining a local estimate of a variational function using an incremental optimization procedure of the first degree is called gradient descent. The idea behind the gradient boosting approach is to use the pattern in the residuals to strengthen a poor forecasting model as much as possible until the residuals are arbitrarily (or possibly random normally) distributed. This is the goal of the algorithm. It consists mostly of the following three stages:

1. A measure of loss that needs to be optimised.
2. An unreliable learner when it comes to making predictions.
3. An incremental model in which weaker learners are added in order to reduce the magnitude of the error function.

3.5.2 Logistic Regression

Logistic regression is a popular technique that can be utilised to predict the chance that a given occurrence corresponds to a specific class. If the predicted probability of an occurrence is higher than fifty percent, then the model suggests that the instance is a member of category 1, and if it is less than fifty percent, then it suggests that the occurrence is not a member of class 1 (Marvin et al., 2021; Midi et al., 2010). This transforms it into a two-way classifier. Analysts devised the logistic function, which is also known as the sigmoid function, in order to characterise the qualities of demographic increase in ecology. These properties include rising swiftly and reaching their maximum at the carrying capacity of the ecosystem. It is an S-shaped arc which can translate any real-valued integer into a value ranging from 0 to 1, but just not directly at those boundaries. It might take any integer and map it to one of those values. Natural logarithms use e as their base, and the quantity X that you want to modify using the logistic function is what you pass into the function.

3.5.3 K-Nearest Neighbors

The K-Nearest Neighbors approach is considered to be a supervised kind of machine learning because the goal variable for the task at hand is already known. A non-parametric method is one that would not establish any assumptions about the input distribution structure that lies under the surface. K is a number that is used to determine neighbouring data points that are comparable to the new information point. It does a head count of how many data items fall under each category across the k surrounding neighbours. The newly added data item will be assigned to the class which contains the most neighbouring points. The Euclidean distance between two clusters is determined by taking the square root of the total of the squared distances between those two points. This determines the length among two neighbours (Chowdhury et al., 2022; Madeti & Singh, 2018). The L2 norm is another name for this concept.

3.5.4 Sequential

The sequential model allows for the input of a list that specifies the different layers that make up the neural network's architecture as illustrated in Fig 3. The data moves in a linear fashion from one layer to the next until it reaches the output layer at the very end. When building a neural network, the input layer is the very first layer to be added. It is not difficult at all when using keras (Gulli & Pal, 2017; Hutter et al., 2011). We begin by generating a layer consisting of 256 neurons, activated by ReLU, with an input size of 4. When compared to the other layers, the output layer depicted in equation (1) is distinguished by the requirement that it accurately reflect the quantity of values that are desired to be produced by the neural network.

$$X = f\_n(f\_\{n-1\}(...f\_2(f\_1(Input))...))$$ (1)

3.5.5 XGBoost

Extreme Gradient Boosting, often known as XGBoost, is a technique that can assist tree-boosting algorithms in making the most of the available memory and hardware resources. It has the advantages of improving algorithms and fine-tuning models, and it can also be implemented in computing environments (Lee et al., 2021; Li et al., 2020). Gradient boosting, regularized boosting, and stochastic boosting are the three primary types of gradient boosting, and XGBoost is capable of implementing all of them. In addition to that, it enables the inclusion of regularization parameters as well as the adjustment of those parameters, setting it apart from other libraries. During development of XGBoost, both extensive considerations in terms of system optimization and fundamentals of machine learning are taken into account.

Figure 3. Sequential model architecture

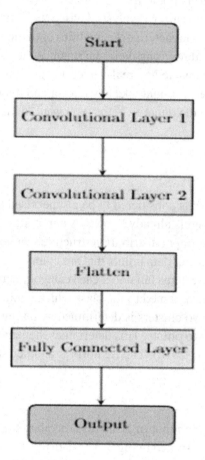

4. RESULTS AND DISCUSSION

The accuracy of the logistic regression model is obtained as 82.45%, which is considered to be a reasonable level of prediction performance. The KNN model is seen to achieve an accuracy of 76.87%, which, despite being significantly lower than the accuracy achieved by logistic regression, is still indicative of satisfactory outcomes. The gradient boosting model attains an accuracy of 72.45%, which indicates that there is room for improvement or that hyper parameters need fine-tuning. On the other hand, the sequential model displays an accuracy that is substantially higher than 95%. This hints that the sequential design, with its layered structure and capability of capturing hierarchical representations, is a good choice for the task of retaining employees. Because of its higher performance, the sequential model demonstrates its capacity to learn intricate patterns and relationships in the data, which ultimately leads to more accurate predictions. In addition, the XGBoost model produces the results with the best accuracy, which is 96.06%. We are able to improve the performance of the XGBoost model by applying the optimized XGBoost library, which results in the XGBoost model being an outstanding option for the employee retention system as indicated in Table. 1. The outstanding predictive capability of the XGBoost algorithm is contributed to by the fact that it could handle complicated relationships and make use of approaches involving boosting. To conclude, in terms of accuracy, the sequential and XGBoost models perform better than the KNN model, the logistic regression model, and the gradient boosting model. Fig 4 shows a comparison between all the models used. These findings highlight how important it is to select the suitable model and make use of optimized libraries in order to attain excellent performance in employee retention systems. The whole system is also hosted as an interface as seen in Fig 5.

Table 1. Accuracy of different models

S:No	Models	Validation Accuracy
1	XGBoost	96.06%
2	Sequential	95.00%
3	Gradient Boosting	72.45%
4	KNN	76.87%
5	Logistic Regression	82.45%

Figure 4. Accuracy comparison of the models used

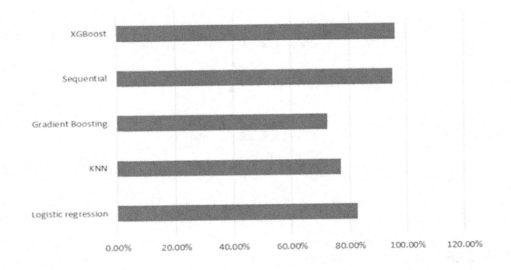

4.1 Directions for Future Works

Our approach for retaining employees has room for improvement in the future, and those improvements can partake a few different forms. To begin, we have the option of investigating additional features or data sources to enhance the strength of prediction. These can include things like employee satisfaction surveys or external economic indicators. Second, there is a possibility of implementing model ensemble techniques, which involve pooling the predictions of numerous models in order to improve the overall performance. Thirdly, improving the results of already-existing models by fine-tuning the hyper parameters of those models using methods such as grid search or Bayesian optimization can produce superior results. In addition, utilizing time-series analysis techniques can allow for the identification of temporal patterns as well as the relationships in the data pertaining to employee retention. By concentrating on these aspects, we can possibly continue to enhance the accuracy, interpretability, and efficiency of our employee retention system, which will be of use to enterprises in reducing the risks associated with employee turnover.

Figure 5. System interface

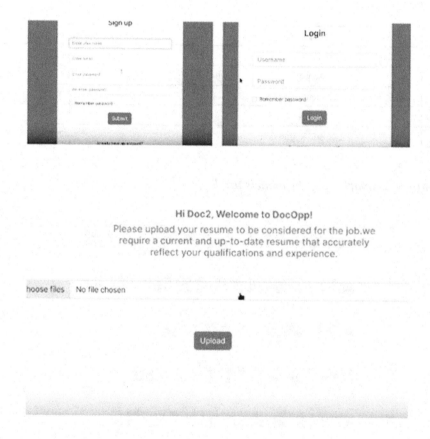

REFERENCES

Bentéjac, C., Csörgő, A., & Martínez-Muñoz, G. (2021). A comparative analysis of gradient boosting algorithms. *Artificial Intelligence Review*, *54*(3), 1937–1967. doi:10.1007/s10462-020-09896-5

Chowdhury, A. H., Malakar, S., Seal, D. B., & Goswami, S. (2022). Understanding employee attrition using machine learning techniques. In *Data Management, Analytics and Innovation: Proceedings of ICDMAI 2021,* Volume 2 (pp. 101-109). Springer Singapore. 10.1007/978-981-16-2937-2_8

Fallucchi, F., Coladangelo, M., Giuliano, R., & William De Luca, E. (2020). Predicting employee attrition using machine learning techniques. *Computers*, *9*(4), 86. doi:10.3390/computers9040086

Gulli, A., & Pal, S. (2017). *Deep learning with Keras*. Packt Publishing Ltd.

Hutter, F., Hoos, H. H., & Leyton-Brown, K. (2011). Sequential model-based optimization for general algorithm configuration. In *Learning and Intelligent Optimization: 5th International Conference, LION 5, Rome, Italy, January 17-21, 2011. Selected Papers 5* (pp. 507-523). Springer Berlin Heidelberg. 10.1007/978-3-642-25566-3_40

Intel. (n.d.). https://www.intel.com/content/www/us/en/developer/tools/oneapi/overview.html

Jarrahi, M. H. (2018). Artificial intelligence and the future of work: Human-AI symbiosis in organizational decision making. *Business Horizons*, *61*(4), 577–586. doi:10.1016/j.bushor.2018.03.007

Johnson, N. E., Ianiuk, O., Cazap, D., Liu, L., Starobin, D., Dobler, G., & Ghandehari, M. (2017). Patterns of waste generation: A gradient boosting model for short-term waste prediction in New York City. *Waste Management (New York, N.Y.)*, *62*, 3–11. doi:10.1016/j.wasman.2017.01.037 PMID:28216080

Lee, J. J., Lee, Y. R., Lim, D. H., & Ahn, H. C. (2021). A Study on the Employee Turnover Prediction using XGBoost and SHAP. *Journal of Information Systems*, *30*(4), 21–42.

Li, H., Cao, Y., Li, S., Zhao, J., & Sun, Y. (2020). XGBoost model and its application to personal credit evaluation. *IEEE Intelligent Systems*, *35*(3), 52–61. doi:10.1109/MIS.2020.2972533

Madeti, S. R., & Singh, S. N. (2018). Modeling of PV system based on experimental data for fault detection using kNN method. *Solar Energy*, *173*, 139–151. doi:10.1016/j.solener.2018.07.038

Marvin, G., Jackson, M., & Alam, M. G. R. (2021, August). A machine learning approach for employee retention prediction. In *2021 IEEE Region 10 Symposium (TENSYMP)* (pp. 1-8). IEEE. 10.1109/TENSYMP52854.2021.9550921

Midi, H., Sarkar, S. K., & Rana, S. (2010). Collinearity diagnostics of binary logistic regression model. *Journal of Interdisciplinary Mathematics*, *13*(3), 253–267. doi:10.1080/09720502.2010.10700699

Najafi-Zangeneh, S., Shams-Gharneh, N., Arjomandi-Nezhad, A., & Hashemkhani Zolfani, S. (2021). An Improved Machine Learning-Based Employees Attrition Prediction Framework with Emphasis on Feature Selection. *Mathematics*, *9*(11), 1226. doi:10.3390/math9111226

Pratt, M., Boudhane, M., & Cakula, S. (2021). Employee attrition estimation using random forest algorithm. *Baltic Journal of Modern Computing*, *9*(1), 49–66. doi:10.22364/bjmc.2021.9.1.04

Raza, A., Munir, K., Almutairi, M., Younas, F., & Fareed, M. M. S. (2022). Predicting Employee Attrition Using Machine Learning Approaches. *Applied Sciences (Basel, Switzerland)*, *12*(13), 6424. doi:10.3390/app12136424

Vardarlier, P., & Zafer, C. (2020). Use of artificial intelligence as business strategy in recruitment process and social perspective. *Digital Business Strategies in Blockchain Ecosystems: Transformational Design and Future of Global Business*, 355-373.

Chapter 14
Research Analysis of Data Exploration and Visualization Dashboard Using Data Science

D. Faridha Banu

Sri Eshwar College of Engineering, Coimbatore, India

C. Keerthi Prashanth

Sri Eshwar College of Engineering, Coimbatore, India

P. T. Kousalya

Sri Eshwar College of Engineering, Coimbatore, India

P. Madhumohan

Sri Eshwar College of Engineering, Coimbatore, India

Kavin Varsha

Sri Eshwar College of Engineering, Coimbatore, India

S. Meivel

https://orcid.org/0000-0002-8717-3881

M. Kumarasamy College of Engineering, Karur, India

ABSTRACT

Any attempt to explain the relevance of data by putting it in a visual context is referred to as data visualization. With the aid of data visualization software, patterns, trends, and correlations that could go unnoticed in text-based data can be exposed and identified more easily. The graphical presentation of quantitative information is known as data visualization. In other words, data visualizations convert big and small data sets into pictures that the human brain can comprehend and digest more readily. In our daily lives, data visualizations are surprisingly prevalent, yet they frequently take the shape of recognizable charts and graphs. It can be applied to find unknown trends and facts. When communication, data science, and design come together, good data visualizations are produced. When done well, data visualizations provide important insights into complex data sets in clear, understandable ways. The authors talk about data visualization, its significance, tools for data visualization, etc. in this chapter.

DOI: 10.4018/979-8-3693-5767-5.ch014

I. INTRODUCTION

Organizations are adopting data visualizations and data technologies to improve their queries and decisions more than ever before. Making better data-driven business decisions has never been easier thanks to developing computer technologies and new, user-friendly software programs (Aparicio & Costa, 2014). Compared to pictures, complicated data is far more difficult for human brains to comprehend when it is stored in numbers and text. Utilizing dataviz approaches, one can convey vast volumes of information in the most effective, visual way possible (Friedman, 2008) (Tukey, 1977). In reality, data that is presented visually is simpler to interpret and evaluate, allowing decision-makers to uncover patterns including new and hidden ones and comprehend even complex ideas more quickly. Data visualization tools like as charts, graphs, maps, and dashboards can be useful in a variety of ways, including identifying problems and inadequacies, selecting the best approach for a product or business operation, predicting sales volume and stock prices, optimizing project management and resource allocation, and many other things. No matter how much data you have, using sophisticated analysis and clear visualizations is one of the greatest ways to identify significant links. The capacity to explore, analyze, and visualize data efficiently is crucial in the age of data-driven decision-making. The Data Exploration and Visualization Dashboard project is an innovative approach created to give users the means to extract useful information from their datasets. Due to the constantly increasing volume of data, this dashboard acts as a portal for both inexperienced and seasoned data aficionados to easily upload, explore, and alter data before turning it into useful visual representations. This project offers users an interactive environment that streamlines data exploration, promotes informed decision-making, and enables a deeper understanding of underlying trends by integrating powerful Python libraries like Pandas, NumPy, Matplotlib, Plotly, Seaborn, and Streamlit. This study examines how the Data Exploration and Visualization Dashboard was developed, its features, and its results, demonstrating how its user-friendly interface closes the gap between data and insights.

II. DATA VISUALIZATION WORKING METHOD

The majority of data visualization tools available today have connectors to widely used data sources, such as Hadoop, the most popular relational databases, and a number of cloud storage platforms. These sources' data are retrieved by the visualization program, which then applies a graphic style to the data. A designer's job is greatly facilitated by automating the process of constructing a visualization, at least in part, when working with data sets that contain hundreds of thousands or millions of data points (Friendly, 2017). Though increasingly technologies automates this process, data visualization software still lets the user choose the best manner to present the data. Some technologies automatically analyze the data's form, find correlations between particular variables, and then insert these findings into the appropriate chart type, as determined by the software (DATA VISUALIZATION FOR HUMAN PERCEPTION). The dashboard feature of data visualization software typically enables users to combine many analysis' visuals into a single interface, which is typically a web portal. A graphical and visual representation of information is a dashboard. Initially, a dashboard was intended to help technology managers of organizations plan, build, execute, and direct applications in real-time while also keeping track of the information that is working. Dashboards are a popular tool for monitoring and analyzing corporate operations nowadays. Your "span of control" over a large amount of business data is improved by cognitive tools like dashboards and visualization. These instruments aid in the visual identification of trends, patterns, and anomalies. They assist people in making sense of what they perceive and direct them toward wise choices. Therefore, these tools must take advantage of users' visual talents. The importance of visual information design has increased due to the prominence of dashboards and the ease with which business users can now review their data .

III. DATA VISUALIZATION LIBRARIES: MATPLOTLIB, PLOTLY, AND SEABORN

The selection of data visualization libraries is essential for a project's success in producing smart and useful representations. The Matplotlib, Plotly, and Seaborn packages are three that are frequently used for data visualization in Python. These libraries are useful resources for various areas of data visualization because each one has particular advantages and use cases. Matplotlib, Plotly, and Seaborn together gave the project the power to produce a wide variety of interactive visualizations. Matplotlib's versatility made it possible to create conventional plots, and Plotly's interactive graphs improved the user experience. The stylistic improvements made by Seaborn improved the outputs of the dashboard. One of the most popular and functional data visualization libraries in Python is Matplotlib. It offers a complete set of tools for producing static charts and figures of publication-caliber. Matplotlib may be used to create a broad variety of visualizations, from straightforward line charts and scatter plots to intricate heatmaps and three-dimensional graphs. It is frequently used for static data visualizations and exploratory data analysis (EDA). Because Matplotlib provides fine-grained control over plot components, users can alter every piece of the visualization. It has a wide range of customization options and is compatible with other libraries.

A strong library for building web-based and interactive data visualizations is Plotly. Building interactive dashboards and web apps is where it shines the most. Plotly excels at producing interactive charts, maps, and dashboards that are dynamic and dynamic. It can be used to create interactive web applications that let users examine data. Plotly creates interactive visuals that are simple to include in online programs. There are many other chart kinds and customization choices available. Python and JavaScript are only two of the many programming languages for which Plotly supports APIs. A high-level data visualization library based on Matplotlib is called Seaborn. It is intended to make it simpler to create visually appealing statistics visuals. For making statistical visualizations like distribution plots, pair plots, and heatmaps, Seaborn is extremely helpful. It is frequently used in data analysis and statistical modeling because it is excellent at displaying relationships between variables. Seaborn offers a high-level interface for writing little to no code while producing complex visuals. It has pre-installed color schemes and themes for visually appealing visualizations. Seaborn works well with Pandas DataFrames as well.

IV. STREAMLIT FRAMEWORK

An open-source Python framework called Streamlit makes it easier to build web applications for data science and visualization projects. Data professionals may quickly and easily convert data scripts into interactive web apps because to its straightforward architecture. The dashboard's user interface was developed with the primary help of Streamlit. Its ease of use and ability to convert data scripts into shareable web applications sped up the dashboard deployment process. User engagement was enabled using Streamlit, enabling for fluid exploration of data and visuals.

V. PANDAS AND NUMPY LIBRARIES

For data handling and mathematical computations, Pandas and NumPy have become essential tools. I successfully organized and worked with the data by utilizing the features of Pandas' DataFrame. Advanced numerical operations were made easier by NumPy, which also provided the structural support for data conversions and statistical calculations. Each of these tools helped students gain a more comprehensive understanding of data manipulation, analysis, and visualization approaches. Additionally, their integration demonstrated how well they worked together, creating a thorough and user-focused Data Exploration and Visualization Dashboard.

VI. ANALYSIS

1. Product Definition

The Data Exploration and Visualization Dashboard project's core objective is to meet the urgent demand for an interactive platform that makes effective data exploration and visualization possible. The project's goal is to develop a user-friendly web application that will enable users to easily upload datasets, receive insights from a variety of visualizations, apply filters, and even alter data without having to have a deep understanding of coding. The key goal is to create a link between data and insights so that decision-makers, analysts, and domain experts may interact with data in a way that promotes quick decision-making and better understanding.

2. Feasibility Analysis

The availability of sophisticated Python modules, which allow for smooth data handling, manipulation, and display, supports the viability of the Data Exploration and display Dashboard project. Pandas, NumPy, Matplotlib, Plotly, Seaborn, and Streamlit integration creates a solid basis for the project's growth. These reputable libraries provide thorough functionality for data analysis and visualization, guaranteeing that the project may successfully achieve its goals. The project's breadth also fits with Streamlit's capabilities, an approachable framework that turns data scripts into web apps. The popularity of data-driven decision-making across numerous industries, which emphasizes the need for easily available data exploration tools, further supports the project's viability.

VII. SOFTWARE REQUIREMENTS

The successful implementation of the Data Exploration and Visualization Dashboard project relies on a comprehensive understanding of the software requirements. This analysis outlines the key components and specifications that the project must adhere to:

1. User Interface

Users should be able to upload datasets in a number of different formats, including CSV and Excel, using the dashboard's straightforward and user-friendly interface. To access data summaries, fundamental statistics, visualizations, data filtering, and modification features, the interface should provide simple 10 navigation. The user interface (UI) should prominently display an easy-to-use data upload button that enables users to pick and upload datasets. In order to provide a smooth import procedure, it should also feature built-in techniques to handle popular data formats like CSV and Excel. Users should be guided through numerous data exploration and visualization tools via the dashboard's clear and organized navigation menu. Users should be able to create and alter several chart kinds, such as bar charts, scatter plots, and histograms, in the dashboard's visualization area. Users should be able to interactively add filters to the dataset using the interface to promote data exploration. By concentrating on particular data subsets, users can improve their analysis and visualization.

2. Data Handling and Manipulation

The project must effectively utilize the Pandas and NumPy libraries for data manipulation. It should provide functionalities to filter data based on numerical ranges and categorical values, as well as the option to add and remove columns seamlessly. The project utilizes the extensive data manipulation and numerical computation capabilities of the Pandas and NumPy libraries through seamless integration. Users can work with datasets effectively thanks to this integration, which makes use of NumPy arrays and Pandas DataFrames. Data may be readily filtered by numerical ranges for users. Users should be able to enter minimum and maximum values for particular numerical columns via input forms on the interface. The data is then appropriately filtered by the system, which only shows the records that satisfy the given requirements. Users should have the option to filter data for categorical data based on particular categorical values or categories. By using user-defined criteria, this functionality streamlines the process of segmenting and evaluating data subsets. The project gives users the ability to clean, modify, and prepare their data for insightful analysis and visualization, ultimately boosting their data exploration experience. This is accomplished by successfully utilizing Pandas and NumPy for data handling and manipulation.

A) Visualization Capabilities

Multiple visualization formats, such as histograms, line charts, scatter plots, bar charts, and pie charts, should be supported by the dashboard. Users should have the ability to examine data patterns dynamically thanks to the seamless integration of the Matplotlib, Plotly, and Seaborn libraries. In order to provide both static and interactive visualizations, the project incorporates Matplotlib, a flexible and well-liked Python charting package. Users can access interactive, web-based visualizations thanks to Plotly. It makes it possible to zoom, pan, hover, and do other things. The quality and selection of accessible plots are improved thanks to Seaborn, which is renowned for its beautiful and instructive statistical visualizations. To see how the distribution of numerical data is distributed, users can generate histograms. The dashboard ought to support customizing the histogram's properties, including bin sizes. The visualization of trends and patterns in time series or sequential data is best accomplished with line charts. Multiple lines can be plotted by users for comparison. Comparing categorical data is effectively done with bar charts. In addition to choosing between vertical and horizontal bars, users can change the labels' colors.

Pie charts are great for showing how something is put together as a whole. Users should have access to interactive zoom and pan features for large datasets so they may concentrate on particular data subsets. To filter the dataset or get more information, users can choose data points in scatter plots or bar charts. For greater context and interpretation, users of the dashboard can add axis labels, titles, and legends to their visualizations. For greater context and interpretation, users of the dashboard can add axis labels, titles, and legends to their visualizations.

B) Real-Time Interactivity

The Streamlit framework should be used for the project to provide real-time interactivity. Without requiring page reloads, users should be able to interact with visualizations, use filters, and see immediate updates. The project is based on the Streamlit framework, a Python toolkit made for quickly and easily building interactive web applications. Data exploration and visualization are made very responsive and dynamic thanks to Streamlit's integration, making it suitable for both technical and non-technical users. The project's generated visualizations are by nature dynamic and interactive. To get more information, users can zoom, pan, and hover over data points. Users can quickly switch between various chart kinds or setups thanks to real-time interactivity. With the help of the Streamlit framework, users can easily interact with the data and visualizations thanks to a number of widgets including sliders, buttons, text inputs, and choose boxes. Changes are made instantly as users engage with the dashboard, whether by applying filters, changing the data, or choosing data points. Users can now view the results of their activities in real time. For instance, changing the filter range on a histogram immediately changes the plot, enabling users to explore various elements of the data in real-time. The lack of page reloads in Streamlit is a key benefit of real-time engagement. Users can continuously examine data and visualize trends, improving workflow effectiveness and user experience.

C) Data Export and Saving

The capability to save processed data in CSV format must be offered by the dashboard. Users should also be able to download the raw data for additional research and save visualizations as image files. The processed and changed data can be readily exported in CSV format by users. For users who want to save their cleaned and modified datasets for use in outside programs or for sharing with coworkers, this functionality is essential. A clear button or option to start the CSV export should be provided in the export procedure. The export and save options are user-configurable. For stored visualizations, they can select the resolution, file type, and file name. Users can choose several CSV file delimiter options to match the specifications of their analysis tools when exporting data. Users can keep a record of their data changes and analyses by using the versioning or timestamping options the dashboard may implement for data exports.

D) Compatibility and Deployment

The project ought to work with a variety of browsers and operating systems. Users should be able to access it using web browsers as an independent web application that may be deployed. The project's platform independence ensures interoperability with well-known operating systems including Windows, macOS, and Linux. Regardless of the OS they want, users can access the dashboard. The user interface and capabilities have undergone extensive testing and optimization for cross-browser compatibility. This includes, but is not limited to, Safari, Opera, Microsoft Edge, Mozilla Firefox, and Google Chrome. Different browser versions should be taken into consideration during compatibility testing to guarantee a consistent user experience. The project can be used to deploy a standalone web application that is hosted on a web server and reachable by a URL. Users are not required to locally install any dependencies or software. The program hosting on a web server should be done according to simple deployment instructions.

E) Security and Privacy

Data security and privacy must be given top priority in the dashboard, ensuring that uploaded datasets are handled securely. Additionally, it must to provide options for erasing user inputs and private data after use. The Data Exploration and Visualization Dashboard project can be created to effectively satisfy the expectations and needs of its consumers by considering these software requirements. User data and uploaded datasets are safely saved on the server. Access control and encryption at rest are just two examples of appropriate safeguards that are used to protect data privacy and integrity. The dashboard's user access is restricted via procedures for authentication and authorisation. Role-based access control ensures that users can only access information and functionality that are pertinent to their responsibilities. Users must log in using secure credentials. Sensitive data in datasets is if necessary anonymised or hidden to preserve user privacy. This stops personally identifiable information (PII) from being displayed in visualizations or exported.

VIII. DESIGN

Table 1. Design

Upload Dataset	Allows users to upload CSV or Excel datasets for analysis.
Display Data Summary	Presents a preview of the uploaded dataset's initial rows.
Display Basic Statistics	Provides basic statistical measures (mean, median, etc.) for numerical columns.
Visualize Numerical Columns	Enables users to select a numerical column and visualize it using various chart types (Histogram, Line Chart, Scatter Plot).
Visualize Categorical Columns	Allows users to select a categorical column and visualize it using chart types (Bar Chart, Pie Chart).
Data Filtering by Numerical	Provides the option to filter data based on a numerical column by specifying a numerical range.
Data Filtering by Categorical	Enables data filtering based on a categorical column by selecting specific categorical values.
Add Column	Allows users to add a new column to the dataset with a user-defined name and default value.
Remove Column	Permits users to remove a selected column from the dataset.
Save Processed Data	Enables users to save the processed dataset after filtering and manipulation in CSV format.
Save Visualization	Allows users to save visualizations generated from selected columns as image files (e.g., PNG).

IX. OUTPUT

The dashboard gives a succinct overview of the uploaded dataset along with important facts and details. Users can rapidly understand the most important details of their data.

The robust data filtering tool of our Data Exploration and Visualization Dashboard is one of its core components, enabling you to precisely segment and examine your collection. You can concentrate on particular subsets of data that are important to your research or analytical goals by filtering the data. By defining a range of values, you can filter data based on columns with numerical values. You can specify minimum and maximum values in the filtering tool to make sure the data provided satisfies your unique requirements. A key technique for honing your data analysis and obtaining accurate insights is filtering. Our dashboard's filtering features provide you control over your data exploration trip whether you're performing exploratory data analysis, market segmentation, or any other data-driven work.

Figure 1. Dashboard

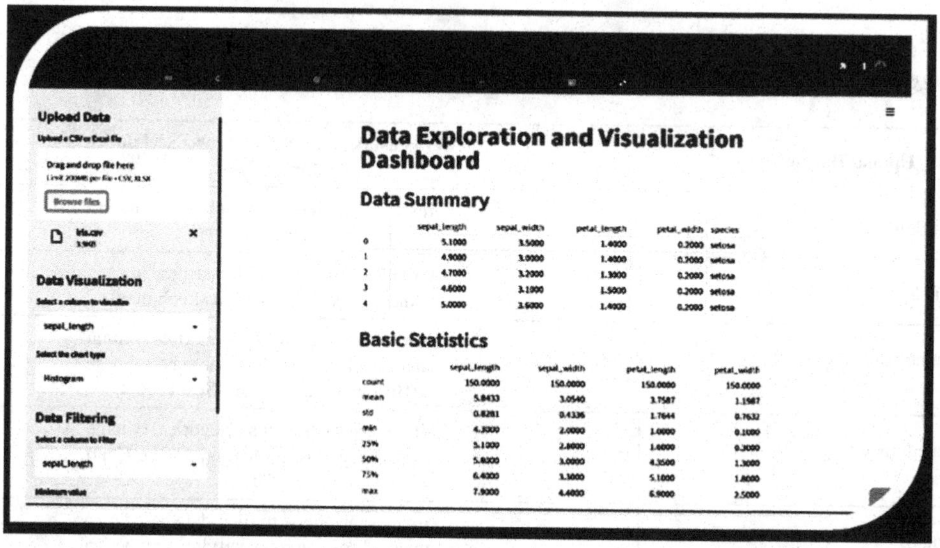

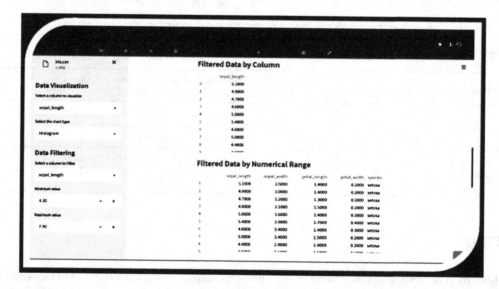

Figure 2. Filtering data

By converting raw data into visual representations, data visualization is a potent tool that makes it simpler to comprehend patterns, trends, and insights using a range of graphs and charts.

Data grouping based on particular attributes or columns is one of the basic data analysis operations. By combining data points with similar qualities, we can acquire insights into the patterns and relationships present in our collection. We include a strong grouping tool in this dashboard that enables you to dynamically group your data. By grouping your data, you can find patterns or trends that might not be visible when viewing the collection as a whole. You may produce insightful summaries by aggregating data within groups, which helps you make complex datasets easier to grasp.

Figure 3. Data visualization

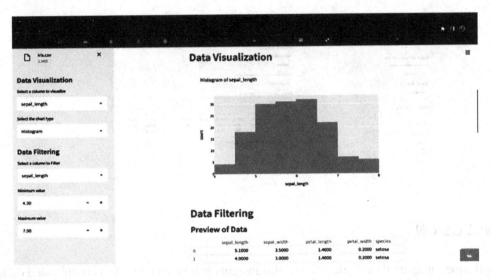

Figure 4. Grouping

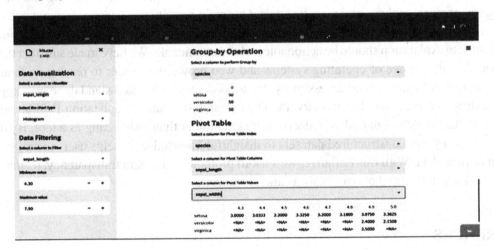

With the organized and dynamic perspective of your dataset that the Pivot Table analysis offers, you can quickly extract insightful information and investigate data correlations. This tool is essential for flexible and interactive data summarization, aggregation, and visualization. You can select the columns for the rows and columns in the pivot table as well as the aggregated data. The summary statistics and calculations performed on your data are completely under your control. By pivoting and displaying data in real time, it is simple to discover links and patterns in the data. Produce individualized reports and visualizations for presentations or additional research. Gaining a deeper grasp of your data will help you make decisions that are more based on data.

Figure 5. Pivot table

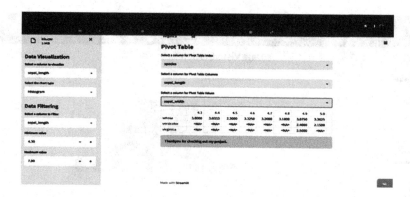

X. CONCLUSION

Making informed judgments and gleaning useful insights from enormous and complicated datasets is a crucial problem in the age of big data. In order to help users overcome this obstacle, the Data Exploration and Visualization Dashboard project was designed and created. It gives users a strong and simple tool to efficiently explore, analyze, and visualize data. Our main priority has been to design a user-friendly interface that is both technical and non-technical people can utilize. We are aware that the conclusions drawn from data exploration should be actionable and communicable. We have made sure that the project is compatible with a range of operating systems and web browsers in order to reach a wide audience. Users benefit from flexibility and accessibility due to its deployability as a standalone web application hosted on different platforms. In summary, the Data Exploration and Visualization Dashboard project is a platform that facilitates data-driven decision-making rather than only acting as a tool. By directing users on a journey from unstructured datasets to insightful knowledge, it helps users to realize the full potential of their data. With this initiative, we seek to promote better data transparency, teamwork, and decision-making in the field of data analysis and visualization.

REFERENCES

Ahmed, L. J., Anish Fathima, B., Mahaboob, M., & Gokulavasan, B. (2021). Biomedical Image Processing with Improved SPIHT Algorithm and optimized Curvelet Transform Technique. *2021 7th International Conference on Advanced Computing and Communication Systems, ICACCS 2021.* 10.1109/ICACCS51430.2021.9441832

Ahmed, L. J., Bruntha, P. M., Dhanasekar, S., Chitra, V., Balaji, D., & Senathipathi, N. (2022). An Improvised Image Registration Technique for Brain Tumor Identification and Segmentation Using ANN Approach. *ICDCS 2022 - 2022 6th International Conference on Devices, Circuits and Systems,* 80-84. 10.1109/ICDCS54290.2022.9780846

Dency Flora, G., Sekar, G., Nivetha, R., Thirukkumaran, R., Silambarasan, D., & Jeevanantham, V. (2022). An Optimized Neural Network for Content Based Image Retrieval in Medical Applications. *8th International Conference on Advanced Computing and Communication Systems, ICACCS 2022*, 1560-1563. 10.1109/ICACCS54159.2022.9785151

Faridha Banu, D., Sindhwani, N. S., G, K. R. A., & M, S. (2022). Fuzzy acceptance Analysis of Impact of Glaucoma and Diabetic Retinopathy using Confusion Matrix. *2022 10th International Conference on Reliability, Infocom Technologies and Optimization (Trends and Future Directions) (ICRITO)*, 1-5. 10.1109/ICRITO56286.2022.9964858

Friedman, V. (2008). Data visualization and infographics. Graphics, Monday Inspiration.

Friendly, M. (2017). *A brief history of data visualization*. Springer-Verlag.

Manuela & Costa. (2014, November). Data Visualization. *Communication Design Quarterly Review*.

Meivel, S., Maheswari, S., & Faridha Banu, D. (2023). Design and Method of an Agricultural Drone System Using Biomass Vegetation Indices and Multispectral Images. In *Proceedings of UASG 2021: Wings 4 Sustainability. UASG 2021. Lecture Notes in Civil Engineering* (vol. 304). Springer. 10.1007/978-3-031-19309-5_25

Mercy, J., Lawanya, R., Nandhini, S., & Saravanan, M. (2022). Effective Image Deblurring Based on Improved Image Edge Information and Blur Kernel Estimation. *8th International Conference on Advanced Computing and Communication Systems, ICACCS 2022*, 855-859.

Tukey, J. (1977). *Exploratory Data Analysis*. Addison-Wesley.

Chapter 15
Research Directions and Challenges in the Deployment of Explainable AI in Human–Robot Interactions for Healthcare 4.0/5.0

A. Kishore Kumar

ⒾⒹ https://orcid.org/0000-0003-4876-319X

Sri Ramakrishna Engineering College, India

S. Sarveswaran

Sri Ramakrishna Engineering College, India

N. Dheerthi

Sri Ramakrishna Engineering College, India

A. Murugarajan

Sri Ramakrishna Engineering College, India

ABSTRACT

In the rapidly evolving landscape of Healthcare 4.0/5.0, the integration of artificial intelligence (AI) and robotics has shown immense potential in transforming patient care. However, the deployment of these technologies in human-robot interactions (HRI) demands a delicate balance between efficiency and transparency. This chapter explores the research directions and challenges associated with the implementation of Explainable AI (XAI) in the context of HRI for the advancement of healthcare services. The authors delve into the critical aspects of ensuring transparency and interpretability in AI-driven robotic systems, emphasizing the need for explainability to foster trust and collaboration between healthcare professionals, patients, and intelligent robotic entities. The chapter highlights key challenges, proposes potential research directions, and suggests methodologies to address the complexities in deploying XAI within the healthcare ecosystem.

DOI: 10.4018/979-8-3693-5767-5.ch015

INTRODUCTION

Central to Healthcare 4.0/5.0 is the reliance on data analytics and artificial intelligence for informed decision-making. The integration of big data, machine learning, and predictive analytics empowers healthcare professionals with actionable insights, optimizing treatment plans and resource allocation. There is a notable shift towards personalized and patient-centric care in Healthcare 4.0/5.0. AI-driven technologies enable the tailoring of treatments based on individual patient data, preferences, and real-time health monitoring, fostering a more holistic and effective approach to healthcare. Robotics, powered by AI, plays a pivotal role in automating routine tasks, such as diagnostics, surgeries, and medication dispensing. This not only enhances efficiency but also minimizes errors, thereby improving overall patient safety (Vale et al., 2022).

The Transformative Role of AI and Robotics

AI algorithms analyze vast datasets, including genetic information, to identify patterns and predict patient responses to specific treatments. This enables the customization of medical interventions, moving towards a precision medicine approach. AI and robotics facilitate remote monitoring of patients, enabling healthcare professionals to track vital signs and disease progression in real-time. This not only enhances patient comfort but also allows for early intervention and prevention. Robotics in surgery, guided by AI, enhances precision and minimizes invasiveness. Surgeons can now perform complex procedures with greater accuracy, reducing recovery times and improving overall patient outcomes. AI's ability to analyze historical data aids in predicting potential health risks and complications. This proactive approach enables healthcare providers to implement preventive measures, ultimately reducing the burden on healthcare systems (Gupta et al., 2022).

The integration of AI and robotics in modern healthcare systems has revolutionized the way medical services are delivered, enhancing efficiency, accuracy, and patient outcomes.

AI-driven robotic systems, such as the da Vinci Surgical System, assist surgeons in performing minimally invasive procedures with greater precision and control. These robots can execute complex maneuvers with high dexterity, leading to reduced recovery times, less scarring, and improved patient safety. AI algorithms analyze medical images such as X-rays, MRIs, and CT scans to detect abnormalities and assist radiologists in diagnosis. Deep learning algorithms can identify patterns and anomalies in medical images more accurately and efficiently than human experts, leading to earlier detection of diseases like cancer and faster treatment decisions. AI-powered devices and wearable sensors enable continuous monitoring of patients' vital signs and health metrics outside of traditional clinical settings. These systems can alert healthcare providers to potential issues in real-time, allowing for timely interventions and proactive management of chronic conditions. AI algorithms analyze vast amounts of patient data, including genetic information, medical history, and treatment outcomes, to tailor treatment plans to individual patients. This personalized approach improves treatment efficacy and reduces the risk of adverse reactions by considering each patient's unique characteristics and needs. AI accelerates the drug discovery process by predicting the efficacy and safety of potential drug candidates and identifying new therapeutic targets. Machine learning algorithms analyze large datasets to identify patterns in biological data, leading to the development of novel treatments for various diseases (Mbunge et al., 2021).

Robotic exoskeletons and rehabilitation devices aid patients in recovering from injuries or surgeries by providing targeted physical therapy and assistance with mobility. These devices can adapt their assistance levels based on patients' progress and feedback, facilitating faster recovery and improved functional outcomes. AI-powered chatbots and virtual assistants help patients schedule appointments, access medical information, and receive personalized health recommendations.

These virtual agents can triage patients, provide basic medical advice, and offer support for managing chronic conditions, relieving the burden on healthcare providers and improving patient access to care. AI algorithms analyze healthcare data to predict disease outbreaks, optimize resource allocation, and improve operational efficiency within healthcare systems. By identifying trends and patterns in data, these predictive models enable proactive decision-making and resource planning to better meet patient needs. Overall, the integration of AI and robotics in modern healthcare systems holds immense promise for improving patient outcomes, enhancing operational efficiency, and advancing medical research and innovation. However, it also raises important ethical and regulatory considerations related to data privacy, algorithm transparency, and the equitable distribution of healthcare resources (Mohanta et al., 2019). Healthcare 4.0/5.0 represents a transformative era where AI and robotics synergistically contribute to the advancement of healthcare delivery. As we navigate this new frontier, it is essential to explore the challenges and opportunities posed by these technologies, ensuring that the benefits of innovation are harnessed responsibly for the betterment of global healthcare.

EXPLAINABILITY IN HUMAN-ROBOT INTERACTIONS

Explainability in human-robot interactions refers to the ability of an AI-driven system to transparently communicate its reasoning and decision-making processes to humans in a way that is understandable and interpretable. Human users are more likely to trust and accept the decisions made by AI-driven robots if they understand the rationale behind those decisions. When robots can explain their actions in a clear and transparent manner, users are more likely to feel comfortable interacting with them and relying on their assistance. In safety-critical applications such as healthcare, manufacturing, and autonomous vehicles, it's essential for humans to understand why a robot made a particular decision or took a specific action.

Explainable AI enables users to identify potential errors or biases in the system's reasoning and intervene if necessary to prevent accidents or malfunctions (Pawar et al., 2020).

By providing explanations for their actions, robots can help humans identify and correct errors in the system's behaviour more effectively. If a robot makes a mistake or encounters a situation it doesn't understand, an explanation of its decision-making process can facilitate troubleshooting and problem-solving. Explainable AI can serve as a valuable educational tool by helping users, particularly students and novice users, understand complex concepts and processes. By explaining how they reach decisions or solve problems, AI-driven systems can enhance users' understanding of the underlying principles and encourage learning and skill development. In regulated industries such as finance and healthcare, there may be legal and ethical requirements mandating transparency and accountability in AI-driven decision-making. Explainable AI enables organizations to demonstrate compliance with these regulations by providing clear explanations for their actions and decisions. Explainable AI can enhance the overall user experience by making interactions with robots more intuitive and meaningful. When users understand why a robot behaves in a certain way or makes a particular recommendation, they are more likely to engage with the system in a positive and productive manner. Transparent explanations can help users identify and mitigate biases in AI-driven systems, thereby promoting fairness and equity in decision-making processes. By revealing the factors influencing their decisions, robots can enable users to assess whether those decisions are fair and unbiased across different demographic groups. Thus, explainability is essential for fostering trust, ensuring safety, promoting learning, and enhancing the overall user experience in human-robot interactions (Shaban Nejad et al., 2020). By providing transparent explanations for their actions and decisions, AI-driven robots can empower users to understand, trust, and effectively collaborate with intelligent systems in various domains.

Explainable AI facilitates communication between humans and robots by enabling robots to provide clear and interpretable explanations for their actions. This communication is essential for effective collaboration, as it allows humans to understand the intentions and goals of robots and coordinate their efforts accordingly. Transparent explanations create a feedback loop between humans and robots, enabling mutual learning and adaptation. When robots explain their decisions, humans can provide feedback and guidance to improve the performance and behaviour of the system, leading to more productive collaboration over time. Overall, explainability plays a crucial role in shaping the dynamics of trust, acceptance, and collaboration in HRI. By providing transparent explanations for their actions and decisions, robots can establish trust with users, increase acceptance of their presence and role, and facilitate effective collaboration in various domains ranging from healthcare and manufacturing to education and entertainment.

Overview of Explainable AI Techniques

Explainable AI (XAI) techniques aim to provide insights into the decision-making processes of AI models, making their outputs understandable to humans. Here's an overview of some common XAI techniques:

Feature Importance Methods:
- Feature Attribution: Techniques such as LIME (Local Interpretable Model-agnostic Explanations) and SHAP (SHapley Additive exPlanations) attribute importance to input features by analysing how changes in those features affect the model's output.
- Saliency Maps: Saliency methods visualize the regions of input data that are most influential in the model's decision, providing insights into which features the model is focusing on.

Model-Specific Interpretation:
- Decision Trees: Decision trees provide a transparent representation of the decision-making process by partitioning the input space based on features and assigning decision rules to each partition.
- Rule-based Models: Rule-based models directly express decision rules in the form of human-interpretable if-then statements, making them easy to understand and interpret.

Local Explanations:
- Local Models: Techniques such as surrogate models and interpretable neural networks learn simpler models to approximate the behavior of complex black-box models locally around specific instances, providing explanations at the individual prediction level.
- Counterfactual Explanations: Counterfactual explanations generate alternative scenarios that could have led to a different prediction, helping users understand why a particular outcome was predicted.

Global Explanations:
- Model Summarization: Methods like global surrogate models and prototype-based explanations summarize the behavior of complex models across the entire dataset, providing insights into general trends and patterns.

    ◦   Partial Dependence Plots: Partial dependence plots visualize the relationship between a specific feature and the model's output while marginalizing over the effects of other features, allowing users to understand how the model's predictions change with variations in individual features.

Natural Language Explanations:

    ◦   Text Explanations: Techniques generate natural language explanations to describe the rationale behind model predictions in a human-readable format, making them more accessible to users who may not have technical expertise.

    ◦   Interactive Explanations: Interactive interfaces allow users to explore and interact with explanations, enabling a more intuitive understanding of the model's behaviour.

Certainty and Uncertainty Estimation:

    ◦   Confidence Intervals: Estimating uncertainty provides users with information about the model's confidence in its predictions, helping users gauge the reliability of the model's outputs.

    ◦   Probabilistic Models: Probabilistic models quantify uncertainty by providing probability distributions over predictions, allowing users to assess the likelihood of different outcomes.

These XAI techniques offer various approaches to elucidate the inner workings of AI models, promoting transparency, trust, and interpretability in AI-driven decision-making processes across different application domains (Islam et al., 2022).

Real-Time Explainability

Real-time explanations play a critical role in enhancing the transparency, trustworthiness, and utility of AI-driven healthcare systems, ultimately improving patient care, safety, and outcomes. By integrating real-time explanation capabilities into AI-driven clinical decision support systems and patient-facing applications, healthcare organizations can leverage the full potential of AI technologies while ensuring ethical, safe, and patient-centered care. Real-time explanations provide healthcare professionals and patients with transparency into the decision-making process of AI-driven systems. Understanding why a particular diagnosis, treatment recommendation, or decision was made is essential for building trust and confidence in AI technologies. Real-time explanations help build trust between healthcare professionals and AI systems by enabling them to understand the rationale behind AI-generated recommendations or decisions. When healthcare professionals and patients trust AI systems, they are more likely to accept and adopt them as valuable tools in clinical practice (Han and Liu, 2022).

Real-time explanations enhance the utility of AI-driven clinical decision support systems by enabling healthcare professionals to interpret and contextualize AI-generated recommendations in the context of individual patient cases. Real-time explanations help healthcare professionals make informed decisions by providing additional insights and justifications for AI recommendations. Real-time explanations serve as valuable educational tools for healthcare professionals, enabling them to learn from AI-driven systems and improve their understanding of complex medical concepts and decision-making processes. By providing real-time insights into AI algorithms' reasoning, healthcare professionals can enhance their clinical knowledge and skills. Thus real-time explanations play a critical role in enhancing the transparency, trustworthiness, and utility of AI-driven healthcare systems, ultimately improving patient care, safety, and outcomes. By integrating real-time explanation capabilities into AI-driven clinical decision support systems and patient-facing applications, healthcare organizations can leverage the full potential of AI technologies while ensuring ethical, safe, and patient-centered care.

CHALLENGES IN HUMAN-ROBOT INTERACTIONS

Human-robot interactions (HRI) present a range of challenges, stemming from technical, social, ethical, and psychological factors. Here are some of the key challenges in HRI:

- Ensuring the safety of humans when interacting with robots is paramount. Robots need to be designed with robust collision detection and avoidance mechanisms to prevent accidents and injuries. Additionally, as robots become more autonomous, ensuring that they adhere to safety standards and regulations is essential.

- Building trust between humans and robots is crucial for effective collaboration. However, trust can be challenging to establish, particularly if humans perceive robots as unpredictable or unreliable. Designing robots that are transparent, consistent, and capable of explaining their actions can help foster trust and acceptance.

- Achieving natural and intuitive communication between humans and robots remains a significant challenge. Robots need to understand and interpret human gestures, facial expressions, speech, and other non-verbal cues accurately. Developing natural language processing algorithms and multimodal interfaces can improve communication effectiveness.

- Balancing autonomy and human control in HRI is complex. While autonomous robots can improve efficiency and productivity, humans may feel uncomfortable relinquishing control, particularly in safety-critical environments. Designing interfaces that allow humans to intervene and override autonomous behaviours when necessary is essential.

- As robots collect and process sensitive information about humans, ensuring privacy and data security is paramount. Robots must adhere to data protection regulations and employ robust encryption and authentication mechanisms to safeguard sensitive data from unauthorized access or misuse.

- Ethical dilemmas arise in HRI, particularly concerning the impact of robots on human employment, autonomy, and dignity. Ensuring that robots adhere to ethical principles such as beneficence, non-maleficence, and respect for human rights is essential. Additionally, addressing issues of robot bias, discrimination, and fairness is critical to promoting ethical HRI.

- Cultural differences and social norms influence human perceptions and interactions with robots. Designing culturally sensitive robots that respect diverse values and customs is essential for promoting acceptance and adoption across different societies and demographics.

- Maintaining user engagement and satisfaction over extended periods is challenging. Users may become bored or disinterested in interacting with robots if they perceive them as repetitive or unresponsive. Incorporating personalized experiences, adaptive behaviors, and emotional intelligence into robots can enhance long-term user engagement.

- Designing robots that accommodate the physical and cognitive abilities of humans is crucial for ensuring usability and accessibility. Considering factors such as anthropometry, reachability, visibility, and cognitive load can improve the user experience and prevent user fatigue or discomfort.

Addressing these challenges requires interdisciplinary collaboration between robotics engineers, human-computer interaction specialists, ethicists, psychologists, and other stakeholders. By considering the technical, social, ethical, and psychological dimensions of HRI, researchers and practitioners can develop robots that are safe, trustworthy, and socially acceptable in diverse human environments (Folke et al., 2021).

Addressing Biases in AI Algorithms in the Healthcare Domain

Addressing biases in AI algorithms in the healthcare domain is crucial to ensure fairness, equity, and accuracy in medical decision-making. Here are several strategies to mitigate biases in AI algorithms used in healthcare:

- Ensuring that AI algorithms are trained on diverse and representative datasets is essential to mitigate biases. This involves collecting data from diverse populations, including individuals of different races, genders, ages, socioeconomic backgrounds, and geographic locations. By incorporating a wide range of data sources, AI algorithms can learn more comprehensive and unbiased representations of the underlying phenomena.
- Implementing rigorous methods to detect and evaluate biases in AI algorithms is essential. This involves analyzing the data used to train the algorithms and identifying any patterns or disparities that may indicate bias. Techniques such as fairness metrics, disparity analysis, and sensitivity analysis can help quantify and assess the presence of biases in AI models.
- Interpretability and Transparency: Enhancing the interpretability and transparency of AI algorithms can help identify and address biases more effectively. By providing explanations for their predictions or decisions, AI models enable users to understand the underlying factors influencing the outcomes and detect potential biases more easily.
- Human Oversight and Validation: Incorporating human oversight and validation into the AI decision-making process can help mitigate biases and ensure accountability. Human experts can review AI predictions, evaluate their fairness and accuracy, and intervene when biases are detected to ensure equitable outcomes.
- Continuous Monitoring and Evaluation: Implementing mechanisms for continuous monitoring and evaluation of AI algorithms in real-world settings is essential to identify and address biases over time. By regularly assessing the performance of AI models and their impact on different population groups, healthcare organizations can proactively mitigate biases and improve algorithmic fairness.

BIAS MITIGATION TECHNIQUES

Bias mitigation techniques aim to reduce or eliminate biases present in AI algorithms, ensuring fair and equitable outcomes. Here are several commonly employed techniques:

- Data Preprocessing:
  Data Cleaning: Identify and correct errors, inconsistencies, and outliers in the training data to ensure its quality and integrity.

Data Augmentation: Increase the diversity of the training dataset by synthesizing new data samples or perturbing existing ones, helping to mitigate biases caused by underrepresentation or imbalances.

Data Balancing: Adjust the class distribution in the training data to mitigate biases caused by skewed or unbalanced datasets, ensuring that all classes are represented equally.

- Algorithmic Fairness:

Fairness Constraints: Incorporate fairness constraints into the learning process to ensure that the model's predictions or decisions satisfy predefined fairness criteria, such as demographic parity or equalized odds.

Fair Loss Functions: Design loss functions that penalize discriminatory behavior and encourage fairness in model predictions, helping to mitigate biases against certain demographic groups.

Fairness Regularization: Add regularization terms to the model's objective function to penalize discriminatory patterns and encourage fairness in the learned representations.

- Bias Detection and Mitigation:

Fairness Metrics: Define and compute fairness metrics to quantify biases in model predictions across different demographic groups, enabling the identification of discriminatory patterns.

Bias Correction: Develop post-processing techniques to adjust model predictions or decisions to mitigate biases identified during model evaluation, ensuring fair outcomes for all individuals.

Counterfactual Analysis: Generate counterfactual explanations to identify alternative scenarios that could mitigate biases in model predictions, enabling targeted interventions to address discriminatory behavior.

- Model Interpretability:

Interpretable Models: Use transparent and interpretable machine learning models, such as decision trees or linear models, to facilitate the inspection and understanding of model predictions, helping to identify and mitigate biases.

Feature Importance Analysis: Analyze the importance of input features in model predictions to identify potential sources of bias and assess their impact on model outputs.

- Human Oversight and Intervention:

Human-in-the-Loop Systems: Incorporate human oversight and intervention mechanisms into AI systems to review and validate model predictions, enabling human experts to detect and mitigate biases as necessary.

Ethical Review Boards: Establish ethical review boards or committees to evaluate the potential societal impact of AI systems and ensure that they adhere to ethical principles, including fairness and non-discrimination.

- Regular Monitoring and Evaluation:

Continuous Evaluation: Continuously monitor and evaluate the performance of AI systems in real-world settings to detect and address biases as they emerge, ensuring that models remain fair and equitable over time.

Feedback Mechanisms: Collect feedback from users and stakeholders to identify potential biases or discriminatory behaviors in AI systems and take corrective actions to mitigate them.

By implementing these strategies and adopting a multi-faceted approach to addressing biases in AI algorithms, healthcare organizations can improve the fairness, accuracy, and equity of AI-driven decision-making processes, ultimately enhancing patient outcomes and healthcare delivery.

EMERGING TRENDS AND FUTURE DIRECTIONS IN THE DEPLOYMENT OF EXPLAINABLE AI IN HRI FOR HEALTHCARE

Emerging trends and future directions in the deployment of explainable AI (XAI) in Human-Robot Interaction (HRI) for healthcare are shaping the development of more transparent, trustworthy, and effective AI-driven healthcare systems. Here are some key trends and directions:

- Interpretability in Robotic Assistance: As robots play increasingly active roles in healthcare settings, there is a growing need for interpretable AI models that can explain their decisions and actions to healthcare professionals and patients. Future research may focus on developing robot-assisted systems that provide real-time explanations for their behavior during medical procedures, rehabilitation exercises, or patient care tasks.
- Patient-Centric Explainability: Future deployments of XAI in HRI for healthcare will prioritize patient-centric explanations, tailoring the level and format of explanations to individual patient preferences, health literacy levels, and cultural backgrounds. Personalized explanations will enhance patient understanding, engagement, and trust in AI-driven healthcare systems.
- Collaborative Decision-Making: XAI techniques will enable collaborative decision-making between healthcare professionals, patients, and AI-driven systems in real-time clinical settings. Future research may explore interactive interfaces that facilitate dialogue and negotiation between human users and AI models, allowing for shared decision-making processes that incorporate both clinical expertise and patient preferences.
- Human-Robot Trust and Acceptance: Addressing trust and acceptance issues will be a key focus of future research on XAI in HRI for healthcare. Future deployments may incorporate explainability features into healthcare robots to enhance human-robot trust through transparent communication, reliable decision-making, and accountability for actions taken in clinical settings.
- Adaptive Explainability: Future XAI systems deployed in HRI for healthcare will be adaptive and context-aware, providing explanations that are tailored to specific interaction contexts, user preferences, and task requirements. Adaptive explainability will enhance the relevance, clarity, and effectiveness of explanations provided by healthcare robots in diverse clinical scenarios.
- Integration with Clinical Workflows: XAI techniques will be integrated seamlessly into clinical workflows, supporting healthcare professionals in decision-making processes without disrupting existing practices or workflows. Future deployments may involve embedding XAI capabilities into electronic health record systems, medical devices, and robotic platforms to provide transparent decision support and enhance clinical decision-making.

Overall, the deployment of XAI in HRI for healthcare is poised to transform the way AI-driven systems interact with healthcare professionals and patients, promoting transparency, trust, and collaboration in clinical settings. By addressing emerging trends and future directions, researchers and practitioners can harness the full potential of XAI to improve patient care, safety, and outcomes in healthcare environments (Guo, 2020).

Training Healthcare Professionals

Training healthcare professionals and end-users on understanding AI decisions is crucial for fostering trust, promoting acceptance, and facilitating effective collaboration between humans and AI-driven systems in healthcare settings. Here are some recommendations for training programs:

Foundational Knowledge: Provide healthcare professionals and end-users with foundational knowledge about AI technologies, including machine learning algorithms, deep learning architectures, and explainable AI techniques. Ensure that participants understand the basic principles and concepts underlying AI-driven decision-making processes.

Clinical Relevance: Emphasize the clinical relevance of AI-driven decision-making in healthcare by illustrating real-world examples and case studies. Demonstrate how AI algorithms are used to analyze medical data, diagnose diseases, recommend treatments, and optimize patient care pathways (Kwong et al., 2022).

Explainability Techniques: Educate participants about various explainability techniques used in AI, such as feature importance analysis, local interpretations, model-agnostic explanations, and natural language explanations. Teach participants how to interpret and understand the explanations provided by AI-driven systems in healthcare contexts.

Interpretation Skills: Develop participants' interpretation skills to enable them to assess the reliability, validity, and relevance of AI-generated predictions or recommendations. Teach participants how to critically evaluate AI outputs, identify potential biases or errors, and make informed decisions based on AI insights.

Transparency and Trust: Highlight the importance of transparency and trust in AI-driven decision-making by discussing ethical principles, regulatory requirements, and best practices for responsible AI deployment in healthcare. Foster a culture of transparency and openness to encourage dialogue and collaboration between healthcare professionals and AI systems.

Interactive Learning: Incorporate interactive learning activities, such as case studies, role-playing exercises, and hands-on demonstrations, to engage participants and reinforce learning objectives. Encourage active participation and discussion to facilitate knowledge sharing and peer learning among participants.

Continuing Education: Offer continuing education and professional development opportunities to healthcare professionals to keep them updated on advances in AI technologies and best practices for AI-driven decision-making in healthcare. Provide access to online courses, workshops, webinars, and conferences focused on AI in healthcare.

Multidisciplinary Collaboration: Facilitate multidisciplinary collaboration and knowledge exchange between healthcare professionals, data scientists, AI engineers, ethicists, and other stakeholders involved in AI-driven healthcare initiatives. Foster interdisciplinary teamwork to leverage diverse expertise and perspectives in addressing complex challenges related to AI decision-making.

By implementing comprehensive training programs that cover foundational knowledge, explainability techniques, interpretation skills, and ethical considerations, healthcare organizations can empower healthcare professionals and end-users to understand AI decisions effectively and leverage AI technologies to improve patient care and outcomes.

Identifying Best Practices for Deploying and Maintaining XAI Systems

Deploying and maintaining Explainable AI (XAI) systems effectively involves a combination of technical considerations, organizational strategies, and ethical principles (Amann et al., 2020). Here are some best practices for deploying and maintaining XAI systems:

Understand the Context: Before deploying an XAI system, it's crucial to understand the specific context in which it will be used. Consider the stakeholders involved, the domain of application, regulatory requirements, and potential ethical implications.

Select Appropriate XAI Techniques: Choose XAI techniques that are suitable for the specific needs of the application and align with the level of transparency required. Techniques such as LIME (Local Interpretable Model-agnostic Explanations), SHAP (SHapley Additive exPlanations), and decision trees are commonly used for providing explanations in machine learning models. The importance of XAI in healthcare applications in trainable attention refers to a set of methodologies aimed at directing focus towards significant content within digital multimedia data, including images, videos, audio, and text. One illustrative application involves integrating text and image data within a hidden layer, enabling clinicians to concentrate on pertinent information relating to regions of interest (ROIs) in images alongside electronic health records. This technique facilitates the alignment of relevant textual and visual cues, enhancing the ability of healthcare professionals to discern critical information in medical data (Chaddad et al., 2023). Integrate XAI into Development Pipeline: Integrate XAI considerations into the entire development pipeline, from data collection and model training to deployment and monitoring. Ensure that XAI techniques are incorporated early in the process to facilitate transparency and interpretability throughout the system's lifecycle.

Provide Transparent Explanations: Ensure that the explanations provided by the XAI system are clear, understandable, and relevant to the end-users. Use visualizations, natural language explanations, or interactive interfaces to convey the reasoning behind AI decisions effectively.

Validate and Evaluate XAI Systems: Conduct rigorous validation and evaluation of XAI systems to assess their effectiveness, reliability, and impact on decision-making processes. Use appropriate metrics and methodologies to measure the quality of explanations and user satisfaction.

Address Bias and Fairness: Pay attention to potential biases in the data and algorithms used by the XAI system. Implement techniques to mitigate bias and ensure fairness, such as fairness-aware machine learning algorithms, bias detection methods, and fairness constraints.

Ensure Robustness and Stability: XAI systems should be robust and stable across different datasets, environments, and user interactions. Perform robustness testing and sensitivity analysis to evaluate the system's performance under various conditions and edge cases.

Provide Continuous Monitoring and Feedback: Establish mechanisms for monitoring the performance of XAI systems in real-world settings and collecting feedback from users. Use this feedback to iteratively improve the system's transparency, accuracy, and usability over time.

Educate Users and Stakeholders: Educate users, stakeholders, and decision-makers about the capabilities and limitations of XAI systems. Provide training and resources to help them understand how to interpret and trust the explanations provided by the system.

Adhere to Ethical Guidelines: Ensure that the deployment and maintenance of XAI systems adhere to relevant ethical guidelines, privacy regulations, and industry standards. Consider the potential societal impact of the system and prioritize ethical considerations in decision-making processes. By following these best practices, organizations can deploy and maintain XAI systems effectively, fostering trust, transparency, and accountability in AI-powered decision-making processes (Tjoa & Guan, 2021).

CONCLUSION

Collaborative efforts play a crucial role in addressing challenges and advancing research in the deployment of Explainable AI (XAI) in Human-Robot Interactions (HRI) for Healthcare 4.0/5.0. Collaborative efforts often involve interdisciplinary research teams comprising experts from fields such as artificial intelligence, robotics, human-computer interaction, healthcare, and ethics. These teams bring diverse perspectives and expertise to tackle the multifaceted challenges of deploying XAI in HRI for healthcare. Collaborations between academia and industry facilitate the translation of research findings into practical solutions. Academic researchers contribute theoretical insights and experimental validation, while industry partners provide real-world data, resources, and implementation expertise. These partnerships accelerate the development and deployment of XAI-enabled healthcare robots. Collaborating with healthcare professionals, clinicians, and medical institutions is essential for understanding the specific requirements, workflows, and challenges in clinical settings. Clinical collaborations enable researchers to design XAI systems that are aligned with the needs of healthcare practitioners and patients, ensuring usability, safety, and effectiveness. Involving end-users, including healthcare professionals, patients, and caregivers, in the design and evaluation of XAI-enabled healthcare robots ensures that the technology meets their needs and preferences. User-centered design approaches, such as participatory design workshops, usability studies, and co-creation sessions, facilitate collaboration between researchers and stakeholders to iteratively refine the system. Collaboration with ethicists, policymakers, and stakeholders in the broader society is essential for addressing ethical and societal implications associated with the deployment of XAI in HRI for healthcare (Taimoor & Rehman, 2022). Collaborative efforts focus on developing ethical guidelines, regulatory frameworks, and governance mechanisms to ensure transparency, fairness, accountability, and privacy in AI-powered healthcare systems. Collaborative efforts often extend beyond national boundaries through international partnerships and collaborations. International collaboration enables knowledge sharing, cross-cultural understanding, and benchmarking of XAI technologies across different healthcare systems and cultural contexts. By fostering collaborative efforts among researchers, industry partners, healthcare professionals, policymakers, and stakeholders, the deployment of Explainable AI in Human-Robot Interactions for Healthcare 4.0/5.0 can be accelerated, leading to transformative advancements in patient care, clinical decision-making, and healthcare delivery.

REFERENCES

Amann, J., Blasimme, A., Vayena, E., Frey, D., & Madai, V. I. (2020). Explainability for artificial intelligence in healthcare: A multidisciplinary perspective. *BMC Medical Informatics and Decision Making*, *20*(no. 1), 310. doi:10.1186/s12911-020-01332-6 PMID:33256715

Chaddad, A., Peng, J., Xu, J., & Bouridane, A. (2023). Survey of Explainable AI Techniques in Healthcare. *Sensors (Basel)*, *23*(2), 1–19. doi:10.3390/s23020634 PMID:36679430

Folke, T., Yang, S. C., Anderson, S., & Shafto, P. (2021). Explainable AI for medical imaging: Explaining pneumothorax diagnoses with Bayesian teaching. CoRR, vol. abs/2106.04684.

Guo, W. (2020). Explainable artificial intelligence for 6G: Improving trust between human and machine. *IEEE Communications Magazine, 58*(6), 39–45. doi:10.1109/MCOM.001.2000050

Gupta, R., Shukla, A., & Tanwar, S. (2020). Aayush: A smart contract-based telesurgery system for healthcare 4.0. *2020 IEEE International Conference on Communications Workshops (ICC Workshops),* 1–6. 10.1109/ICCWorkshops49005.2020.9145044

Han, H., & Liu, X. (2022, January). The challenges of explainable ai in biomedical data science. *BMC Bioinformatics, 22*(12), 443. PMID:35057748

Islam, A. M. R., Ahmed, M. U., Barua, S., & Begum, S. (2022). A systematic review of explainable artificial intelligence in terms of different application domains and tasks. *Applied Sciences (Basel, Switzerland), 12*(3), 1353. doi:10.3390/app12031353

Kwong, J. C., Khondker, A., Tran, C., Evans, E., Cozma, A. I., Javidan, A., Ali, A., Jamal, M., Short, T., Papanikolaou, F., Srigley, J. R., Fine, B., & Feifer, A. (2022). Explainable artificial intelligence to predict the risk of side-specific extraprostatic extension in pre-prostatectomy patients. *Canadian Urological Association Journal, 16*(6). Advance online publication. doi:10.5489/cuaj.7473 PMID:35099382

Mbunge, E., Muchemwa, B., Jiyane, S., & Batani, J. (2021). Sensors and healthcare 5.0: Transformative shift in virtual care through emerging digital health technologies. *Global Health Journal (Amsterdam, Netherlands), 5*(4), 169–177. doi:10.1016/j.glohj.2021.11.008

Mohanta, B., Das, P., & Patnaik, S. (2019). Healthcare 5.0: A paradigm shift in digital healthcare system using artificial intelligence, IOT and 5G communication. *2019 International Conference on Applied Machine Learning (ICAML),* 191–196. 10.1109/ICAML48257.2019.00044

Pawar, U., O'Shea, D., Rea, S., & O'Reilly, R. (2020). Incorporating explainable artificial intelligence (XAI) to aid the understanding of machine learning in the healthcare domain. *Irish Conference on Artificial Intelligence and Cognitive Science.*

Shaban Nejad, M. (2020). *Explainable AI in healthcare and medicine: building a culture of transparency and accountability* (Vol. 914). Springer Nature.

Taimoor, N., & Rehman, S. (2022). Reliable and resilient AI and IOT-based personalised healthcare services: A survey. *IEEE Access : Practical Innovations, Open Solutions, 10,* 535–563. doi:10.1109/ACCESS.2021.3137364

Tjoa, E., & Guan, C. (2021). A survey on explainable artificial intelligence (XAI): Toward medical xai. *IEEE Transactions on Neural Networks and Learning Systems, 32*(11), 4793–4813. doi:10.1109/TNNLS.2020.3027314 PMID:33079674

Vale, D., El-Sharif, A., & Ali, M. (2022, March). Explainable artificial intelligence (XAI) post-hoc explainability methods: Risks and limitations in non-discrimination law. *AI and Ethics, 2*(4), 815–826. doi:10.1007/s43681-022-00142-y

Chapter 16
Security System for Smart Homes to Prevent Theft

Anitha Mary
Karunya University, India

P. Kingston Stanley
Karunya Institute of Technology and Sciences, India

V. Evelyn Brindha
Karunya Institute of Technology and Sciences, India

J. Jency Joseph
Krishna College of Technology, India

ABSTRACT

The chapter intends to create a system that alerts the victim of theft in real time. The current system does not distinguish between people and objects; instead, it uses a methodology to identify the burglar after the theft has taken place. The internet of things and advancements in wireless sensor networks make it possible to create a smart, safe home that can detect burglars in real time and notify the homeowner while the theft is occurring. The suggested approach is better than the current ones that use CCTV cameras for surveillance.

1. INTRODUCTION

Security is a significant concern for both homes and offices, especially in light of recent theft activities. There is a clear need for a security system capable of detecting intruders in real-time during theft incidents. The current approach relies on CCTVs and DTRs, but these systems cannot distinguish between humans and objects. Additionally, they can only analyze stored images after an incident has occurred, requiring human intervention. In contrast, the proposed system not only detects intruders live but also promptly alerts homeowners or office occupants via a GSM module, notifying them of the theft immediately. This

DOI: 10.4018/979-8-3693-5767-5.ch016

solution is both cost-effective and efficient, offering a reliable security measure for properties without the need for surveillance cameras.

2. RELATED WORKS

The absence of smart devices and sophisticated facial recognition software at home is contributing to an increase in theft tracking (Ahmed et.al. 2010). The inaccuracy of face detection techniques acquired by CCTV cameras arises from theft victims hiding their faces, either completely or partially, using materials made of leather or cloth. Numerous researchers have directed their efforts towards improving face recognition technology (Jian et.al.2018). For example, Zhiwei et al. (Zhang et.al.2012) introduced a regularization technique aimed at enhancing face identification across different spectrums. Additionally, many researchers have focused on refining face and eye detection using the Haar classifier algorithm. Xin et al. (Xin et.al 2018) proposed a system for person face recognition, claiming it offers superior security compared to unimodal biometric systems. Nguyen et al. Nguyen et.al. 2018 suggested employing deep learning techniques, while Cho et al. Cho et.al 2018 advocated for the use of convolutional neural networks to improve image features in face recognition systems, particularly those equipped with visible light camera sensors. Binary pattern techniques were used in the creation of the face detection system by Alobaidi et al.2018. Zhang et al.'s approach to face detection in a stable environment combines model-driven and data-driven methods. Using support vector machines, Omid et al. created a face recognition system that works in the presence of cosmetics like contact lenses and facial makeup (Sharifi et.al.2018). Door and window security is a major consideration when it comes to house security. These days, digital doors that don't require a physical key are employed because to advancements in IoT technology. Nevertheless, it is simple to damage digital doors, and the owners are only aware of theft once they get home. Huth et al. devised a wirelessly connected security system employing a physical key generation method, demonstrating its application in smart homes (Andreasa et.al.2019). To enhance energy efficiency, WiFi modules, temperature sensors, and door sensors are commonly employed.

Intruder detection systems often utilize laser and LDR sensors to detect movement, with alerts sent to homeowners via SMS (Cristian et.al.2016). However, this method may fail to transmit messages through a GSM module in areas lacking internet coverage. Anitha et al. proposed an artificial intelligence-based home system (Anitha 2016).

Patel et al. introduced a modern door lock system for homes, incorporating a Raspberry Pi system-on-chip (SoC), a camera, and an infrared sensor (Jay Patel et.al.2019). This system operates by granting access to individuals whose images are stored in the cloud. Nivo Suranth et al. implemented a home security system utilizing a PIR sensor, an Arduino microcontroller, and a Raspberry Pi 3 SoC (Nico-surantha 2018). The Raspberry Pi processes images captured by a webcam connected to the SoC, along with sensor data from the Arduino microcontroller. Additionally, intruders can be identified using a support vector algorithm, capable of detecting intruders within 2 seconds.

Beyond home security, several researchers have developed anti-theft systems for vehicles. Kiruthiga et al. devised a system employing a global system and a PIC microcontroller (Kiruthigara et.al 2015). This controller identifies unauthorized access and notifies the owner via SMS. Although GPS technology is utilized for vehicle tracking (win et.al.2011), it may encounter limitations at the receiver end, resulting in inaccurate location data due to limited sky view. Radio frequency identification (RFID) is another method used for theft tracking, allowing access to the card, which poses a risk of theft.

3. METHODOLOGY

Figure 1 illustrates the block diagram of the anti-theft detection system. The controller acts as the master and interfaces with sensors to detect intruders using a PIR sensor and a Reed Switch. These sensors can be affixed to doors or windows in homes and/or offices. When an intruder opens a door or window, the controller detects their presence and promptly sends a message to the owner via a GSM module.

3.1 Sensor Module

The sensor module comprises a Passive Infrared (PIR) sensor and a Piezo vibration sensor. The PIR sensor contains a pyroelectric sensor that produces an electrical signal when there is a change in temperature. This enables detection of human presence within a range of 14 meters. The piezo-vibration sensor generates an electric signal output in response to vibrations, aiding in the identification of intruders, such as when they bore holes in walls.

3.2 Controller

The controller utilized in the system is the Atmel 2560 microcontroller, an 8-bit microcontroller offering several advantages over an Arduino UNO board, including a higher number of analog and digital pins. It operates at a voltage of 3.3V and a frequency of 16MHz, supporting various types of digital and analog sensors.

3.3 SIM 808 Global System for Mobile communication (GSM)

For GSM communication, the system employs the SIM808 GSM Module, which integrates both GPRS and GPS modules. This combination ensures cost-effectiveness and quick response times. Operating at voltages ranging from 3.3V to 4.4V, the GSM module communicates with the controller via UART (Universal Asynchronous Receiver Transmitter), facilitating the transmission of theft-related SMS alerts to the owner.

3.4 Real Time Clock (RTC) Module

The Real-Time Clock (RTC) Module is responsible for maintaining accurate timekeeping, including seconds, minutes, and hours. Additionally, it offers provisions for adjusting the date, month, and year, including corrections for leap years. The module features a square wave output pin (SQW) providing frequencies of 1, 4, 8, or 32 KHz, which can be modified through programming. These features serve as interrupts during theft detection. To mitigate the effects of temperature variations on the circuit, the RTC Module includes an inbuilt temperature-compensated crystal oscillator.

3.5 Radio Frequency (RF) Module (Transmitter and Receiver)

The RF Module uses simplex communication to create a 433MHz frequency. The transmitter and receiver are two separate devices; the transmitter sends information from one end (such as one microcontroller board), while the receiver receives it from another controller board. Amplitude shift keying and an

HT12E encoder are needed for controller interface. Long-distance communication typically uses these kinds of modules.

Figure 1. Methodology-block diagram

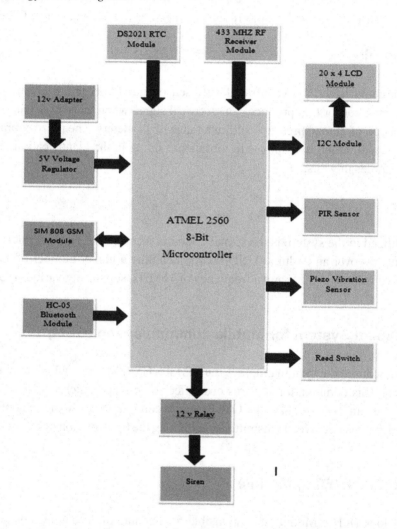

3.6 Li-Ion Battery

With a capacity of providing power up to 1000 mA, the 4300 mAh Li-ion battery sustains the proposed system for nearly 1 hour. Charging occurs in two stages: the constant current stage, where a steady current is applied until reaching the voltage threshold, and the constant voltage stage, where a stable voltage is applied as the current decreases to the threshold limit.

3.7 LCD (Liquid Crystal Display)

The 20x4 LCD, featuring 20 columns and 4 rows, is designed to display large amounts of text with a resolution of 5x8 pixels. Communication with the microcontroller is facilitated through the 2-wire communication I2C protocol.

3.8 Buzzer

The buzzer emits a continuous beep sound upon activation and comes in two types: the simple buzzer emits a continuous sound when supplied with voltage, while the readymade buzzer produces a beeping sound through an oscillating circuit. Commonly used in automobile electronics and portable equipment, it operates on a +5V power supply.

3.9 Siren

A siren serves as a noise-making system typically mounted on rooftops or posts, used to warn of natural disasters or attacks.

3.10 Relay Module

The relay module acts as an electromagnetic switch, toggling ON/OFF when current flows through the circuit. It includes three terminals—normally open (NO), common (COM), and normally closed (NC)—with the NO connecting to the COM and the NC disconnected. When current flows, the NC connects with the COM, returning to its initial position when the current is cut off.

3.11 Bluetooth Module (HC-05)

The HC-05 Bluetooth module utilizes the Serial Port Protocol (SPP) and features Tx and Rx terminals for communication with the microcontroller. Powered by a +5V battery, it enters command mode during power-up and switches to data mode when idle.

3.12 Android Studio

Android Studio, based on Google's Android OS, is the Integrated Development Environment (IDE) utilized for this project. It supports languages such as C, C++, and JAVA, along with their extensions, facilitating software development.

4. WORKFLOW

This suggested work is a self-diagnostic anti-theft system that, without the need for a service technician, examines the state of interfaced sensor modules and notifies the owner appropriately. It is a full security system made for residences and retail spaces. The PIR and vibration sensors are part of the sensor module. PIR sensors are employed to detect theft based on variations in temperature. If a dacoit creates a hole in the wall, vibration sensors are utilised to detect the vibration. The microcontroller is interfaced with these sensors. The microcontroller notifies the owner in real time when the security system is activated, based on the data that the sensor has received.

4.1 Operating Modes

As seen in Figure 2, this system has five modes of operation, each of which has a distinct functional operation. A standard security feature that assists elderly individuals living alone at night is called "stay mode." A siren that uses a PIR sensor and reed switch is activated if an unknown person enters the house.Using the GSM module, an outing mode sends a real-time message to the owner alerting them to the theft. Using the emergency mode makes it easier for the user to message the local police station or ask for assistance from neighbours.

Figure 2. Modes of operation

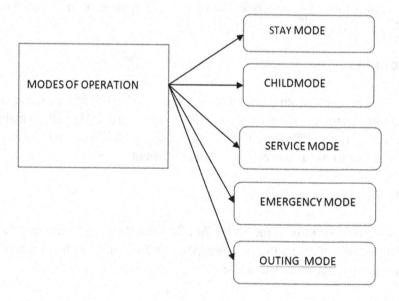

The 500-meter radius around the child is protected when in child mode (Figure 3). This option uses the child's GPS position to follow their whereabouts, updating the parents via a mobile app every ten minutes. The system has a safety button that the child can press if they are in distress. The CPU calls the preprogrammed list of phone numbers when the safety button is pressed. It will call the subsequent number till it reaches the last one if no one answers. This option assists children who are missing, abducted, or involved in accidents.

In addition to the five operating modes listed above, there is one more mode known as service mode. Service mode guarantees that the device is operating properly by allowing it to self-check the system once a month.

Figure 3. Self-protection system for children

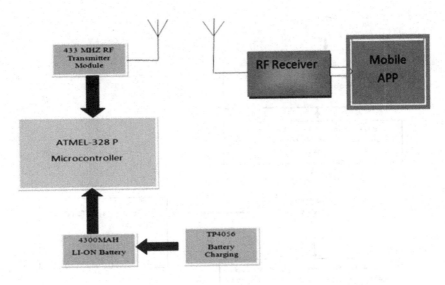

5. RESULTS AND DISCUSSION

A feature of the suggested anti-theft system (Figure 4) allows for voice, GPS, and SMS permissions. Bluetooth is enabled to pair the device with the mobile app after authorization is granted. There are five modes to choose from on the programme (Figure 6), which include stay, outing, emergency, and child mode. Any mode can be used by the user. The owner's phone number and GPS location are forwarded to the local police station when the system is in emergency mode and there are intruders inside the home. The youngster can be viewed up to 500 metres away while the system is in child mode and the RF tag is active. The flowchart for the suggested anti-theft system is displayed in Figure 5. Figure 3: Self Protection system for children

Figure 4. Experimental set-up

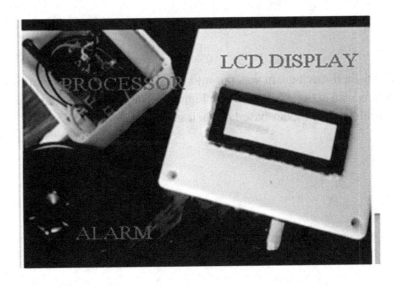

Figure 5. Flowchart for the proposed system

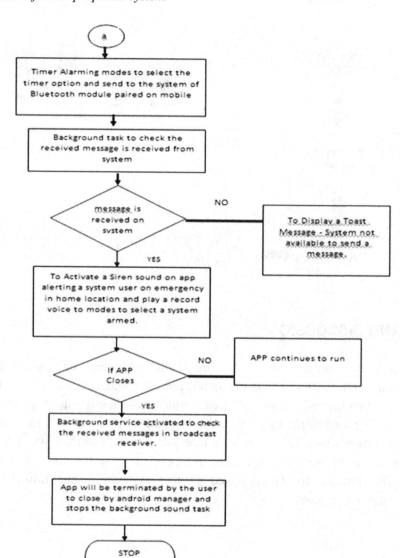

The suggested anti-theft mechanism was put into place. With all of its operating modes displayed, the suggested system is seen at the window side in Figure 7. The red LED on the power on button allows the user to choose from various modes based on their needs.

Figure 6. Mobile app for selecting modes

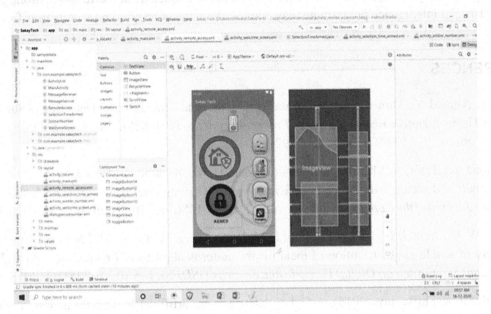

Figure 7. Implementation of proposed system

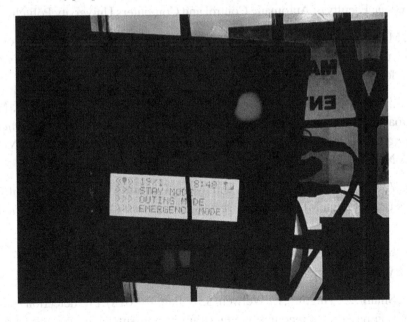

6. CONCLUSION

The suggested system offers superior protection for not only houses but also workplaces, cars, and other spaces. It may be operated by an Android smartphone user through an application. Through a GSM phone and Bluetooth, the user can get the status update. Using kid mode configuration, it is possible to keep an eye on the youngster and prevent them from being abducted. In the future, a drone equipped with an artificial intelligence programme may employ the same concept to protect its user from theft.

REFERENCES

Ahmed, T., Ahmed, S., Ahmed, S., & Motiwala, M. (2010). Real-Time Intruder Detection in Surveillance Networks Using Adaptive Kernel Methods. *Proceedings of the 2010 IEEE International Conference on Communications.* 10.1109/ICC.2010.5502592

Almohamad, A., Tahir, A. M., Al-Kababji, A., Furqan, H. M., Khattab, T., Hasna, M. O., & Arslan, H. (2020). Smart and secure wireless communications via reflecting intelligent surfaces: A short survey. *IEEE Open Journal of the Communications Society, 1*, 1442–1456. doi:10.1109/OJCOMS.2020.3023731

Alobaidi, W. H., Aziz, I. T., Jawad, T., Flaih, F. M. F., & Azeez, A. T. (2018). Face detection based on probability of amplitude distribution of local binary patterns algorithm. *Proceedings of the 2018 6th International Symposium on Digital Forensic and Security (ISDFS)*, 1–5 10.1109/ISDFS.2018.8355319

Anitha, Kalra, & Shrivastav. (2016). A Cyber defence using artificial home automation system using IoT. *International Journal of Pharmacy and Technology, 8*, 25358-64.

Cristian, Ursache, Popa, & Pop. (2016). *Energy efficiency and robustness for IoT: Building a smart home security system.* Faculty of Automatic Control and Computers University Politehnica of Bucharest.

Enoch Sam, M., Misra, S. K., Anitha Mary, X., Karthik, C., & Chowdhury, S. (2023, November). Review of different types of spatial positioning platforms. In AIP Conference Proceedings (Vol. 2878, No. 1). AIP Publishing. doi:10.1063/5.0171256

Evangeline, C. S., Sarah, M., Lenin, A., Reddy, J. H. V., Mary, X. A., & Karthiga, M. (2023, May). Design of On-Board Unit for Vehicular Applications. In *2023 2nd International Conference on Vision Towards Emerging Trends in Communication and Networking Technologies (ViTECoN)* (pp. 1-6). IEEE. 10.1109/ViTECoN58111.2023.10157654

Evelyn Brindha, V., & Anitha Mary, X. (2023). Analysing Control Algorithms for Controlling the Speed of BLDC Motors Using Green IoT. *Power Converters, Drives and Controls for Sustainable Operations*, 779-788.

Huth, C. (2015). Securing systems on the Internet of Things via physical properties of devices and communications. In Systems Conference (SysCon), 9th Annual IEEE International. IEEE.

Jian, Z., Chao, Z., Shunli, Z., Tingting, L., Weiwen, S., & Jian, J. (2018). Pre-detection and dual-dictionary sparse representation based face recognition algorithm in non-sufficient training samples. *Journal of Systems Engineering and Electronics, 29*(1), 196–202. doi:10.21629/JSEE.2018.01.20

Makarfi, A. U., Rabie, K. M., Kaiwartya, O., Li, X., & Kharel, R. (2020, May). Physical layer security in vehicular networks with reconfigurable intelligent surfaces. In *2020 IEEE 91st Vehicular Technology Conference (VTC2020-Spring)* (pp. 1-6). IEEE. 10.1109/VTC2020-Spring48590.2020.9128438

Ramadan, M. N., Al-Khedher, M. A., & Al-Kheder, S. A. (2012). Intelligent anti-theft and tracking system for automobiles. *International Journal of Machine Learning and Computing, 2*(1), 83–88. doi:10.7763/IJMLC.2012.V2.94

Ritharson, P. I., Raimond, K., Mary, X. A., Robert, J. E., & Andrew, J. (2024). DeepRice: A deep learning and deep feature based classification of Rice leaf disease subtypes. *Artificial Intelligence in Agriculture, 11*, 34–49. doi:10.1016/j.aiia.2023.11.001

Sadagopan, V. K., Rajendran, U., & Francis, A. J. (2011, July). Anti theft control system design using embedded system. In *Proceedings of 2011 IEEE International Conference on Vehicular Electronics and Safety* (pp. 1-5). IEEE. 10.1109/ICVES.2011.5983776

Sharifi, O., & Eskandari, M. (2018). Cosmetic Detection Framework for Face and Iris Biometrics. *Sensors (Basel), 10*, 122.

Win, Z. M., & Sein, M. M. (2011). *Fingerprint recognition system for low quality images.* Presented at the SICE Annual Conference, Waseda University, Tokyo, Japan.

Xin, Y., Kong, L., Liu, Z., Wang, C., Zhu, H., Gao, M., Zhao, C., & Xu, X. (2018). Multimodal Feature-Level Fusion for Biometrics Identification System on IoMT Platform. *IEEE Access, 6*, 21418–21426.

Zhang, H., Li, Q., Sun, Z., & Liu, Y. (2018). Combining Data-Driven and Model-Driven Methods for Robust Facial Landmark Detection. *IEEE Transactions on Information Forensics and Security, 13*(10), 2409–2422. doi:10.1109/TIFS.2018.2800901

Zhang, Z., Yi, D., Lei, Z., & Li, S.Z. (2012). Regularized Transfer Boosting for Face Detection Across Spectrum. *IEEE Signal Process. Lett., 19*, 131–134.

Chapter 17
Study on Nano Robotic Systems for Industry 4.0:
Overcoming Challenges and Shaping Future Developments

G. V. Krishna Pradeep

Department of Mechanical Engineering, Aditya Engineering College, Surampalem, India

M. Balaji

Department of Mechanical Engineering, V.R. Siddhartha Engineering College, Vijayawada, India

Vivek Narula

Department of Management, Shri Venkateshwara University, India

V. Nirmala

Department of Mathematics, R.M.K. Engineering College, Chennai, India

I. John Solomon

Department of Mechanical Engineering, Panimalar Engineering College, Chennai, India

M. Sudhakar

Department of Mechanical Engineering, Sri Sai Ram Engineering College, Chennai, India

ABSTRACT

Nano robotics is a rapidly developing technology that operates at microscopic scales, revolutionizing fields like medicine and manufacturing. However, it faces numerous challenges, including technical, ethical, and practical issues. These include precision engineering, control mechanisms, and power sources, as well as ethical concerns about autonomy, safety, and societal impact. The chapter explores the future of nano robotics, highlighting its potential in various fields such as medicine and manufacturing. It highlights the potential of nano robots in enhancing durability and functionality, offering targeted drug delivery, minimally invasive surgeries, and precise diagnostics. The chapter also addresses technical challenges, ethical considerations, and potential developments, aiming to make the seemingly impossible achievable at the tiniest scales, emphasizing the need for further advancements.

DOI: 10.4018/979-8-3693-5767-5.ch017

INTRODUCTION

Nano robotics is a groundbreaking field that combines robotics, nanotechnology, and engineering at an unprecedented scale. These tiny mechanical marvels, operating in the nanometer range, can navigate realms imperceptible to the human eye, offering potential breakthroughs in medicine, manufacturing, and beyond. Originating from the visionary concepts of manipulating matter at molecular scales, pioneers like Richard Feynman have made significant progress in developing nano robots capable of traversing and interacting within the nanoworld's infinitesimally small dimensions (Daudi, 2015).

Nano robotics faces a challenge in precision engineering due to the need for devices at minuscule scales. This requires innovations in materials science, manufacturing techniques, and control mechanisms. Inspired by biological systems, these robots mimic the agility of nature's smallest organisms but require unparalleled complexity. However, their potential applications in medicine are boundless. These machines can revolutionize diagnostics, drug delivery, and surgical procedures with unprecedented precision. Their ability to navigate the human body at cellular or molecular levels offers new avenues for targeted therapies and treatments (Gheorghe et al., 2014).

This chapter delves into the intricacies, challenges, and transformative potential of nano robotics, a cutting-edge field that has the potential to revolutionize manufacturing processes by creating materials and products with unprecedented precision, durability, and functionality. It explores the intricacies and challenges of this cutting-edge field. Understanding operations at the nanoscale is fundamental to grasp the intricacies and challenges of nano robotics. At this minute level, physical laws and behaviors diverge from those in the macroscopic world, demanding a nuanced comprehension of how matter interacts, moves, and functions (Halder & Sun, 2019).

The nanoscale, characterized by dimensions typically between 1 and 100 nanometers, introduces a realm where quantum effects dominate and classical mechanics often lose their relevance. Quantum phenomena such as tunneling, where particles traverse barriers they theoretically shouldn't overcome, become significant at this scale. Moreover, surface forces such as Van der Waals and electrostatic forces become remarkably influential, even overpowering gravitational or inertial forces. Manipulating matter at this scale involves harnessing these unique properties and overcoming the challenges they pose. Precision becomes paramount; slight disturbances or fluctuations that might be inconsequential at larger scales can significantly impact nanoscale operations. Additionally, materials behave differently at the nanoscale, with altered mechanical, electrical, and optical properties, necessitating specialized approaches for design and manipulation (Chen et al., 2022a).

Nano robotics capitalizes on these principles by developing mechanisms and machines tailored to exploit nanoscale phenomena. For instance, nanorobots may leverage molecular motors inspired by biological structures or use specialized materials exhibiting unique properties at the nanoscale to achieve desired functionalities. However, navigating this domain is rife with challenges. Fabricating nanoscale components with precision, ensuring their reliability, and developing control mechanisms that operate effectively in this realm are formidable tasks. Moreover, environmental factors, such as temperature fluctuations or surface interactions, can significantly influence the behavior and stability of nanorobots, necessitating robust solutions (Ghanbarzadeh-Dagheyan et al., 2021).

A comprehensive understanding of nanoscale operations serves as the bedrock for advancements in nano robotics. By delving into the nuances of this scale, researchers and engineers aim to harness its peculiarities to design and create innovative nano robots capable of operating seamlessly in these unique environments, unlocking a world of possibilities across various disciplines. The 1980s saw significant advancements in nano-scale structures, thanks to the development of scanning tunneling and atomic force microscopes. These tools allowed scientists to visualize and manipulate individual atoms and molecules, paving the way for nanotechnology and nano robotics (Sivasankar & Durairaj, 2012).

Advancements in technology led to the development of nano robots capable of performing basic tasks at the nanoscale. These robots often drew inspiration from biological systems, mirroring the functionality of cellular machinery or molecular motors. Advancements in nanomaterials and fabrication techniques, such as self-assembly and DNA origami, expanded the range of nano robotic systems. The convergence of disciplines, including robotics, materials science, biology, and engineering, has led to the development of advanced nano robots. These tiny machines have potential applications in medicine, manufacturing, and environmental remediation. The evolution of nano robotic systems is driven by ongoing research and innovation, with the future promising further advancements and potential industry revolution (Ghanbarzadeh-Dagheyan et al., 2021).

Nano robotics is poised to revolutionize various industries, including healthcare, manufacturing, environmental remediation, and information technology, with its potential impact being far-reaching and promising, marking a transformative era in the field. Nano robotics holds great potential in medicine and healthcare, as they can navigate the human body with precision, enabling targeted drug delivery to specific cells or tissues. These nano-scale instruments can perform minimally invasive surgeries, minimizing trauma and recovery times (Pedram & Nejat Pishkenari, 2017).

They can also facilitate accurate diagnostics at the cellular or molecular level, potentially revolutionizing disease detection and treatment. Beyond healthcare, nano robotics holds promise in manufacturing and materials science, as they can assemble materials with unparalleled precision, creating novel materials with enhanced properties. This could lead to the development of innovative products across various industries. Nano robotics can be used in environmental remediation by precisely targeting pollutants at their source, offering potential solutions for water purification and soil remediation (Daudi, 2015; Pedram & Nejat Pishkenari, 2017).

They can also revolutionize data storage, computing, and communication by packing more information into smaller spaces. This paradigm shift in various industries presents new possibilities and challenges. However, achieving these applications requires technical hurdles, ethical considerations, and ensuring the safety and reliability of nano robotic systems. Despite these challenges, the transformative potential of nano robotics in shaping the future landscape of various industries is undeniable.

The following provides an overview of the background and scope of nano robotics (Puranik et al., 2024a; Srinivas et al., 2023).

- Nano robotics is a rapidly growing field that combines nanotechnology and robotics, focusing on designing and controlling robots at the nanoscale. Initially a futuristic concept, it has become a practical reality due to advancements in materials science, nanofabrication techniques, and control systems.

- Nano robotics has a wide range of applications across various industries, including medicine, manufacturing, electronics, and environmental remediation. In medicine, they could revolutionize diagnostics, drug delivery, and surgical procedures. In manufacturing, they could improve assembly and quality control. Additionally, nanorobots could help tackle environmental pollution at the molecular level.

- Nano robotics faces technical challenges such as precision engineering, robust control mechanisms, and identifying suitable power sources for operation. To overcome these, interdisciplinary collaboration and innovative approaches from fields like materials science, computer science, and bioengineering are needed, leveraging advances in these fields.

- The ethical implications of nano robotics are significant, as they could impact human health, privacy, and societal dynamics. These concerns include autonomy, safety, equitable access, and potential misuse. Balancing technological advancements with ethical responsibilities requires stakeholder engagement, robust regulatory frameworks, and ongoing scientific deliberation.

- Nano robotics holds immense promise despite challenges, with ongoing research aiming to overcome technical hurdles, expand applications, and address ethical concerns. Future developments include nanomaterial advancements for durability and functionality, control algorithm breakthroughs for precise manipulation, and power generation innovations for sustained operation. Interdisciplinary collaboration, ethical foresight, and societal dialogue are crucial for navigating complexities and maximizing benefits.

CHALLENGES IN NANO ROBOTICS

Technical Constraints at Nanoscale

Nano robot manufacturing presents significant technical challenges due to the unique constraints of the nanoscale environment. The principles of physics differ from larger scales, necessitating innovative design, fabrication, and assembly approaches. Precision engineering is required at an unprecedented level, and conventional manufacturing techniques struggle to maintain accuracy and consistency. Novel manufacturing processes, like DNA origami or molecular self-assembly, are crucial for constructing nano robots with precision (Dash & Maiti, 2023; Patwardhan, 2006).

Materials selection at the nanoscale is crucial due to their unique properties like surface area, conductivity, and strength, which differ significantly from bulk materials. Identifying and using these materials is essential for creating robust and functional nano robots. The assembly and manipulation of nano-scale components pose challenges due to the limitations of traditional tools, necessitating the development of specialized nanomanipulation techniques using atomic force microscopes, electron beams, or biological processes for accurate positioning and assembly (Mohanty et al., 2023; Pramila et al., 2023).

Nano robots face challenges in reliability and stability due to environmental factors like temperature fluctuations and surface interactions. Robust design methodologies are crucial for their functionality and longevity. Interdisciplinary collaboration between robotics, materials science, nanotechnology, and engineering is needed to overcome these technical constraints. Advancements in nanofabrication techniques, innovative materials, and precision assembly methods are essential for developing functional and reliable nano robotic systems for real-world applications (Pramila et al., 2023; Rahamathunnisa, Sudhakar, et al., 2023).

Control Mechanisms and Precision Engineering

Nano robotics relies on precise control mechanisms and precision engineering to ensure their functionality and efficacy. Conventional control methods are often insufficient due to the limitations of physics at nanoscale. Innovative control strategies, such as feedback loops, stochastic algorithms, and biological principles, are being explored to enable precise manipulation and navigation of nano robots. Reliability and predictability are crucial for nano robots to perform tasks accurately and consistently, despite inherent uncertainties and fluctuations. This requires robust control systems that can adapt to environmental changes and variations, maintaining stability and reliability in operation (Karimov et al., 2022).

Nano robotics relies on precision engineering to design and fabricate nano-scale components with accuracy and repeatability. Nanorobotics, created using advanced nanofabrication techniques like electron beam lithography and molecular self-assembly, require intricate structures and intricate engineering to integrate various functionalities. These robots require diverse components like sensors, actuators, and power sources, all scaled down to nanometers while maintaining functionality. Interdisciplinary collaboration between robotics, control systems, materials science, and nanotechnology is crucial for developing robust, reliable nano robotic systems (Rahamathunnisa, Sudhakar, et al., 2023; Srinivas et al., 2023).

MATERIALS INNOVATIONS IN NANO ROBOTICS

Advanced Nanomaterials for Construction

Advanced nanomaterials are crucial in the development of nano robotic systems due to their unique properties and functionalities. These materials, engineered and manipulated at atomic or molecular levels, offer exceptional mechanical, electrical, or optical properties. Carbon-based nanomaterials like carbon nanotubes or graphene are ideal for constructing structural elements or providing electrical pathways within nano robots due to their exceptional strength, flexibility, and electrical conductivity (Walsh & Strano, 2018).

Nanoparticles of metals, semiconductors, or magnetic materials offer functionalities like catalytic activity, sensing, and magnetic manipulation, expanding the capabilities of nano robotic systems. These nanoparticles can be integrated into specific components for tasks like targeted drug delivery or precise object manipulation. Biomimetic nanomaterials, inspired by natural structures, offer promising possibilities in nano robotics, such as self-healing polymers and nanoscale structures mimicking biological systems, enabling resilient and adaptable nano robots in dynamic environments (Puranik et al., 2024b; Revathi et al., 2024).

Advanced nanomaterials are produced using techniques like molecular self-assembly, nanolithography, and chemical vapor deposition, allowing precise manipulation of atoms or molecules. However, challenges remain in ensuring scalability, reproducibility, and stability. Quality control and standardization are crucial for nano robotics production to ensure consistent performance and reliability. Advancements in materials science, nanotechnology, and engineering drive the development of novel nanomaterials, offering new possibilities for nano robotic systems with enhanced capabilities and performance at the nanoscale (Pramila et al., 2023; Rahamathunnisa, Sudhakar, et al., 2023).

Functionalities and Durability Enhancements

Nano robotic systems, with their enhanced functionality and durability, can perform complex tasks with precision, adapt to dynamic environments, and offer extended operational capabilities, extending their potential in various fields (Maheswari et al., 2023; Mohanty et al., 2023; Rahamathunnisa, Sudhakar, et al., 2023).

- **Enhanced Sensing Abilities:** Nano robots can be equipped with advanced sensors capable of detecting various physical and chemical parameters at the nanoscale. These sensors enable precise measurements, facilitating tasks such as environmental monitoring, medical diagnostics, or quality control in manufacturing processes (Chen et al., 2022b; Walsh & Strano, 2018).
- **Targeted Drug Delivery:** Nano robots with specialized functionalities can transport drugs or therapeutic agents to specific cells or tissues in the body, minimizing side effects and maximizing treatment efficiency. Functional modifications allow for controlled release mechanisms triggered by specific stimuli.
- **Adaptive and Responsive Behavior:** Incorporating responsive materials and mechanisms enables nano robots to adapt to changing environments. Responsive functionalities can be designed to react to stimuli like pH changes, temperature variations, or specific biomarkers, enhancing their versatility in diverse applications.
- **Self-Healing and Self-Repair:** Integration of self-healing materials or mechanisms within nano robots enhances their durability. These capabilities enable the robots to autonomously repair damage or wear, prolonging their operational lifespan in challenging conditions.
- **Multimodal Capabilities:** Nano robots can possess multiple functionalities, combining tasks such as manipulation, sensing, and drug delivery within a single system. This multimodal approach enhances their utility and efficiency in performing complex tasks.
- **Improved Energy Efficiency:** Enhancements in power sources and energy harvesting technologies lead to more energy-efficient nano robots. Utilizing nanoscale power generation or energy harvesting mechanisms can prolong operational periods and reduce the need for frequent recharging or maintenance.
- **Durability in Harsh Environments:** Nano robots engineered with materials resistant to extreme conditions, such as high temperatures, corrosive environments, or radiation exposure, ensure their functionality and reliability in challenging operational environments.
- **Bio-Inspired Design:** Drawing inspiration from biological systems allows for the development of nano robots with biomimetic functionalities. Mimicking natural structures and processes enhances adaptability, efficiency, and durability in various applications.

Integration of Nanoscale Components

The integration of nanoscale components into nano robotic systems necessitates meticulous engineering and innovative approaches due to the inherent challenges of working at such tiny scales (Indiveri et al., 2013). The figure 1 depicts the integration of nanoscale components into nano robotic systems. The integration of nanoscale components into nano robotic systems requires a comprehensive approach considering functionality, compatibility, interaction, and collective performance within the nano robot's confined space (Boopathi, 2023; Malathi et al., 2024).

- **Miniaturization and Compact Design:** Nano robots demand compact designs that accommodate various functionalities within a limited space. Integrating diverse nanoscale components—sensors, actuators, power sources—requires meticulous design to optimize space and functionality without compromising performance.
- **Nanoscale Assembly Techniques:** Specialized assembly techniques like DNA origami, self-assembly, or bottom-up fabrication methods are employed to precisely position and integrate na-

noscale components. These methods enable the construction of intricate structures with nanometer precision.

- **Multifunctional Platforms:** Designing platforms that can host multiple functionalities is crucial. Nanoscale components with different purposes—such as drug delivery, sensing, or manipulation—need to seamlessly integrate and work in tandem within the confined space of a nano robot.

Figure 1. Integration of nanoscale components into nano robotic systems

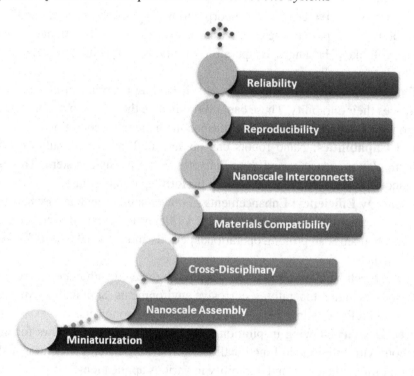

- **Cross-Disciplinary Collaboration:** Integrating nanoscale components necessitates collaboration across various disciplines, including materials science, nanotechnology, robotics, and biology. Experts from these fields collaborate to design and integrate components effectively.
- **Materials Compatibility and Interface Engineering:** Ensuring compatibility between different nanoscale components is crucial. Materials used in different components should exhibit compatibility to avoid adverse reactions or functionality issues. Interface engineering is essential to promote interaction between components while maintaining stability.
- **Nanoscale Interconnects and Communication:** Establishing communication and interconnectivity between nanoscale components is vital for their coordinated operation. Developing nanoscale communication protocols or methods for transmitting signals within nano robots is essential for seamless functionality.
- **Scalability and Reproducibility:** Methods for integrating nanoscale components should be scalable and reproducible. Consistent fabrication processes enable the mass production of reliable nano robots, critical for practical applications.

- **Reliability and Robustness Testing:** Stress tests and simulations are crucial for rigorous testing and validation of integrated nanoscale components, identifying potential weaknesses and optimizing design for enhanced performance and durability.

ADVANCEMENTS IN AI AND MACHINE LEARNING

Empowering Nano Robots With AI Capabilities

AI and machine learning advancements are revolutionizing nano robotics, providing autonomy, adaptability, and enhanced functionality. Important factors include AI integration into nano robots (Pugliese & Regondi, 2022). The integration of AI and machine learning in nano robots enhances their adaptability, intelligence, and versatility, enabling autonomous operation in complex environments, paving the way for applications in medicine, manufacturing, and environmental monitoring.

- **Autonomous Decision-Making:** AI algorithms enable nano robots to make decisions autonomously in dynamic environments. These algorithms process sensory data, analyze patterns, and make real-time decisions, allowing nano robots to adapt their actions without human intervention.
- **Adaptive Behavior:** Machine learning algorithms equipped within nano robots enable adaptive behavior. They learn from experience, adjusting their actions based on feedback, optimizing performance, and enhancing efficiency in completing tasks.
- **Sensory Processing and Analysis:** AI enhances the processing and analysis of sensory information gathered by nano robots. Machine learning algorithms can interpret complex data gathered at the nanoscale, enabling precise and efficient decision-making based on environmental cues.
- **Navigation and Control:** AI algorithms aid in navigation and control at the nanoscale. They assist in precise maneuvering within intricate environments, ensuring accurate positioning and manipulation of nano-scale objects or within biological systems.
- **Fault Detection and Self-Repair:** AI facilitates fault detection and self-repair mechanisms in nano robots. Machine learning algorithms can identify anomalies or malfunctions, triggering corrective actions or self-repair processes to maintain functionality.
- **Task Optimization and Learning:** Nano robots equipped with AI continuously optimize their tasks and operations. They learn from previous experiences, refining strategies and techniques to improve performance and achieve better outcomes.
- **Energy Efficiency and Resource Management:** AI enables efficient resource utilization and energy management within nano robots. Algorithms optimize power consumption, ensuring prolonged operational periods or the ability to harvest energy from the environment.
- **Security and Redundancy:** AI-driven security protocols enhance the robustness of nano robots against external threats or potential malfunctions. Redundancy and fail-safe mechanisms guided by AI algorithms ensure continued operation even in challenging scenarios.

Autonomous Decision-Making at Microscopic Levels

Advancements in AI and machine learning have enabled microscopic autonomous decision-making, a transformative capability in nano robotics (Vigelius et al., 2014). Nano robots with autonomous decision-making capabilities are revolutionizing applications in precision medicine, targeted therapies, environmental monitoring, and nanoscale manufacturing. These advancements revolutionize the operation, navigation, and interaction of these miniature machines, offering unprecedented possibilities for innovation and impact (Sampath et al., 2022).

- **Sensory Perception:** Nano robots equipped with sophisticated sensors capture and interpret data at the microscopic scale. These sensors detect environmental cues, molecular interactions, or cellular responses, providing essential input for decision-making.
- **AI Algorithms for Nanoscale Environments:** AI algorithms process vast amounts of sensory data collected by nano robots. These algorithms are designed to interpret and make decisions based on nanoscale phenomena, leveraging pattern recognition and analysis techniques specific to this level.
- **Real-time Adaptation:** Autonomous decision-making in nano robots involves real-time adaptation to dynamic environments. AI algorithms continuously assess and respond to changes at the microscopic level, enabling swift adjustments in operations or tasks.
- **Environmental Sensing and Response:** Nano robots autonomously sense and respond to variations in their environment. They detect factors like temperature shifts, chemical gradients, or cellular behaviors, altering their actions accordingly for specific tasks or interventions.
- **Decision Trees and Predictive Modeling:** AI-driven decision-making involves the creation of decision trees and predictive models at the nanoscale. These models enable nano robots to anticipate outcomes or responses based on various sensory inputs, optimizing their actions.
- **Learning from Interactions:** Machine learning capabilities within nano robots allow them to learn from their interactions. They adapt their decision-making processes based on past experiences, refining strategies for improved outcomes.
- **Adaptive Control Systems:** Autonomous decision-making is complemented by adaptive control systems within nano robots. These systems modulate actions, movements, or functions in response to the information processed by AI algorithms.
- **Safety Protocols and Redundancy:** Autonomous decision-making in nano robots incorporates safety protocols and redundancy measures. AI algorithms ensure that decisions prioritize safety and reliability, while redundant systems act as fail-safes in case of unexpected circumstances.

Applications in Complex Environments

Nano robots, equipped with advanced capabilities, find diverse applications in complex environments, navigating unique challenges in these intricate settings (Balaguer et al., 2000). The figure 2 depicts the applications of nano robots in complex environments. Nano robots' adaptability, precision, and versatility in complex environments reveal their potential in various fields, addressing challenges that traditional technology cannot overcome due to their ability to navigate and perform tasks at the smallest scales.

- **Biomedical Applications:** Nano robots excel in navigating complex biological environments. They can perform precise tasks such as targeted drug delivery within the human body, maneuvering through intricate networks of cells and tissues to reach specific targets.

Figure 2. Nano robots applications in complex environments

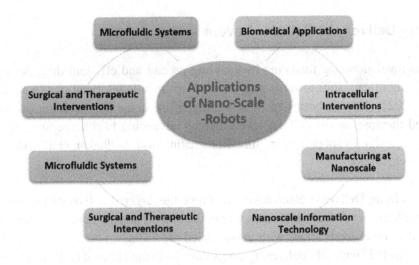

- **Intracellular Interventions:** These tiny machines have the potential to operate within cells, performing interventions or carrying out repairs at the molecular level. They navigate within cellular structures, aiding in diagnostics or therapeutic interventions.
- **Environmental Sensing and Remediation:** Nano robots are deployed in complex environmental settings for tasks like pollution monitoring or remediation. They navigate diverse and challenging terrains, detecting and neutralizing pollutants or assisting in environmental cleanup.
- **Manufacturing at Nanoscale:** In complex manufacturing processes, nano robots facilitate precision assembly and manipulation of materials at the atomic or molecular level. They operate in intricate setups, enabling the creation of advanced materials or devices.
- **Nanoscale Information Technology:** Nano robots contribute to advancements in information technology by enabling data storage at the molecular level or facilitating nanoscale computing. They navigate and manipulate information-bearing structures at this minuscule scale.
- **Surgical and Therapeutic Interventions:** Within the medical field, nano robots perform minimally invasive surgeries, offering precise interventions with minimal disruption. They navigate complex anatomical structures, enabling highly targeted treatments.
- **Exploration in Extreme Environments:** Nano robots navigate and explore extreme environments, such as deep-sea or outer space, where conventional technology faces challenges. They collect data, perform tasks, or assist in scientific exploration in these harsh settings.
- **Microfluidic Systems:** Nano robots contribute to the development of microfluidic systems by navigating and manipulating fluids at the microscale. They assist in tasks like controlled mixing, sorting, or analysis within microfluidic devices.

NANO ROBOTICS IN MEDICINE

Targeted Drug Delivery at Cellular Levels

Nano robotics is revolutionizing medicine by enabling precise and efficient drug delivery at cellular levels, offering a significant shift towards personalized therapeutic interventions (Karan & Majumder, 2011; Xu et al., 2022). Nano robotics are revolutionizing medicine by providing precision, efficacy, and personalized therapies at the cellular level, thereby advancing healthcare and improving patient outcomes, marking a significant paradigm shift (Rahamathunnisa, Sudhakar, et al., 2023; Senthil et al., 2023; Srinivas et al., 2023).

- **Precision in Drug Delivery:** Nano robots facilitate the targeted delivery of drugs or therapeutic agents directly to specific cells or tissues within the body. They navigate complex biological environments, delivering medications with unparalleled precision.
- **Minimizing Side Effects:** By delivering drugs directly to the affected cells or tissues, nano robots minimize systemic exposure and reduce side effects commonly associated with conventional drug delivery methods.
- **Enhanced Efficacy:** Targeted drug delivery at the cellular level enhances the efficacy of treatments. Nano robots ensure that therapeutic agents reach their intended targets, improving the effectiveness of medications in treating diseases.
- **Navigating Biological Barriers:** Nano robots are designed to navigate through biological barriers such as the blood-brain barrier or cellular membranes. They can transport drugs across these barriers, accessing areas that were previously challenging to reach.
- **Site-Specific Treatments:** Nano robots enable site-specific treatments, addressing diseases at their origin. They can target cancerous cells, infectious agents, or specific diseased tissues, offering tailored and precise interventions.
- **Controlled Release Mechanisms:** Nano robots can be engineered to carry payloads of drugs or therapeutic compounds and release them in a controlled manner. This controlled release ensures a sustained therapeutic effect at the cellular level.
- **Real-time Monitoring:** Some nano robots are equipped with sensors for real-time monitoring of cellular responses or drug efficacy. They gather feedback on treatment effectiveness, enabling adjustments or modifications in drug delivery strategies.
- **Potential for Personalized Medicine:** Nano robotics in targeted drug delivery holds promise for personalized medicine. These technologies can be tailored to individual patient needs, offering customized treatments based on specific cellular characteristics or conditions.

Minimally Invasive Surgeries and Nano Surgical Tools

Nano robotics has brought forth a new era in minimally invasive surgeries by introducing advanced nano surgical tools that enable precise interventions with minimal tissue disruption (Xu et al., 2022). Nano robotics and nano surgical tools are revolutionizing minimally invasive surgeries, providing precision, reduced risks, and faster recovery times. These advancements enhance patient outcomes and pave the way for future surgical interventions.

- **Microscale Precision:** Nano surgical tools operate at the micro and nanoscale, allowing surgeons to perform extremely precise interventions. These tools navigate intricate anatomical structures with high precision.

- **Reduced Tissue Trauma:** Minimally invasive surgeries utilizing nano surgical tools result in minimal tissue damage compared to traditional surgical procedures. Nano robots perform delicate procedures with minimal disruption to surrounding tissues.

- **Enhanced Dexterity:** Nano surgical tools offer enhanced dexterity, enabling surgeons to manipulate tissues or perform intricate tasks with precision that surpasses the capabilities of human hands or conventional surgical instruments.

- **Access to Challenging Areas:** Nano robots provide access to anatomical areas that are hard to reach using traditional surgical methods. They can navigate within confined spaces or delicate structures within the body.

- **Remote Surgical Procedures:** Some nano surgical tools enable remote surgical procedures. Surgeons can control these tools from a distance, offering the potential for remote surgeries, especially in scenarios where physical access is limited (Boopathi, 2023; Rahamathunnisa, Subhashini, et al., 2023).

- **Intraoperative Imaging and Sensing:** Nano robots equipped with imaging or sensing capabilities provide real-time information to surgeons during procedures. This aids in navigation, decision-making, and ensuring precision during surgery.

- **Faster Recovery and Reduced Risks:** Minimally invasive surgeries using nano surgical tools result in faster recovery times for patients due to reduced trauma. Additionally, there is a lower risk of complications and infections associated with these procedures.

- **Future Integration with AI:** Integration with AI and machine learning could further enhance the capabilities of nano surgical tools. AI algorithms could aid in real-time decision-making, optimizing surgical techniques, and providing additional guidance to surgeons.

Diagnostic Innovations With Nano Robots

Nano robots provide precise and efficient methods for disease detection and diagnostic imaging at the cellular or molecular level (S. Kumar et al., 2018). The figure 3 depicts nano robots being utilized for precise and efficient disease detection. Nano robots offer precise, accurate, and minimally invasive diagnostic techniques, enhancing patient outcomes and facilitating more effective treatments through early detection and precise imaging, thus advancing healthcare.

- **Early Disease Detection:** Nano robots equipped with sensors can detect biomarkers or abnormalities at early stages of diseases. They enable early detection of conditions such as cancer, infections, or other illnesses before symptoms manifest.

- **High Precision Imaging:** Nano robots facilitate high-resolution imaging at the cellular or molecular level. They can provide detailed images of tissues, organs, or cellular structures, aiding in accurate diagnosis and treatment planning.

- **Targeted Biopsy and Sampling:** Nano robots enable targeted biopsies or sampling of specific cells or tissues. They extract samples precisely, minimizing invasiveness and reducing the need for extensive tissue sampling.

- **In Vivo Imaging and Monitoring:** Some nano robots operate within the body, providing real-time in vivo imaging and monitoring of physiological processes. They offer continuous observation of cellular activities or responses during treatments.
- **Nanoscale Sensors for Diagnostics:** Nano robots carry nanoscale sensors capable of detecting minute changes or biomarkers indicative of diseases. These sensors provide sensitive and specific diagnostic information, improving accuracy.
- **Multiplexed Diagnostics:** Nano robots with multiplexed diagnostic capabilities can simultaneously detect multiple biomarkers or indicators within a single sample. This capability enhances efficiency and accuracy in diagnostics.
- **Point-of-Care Diagnostics:** Nano robots contribute to the development of point-of-care diagnostic tools. They enable rapid and accurate diagnostics at the patient's bedside or in resource-limited settings, improving access to healthcare.
- **Customized and Personalized Diagnostics:** Nano robotics allows for personalized diagnostics based on individual patient profiles. The ability to tailor diagnostic tests to specific cellular or molecular characteristics offers personalized and precise treatment strategies.

Figure 3. Nano robots provide precise and efficient methods for disease detection

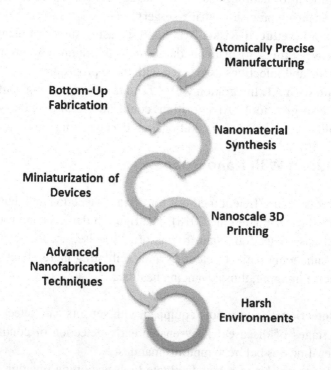

NANO ROBOTICS IN MANUFACTURING

Precision Assembly and Manipulation at Nanoscale

Nano robotics in manufacturing is revolutionizing processes by enabling precision assembly and manipulation at the nanoscale, paving the way for innovative material creation and device fabrication (Dash & Maiti, 2023). Nano robotics in manufacturing offers precision-driven creation of advanced materials, devices, and structures, transforming industries by enabling the development of novel products with enhanced functionalities and tailored properties.

- **Atomically Precise Manufacturing:** Nano robots facilitate manufacturing processes with atomic precision, allowing the assembly of materials at the molecular or atomic level. This precision leads to the creation of materials with specific properties and functionalities.
- **Bottom-Up Fabrication:** Nano robotics enables bottom-up fabrication techniques, where materials are built atom by atom or molecule by molecule. This approach allows for precise control over material structure and properties.
- **Nanomaterial Synthesis:** Nano robots contribute to the synthesis of advanced nanomaterials. They manipulate nanoparticles or nanostructures, assembling them into desired configurations, leading to the development of novel materials with unique properties.
- **Miniaturization of Devices:** Nano robots aid in the miniaturization of devices by assembling components at the nanoscale. They contribute to the production of smaller, more efficient devices with enhanced functionalities.
- **Nanoscale 3D Printing:** Nano robotics enables 3D printing at the nanoscale, allowing the creation of intricate structures with precise dimensions. This capability opens avenues for the production of nanoscale devices or components (Boopathi, Khare, et al., 2023; Palaniappan et al., 2023; Senthil et al., 2023).
- **Advanced Nanofabrication Techniques:** Nano robots employ advanced nanofabrication techniques such as molecular self-assembly, nanolithography, or nanoimprinting. These techniques enable precise manipulation of materials and structures.
- **Precision Manipulation in Harsh Environments:** Nano robots can operate in challenging environments such as extreme temperatures or pressures, facilitating precision manipulation of materials under conditions unsuitable for traditional manufacturing methods (Boopathi, 2022; Gowri et al., 2023).
- **Tailored Material Properties:** Manufacturing with nano robots allows for tailoring material properties at the nanoscale. By controlling material composition and structure, unique mechanical, electrical, or optical properties can be achieved.

Impact on Production Processes

Nano robotics is revolutionizing various industries by enhancing efficiency, precision, and novel capabilities, with significant impacts on production processes highlighted in important factors (Dash & Maiti, 2023). The integration of nano robotics into production processes redefines manufacturing paradigms, offering unprecedented precision, efficiency, and the ability to create materials and devices with unique properties. This transformative impact extends across industries, driving innovation and shaping the future of production.

- **Increased Precision and Accuracy:** Nano robotics elevates precision in production processes to the atomic or molecular level. This precision ensures unparalleled accuracy in manufacturing components or structures, reducing errors and enhancing quality.
- **Miniaturization and Efficiency:** Nano robots contribute to the miniaturization of components and devices, improving efficiency by reducing material wastage and optimizing resource utilization.
- **Novel Material Development:** Nano robotics enables the creation of novel materials with unique properties and functionalities. These materials find applications in diverse industries, offering innovative solutions and improved performance.
- **Customization and Personalization:** The precision afforded by nano robotics allows for customization and personalization of products. Tailored components or devices can be manufactured to meet specific customer requirements.
- **Advanced Manufacturing Techniques:** Nano robots drive the development of advanced manufacturing techniques such as nanolithography, molecular self-assembly, or nanoscale 3D printing. These techniques revolutionize how materials are manipulated and structured (Revathi et al., 2024).
- **Speed and Efficiency in Prototyping:** Nano robotics accelerates prototyping processes by enabling rapid and precise fabrication of prototypes at the nanoscale. This speed expedites innovation and product development.
- **Resource Conservation:** Nano robotics minimizes material wastage and energy consumption in manufacturing processes. The precision and efficiency of these technologies reduce resource utilization, contributing to sustainable production practices.
- **Cross-Industry Applications:** Nano robotics transcends industry boundaries, impacting sectors ranging from electronics and healthcare to materials science and environmental remediation. Its versatility opens doors to diverse applications.
- **Emergence of New Industries:** The capabilities unlocked by nano robotics pave the way for the emergence of entirely new industries and markets. Novel materials, devices, and production methods spur innovation and economic growth (Ravisankar et al., 2024; Vijayakumar et al., 2024).

Future of Nanoscale Manufacturing

The future of nanoscale manufacturing holds immense potential for industry transformation, groundbreaking advancements, and the creation of novel materials, devices, and applications (S. Kumar et al., 2018).

- **Nanoscale Additive Manufacturing:** Advancements in nanoscale 3D printing technologies will enable the precise fabrication of intricate structures at the molecular or atomic level. This will revolutionize manufacturing by allowing the creation of custom-designed nanoscale components and devices.
- **Nanomaterials with Tailored Properties:** Future nanoscale manufacturing techniques will focus on engineering materials with precisely tailored properties. This includes materials with superior strength, conductivity, flexibility, or other desired characteristics for various applications.

- **On-Demand Nanofabrication:** The future of nanoscale manufacturing envisions on-demand and decentralized fabrication capabilities. This may involve the development of portable or desktop-scale nanofabrication devices, enabling rapid prototyping and customization.
- **Nanorobot Swarms for Manufacturing:** The use of swarms of autonomous nanorobots working collaboratively is anticipated. These nanorobot swarms will perform complex tasks, such as assembly, manufacturing, or environmental remediation, in a coordinated manner.
- **AI-Driven Nanoscale Production:** Integration of AI and machine learning into nanoscale manufacturing will optimize processes, predict material behaviors, and enhance decision-making in real time. AI algorithms will enable autonomous control and optimization of nanoscale manufacturing systems (Boopathi & Khang, 2023; Koshariya, Kalaiyarasi, et al., 2023; Koshariya, Khatoon, et al., 2023; Maguluri et al., 2023).
- **Nanoscale Biomanufacturing:** Future developments might explore nanoscale biomanufacturing, leveraging biological systems or biomimetic approaches to produce functional nanomaterials or devices inspired by natural structures.
- **Nanomedicine and Therapeutics:** Nanoscale manufacturing will continue to advance in the medical field, facilitating the creation of targeted drug delivery systems, nanoscale implants, or diagnostic devices for personalized medicine.
- **Nanotechnology Integration in Everyday Products:** Nanoscale manufacturing will lead to the integration of nanotechnology in everyday products, enhancing their functionalities, durability, and performance (Boopathi, Umareddy, et al., 2023; Fowziya et al., 2023; Vijayakumar et al., 2024).
- **Sustainable Nanomanufacturing:** There will be a focus on developing sustainable and environmentally friendly nanomanufacturing techniques. Innovations will aim to reduce energy consumption, minimize waste, and employ eco-friendly materials.

Nanoscale manufacturing holds immense potential for innovation across industries, shaping technological progress and addressing global challenges as advancements continue.

ENVISIONING THE FUTURE OF NANO ROBOTICS

Overcoming Current Limitations

The future of nano robotics is predicted to be significantly enhanced through innovative methods and technological advancements. The following strategies can be employed to tackle obstacles (Pandya & Auner, 2004):

- **Enhanced Control and Manipulation:** Developing more precise and adaptable control mechanisms at the nanoscale is critical. Advancements in nanomanipulation techniques, such as improved nano grippers or nanomanipulators, will enable more accurate manipulation of nano-scale objects.
- **Improved Energy Sources:** Overcoming limitations in power sources for nano robots is crucial. Research into nanoscale power generation or energy harvesting methods could provide sustainable and efficient energy sources for these miniature machines.

- **Reliability and Durability:** Enhancing the reliability and durability of nano robots remains a challenge. Future advancements in materials science may yield more robust materials suitable for nanoscale applications, ensuring longevity and stability in various environments.
- **Miniaturization of Components:** Shrinking components without sacrificing functionality is essential. Advances in nanofabrication techniques and the development of nanoscale components, sensors, and actuators will further miniaturize nano robots, improving their capabilities.
- **Biocompatibility and Safety:** Ensuring the biocompatibility and safety of nano robots used in medical applications is critical. Continued research into materials that are safe for biological systems and the development of biodegradable nanomaterials will address these concerns.
- **Scalability and Mass Production:** Scaling up nanoscale manufacturing processes for mass production is challenging. Innovations in scalable nanofabrication techniques and automated assembly methods will enable the large-scale production of nano robotic systems.
- **Interdisciplinary Collaboration:** Collaborations between various scientific disciplines are vital for overcoming limitations. Continued cooperation between robotics, nanotechnology, materials science, and biomedicine will drive innovation in nano robotics.
- **Integration with AI and Machine Learning:** Further integration of AI and machine learning will enhance the capabilities of nano robots. AI algorithms can optimize decision-making, improve autonomous functionalities, and enable adaptive behavior in these miniature systems (Hussain et al., 2023; M. Kumar et al., 2023).

Nano robotics advancements will revolutionize fields like medicine, manufacturing, and environmental remediation, necessitating collaboration among researchers, engineers, policymakers, and ethicists. The advancement and widespread adoption of nano robotics are poised to bring about substantial societal and economic impacts across various domains. Here are potential impacts in both societal and economic spheres:

Societal Impacts

- **Healthcare Transformation:** Nano robotics in medicine could revolutionize healthcare by offering personalized treatments, precise diagnostics, and targeted drug delivery. This could lead to improved patient outcomes, reduced treatment costs, and better overall health for individuals (Kushwah et al., 2024; Ramudu et al., 2023; Satav et al., 2023).
- **Enhanced Quality of Life:** Nano robotics might lead to the development of assistive devices or technologies that enhance the quality of life for individuals with disabilities or age-related limitations, enabling greater independence.
- **Environmental Remediation:** Nano robots could be utilized for environmental cleanup, pollution monitoring, or remediation, contributing to cleaner ecosystems and sustainable environmental practices (Boopathi, Alqahtani, et al., 2023; Gowri et al., 2023; Hanumanthakari et al., 2023).
- **Access to Advanced Technologies:** Advancements in nano robotics might offer access to advanced technologies in developing regions, bridging technological gaps and providing innovative solutions in healthcare, agriculture, and infrastructure.
- **Ethical Considerations:** Societal impacts also involve addressing ethical concerns related to the use of nano robotics, including privacy, autonomy, and equitable access to these technologies across different socio-economic groups.

Economic Impacts

Understanding and harnessing the potential societal and economic impacts of nano robotics will require collaboration among policymakers, researchers, industries, and communities to ensure responsible and inclusive development, maximizing the benefits for society while addressing any associated challenges or risks (Babu et al., 2022; Dhanya et al., 2023; Hussain et al., 2023; Ravisankar et al., 2023).

- **Innovation and New Industries:** Nano robotics will spur innovation, leading to the emergence of new industries and markets. Companies investing in nano robotics could drive economic growth and create job opportunities in research, development, and manufacturing sectors.
- **Increased Productivity:** Adoption of nano robotic technologies in manufacturing processes could enhance productivity and efficiency. Precise manufacturing capabilities may lead to cost savings, reduced waste, and higher-quality products.
- **Healthcare Expenditure Reduction:** Improved healthcare through nano robotics may lower healthcare costs by providing targeted treatments and early disease detection, potentially reducing long-term healthcare expenditures.
- **Global Competitiveness:** Nations investing in nano robotics research and development may enhance their global competitiveness. Leading the way in advanced technologies could strengthen a country's position in the global economy.
- **Resource Optimization:** Nano robotics may lead to better utilization of resources, whether in manufacturing, energy production, or environmental management, contributing to sustainable economic practices.

FUTURE POSSIBILITIES AND TRANSFORMATIONS

Nano robotics holds immense potential for transformative changes in industries, healthcare, sustainability, and technology interaction. However, careful consideration of ethical, societal, and environmental implications is crucial for harnessing these advancements for the greater good (Mbunge et al., 2021; Pokrajac et al., 2021). The figure 4 depicts the potential for future possibilities and transformations.

Figure 4. Future possibilities and transformations

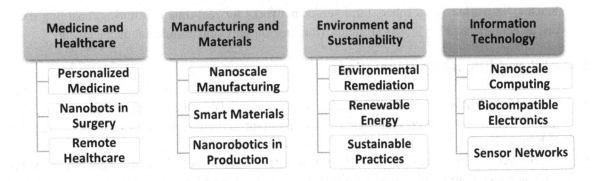

Medicine and Healthcare

- **Personalized Medicine:** Nano robots enable precise targeted therapies, personalized drug delivery, and diagnostics tailored to individual genetic profiles, revolutionizing healthcare.
- **Nanobots in Surgery:** Surgical procedures become minimally invasive, with nanobots performing precise interventions within the body, reducing risks and improving recovery times.
- **Remote Healthcare:** Nanobots equipped with sensors could provide remote monitoring and treatments, extending healthcare access to remote or underserved areas (Malathi et al., 2024).

Manufacturing and Materials

- **Nanoscale Manufacturing:** Advanced nanofabrication techniques enable the production of novel materials, leading to the development of highly efficient devices and structures.
- **Smart Materials:** Nano robotics contribute to the creation of smart materials with adaptive functionalities, finding applications in electronics, construction, and energy.
- **Nanorobotics in Production:** Nanobots aid in advanced manufacturing processes, optimizing production efficiency, reducing waste, and enabling rapid prototyping.

Environment and Sustainability

- **Environmental Remediation:** Nano robots assist in cleaning pollutants, monitoring ecosystems, and remediating environmental damage in a targeted and efficient manner.
- **Renewable Energy:** Nanotechnology contributes to the development of advanced materials for energy harvesting, storage, and more efficient renewable energy devices.
- **Sustainable Practices:** Nanoscale technologies promote sustainability by optimizing resource utilization, reducing energy consumption, and fostering eco-friendly manufacturing processes (Boopathi, Umareddy, et al., 2023; Boopathi & Davim, 2023b, 2023a).

Information Technology

- **Nanoscale Computing:** Advances in nanoscale computing and data storage could revolutionize information technology, leading to faster and more efficient devices.
- **Biocompatible Electronics:** Nanoscale electronics enable the development of biocompatible devices, potentially integrating technology with biology for various applications.
- **Sensor Networks:** Nano robots form sensor networks for real-time monitoring, enabling data-driven decision-making in various fields, from healthcare to environmental management.

Societal and Ethical Considerations

- **Ethical Implications:** Discussions around ethical use, privacy concerns, regulation, and societal impact of nano robotics become increasingly important as these technologies advance (Boopathi & Khang, 2023).
- **Workforce Adaptation:** Transformations in industries due to nanotechnology might require workforce adaptation, training, and new skillsets to integrate these advanced technologies effectively.

- **Global Collaboration:** International collaboration and ethical frameworks become crucial in ensuring responsible development and equitable access to nano robotics across regions.

CONCLUSION

The advancement of nano robotics is a significant technological advancement, poised to revolutionize various fields with its precision, adaptability, and transformative capabilities, marking a new era of advancements and possibilities in various fields. Nanotechnology and robotics are combining to create precision at atomic and molecular levels, enabling improved treatments and outcomes in medicine and manufacturing. Nano robots offer targeted drug delivery, minimally invasive surgeries, and personalized healthcare, while manufacturing machines enable the creation of novel materials and efficient production processes.

Nano robotics have significant societal and economic impacts, including healthcare transformation, environmental remediation, and increased accessibility to advanced technologies. However, ethical considerations and regulatory frameworks are crucial for responsible development and equitable access to these technologies. Nano robotics holds immense potential for personalized medicine, smart materials, and renewable energy. However, navigating this transformative landscape requires collaboration, ethical discourse, and a proactive approach to societal and environmental implications to harness its potential for humanity's betterment.

Nano robotics is a cutting-edge technology that is poised to revolutionize industries, redefine possibilities, and positively impact our lives, presenting a promising future.

Abbreviations

AI: Artificial Intelligence
   3D: Three-Dimensional

REFERENCES

Babu, B. S., Kamalakannan, J., Meenatchi, N., Karthik, S., & Boopathi, S. (2022). Economic impacts and reliability evaluation of battery by adopting Electric Vehicle. *IEEE Explore*, 1–6.

Balaguer, C., Giménez, A., Pastor, J. M., Padron, V., & Abderrahim, M. (2000). A climbing autonomous robot for inspection applications in 3d complex environments. *Robotica*, *18*(3), 287–297. doi:10.1017/S0263574799002258

Boopathi, S. (2022). Performance Improvement of Eco-Friendly Near-Dry wire-Cut Electrical Discharge Machining Process Using Coconut Oil-Mist Dielectric Fluid. *World Scientific: Journal of Advanced Manufacturing Systems*.

Boopathi, S. (2023). Internet of Things-Integrated Remote Patient Monitoring System: Healthcare Application. In *Dynamics of Swarm Intelligence Health Analysis for the Next Generation* (pp. 137–161). IGI Global. doi:10.4018/978-1-6684-6894-4.ch008

Boopathi, S., Alqahtani, A. S., Mubarakali, A., & Panchatcharam, P. (2023). Sustainable developments in near-dry electrical discharge machining process using sunflower oil-mist dielectric fluid. *Environmental Science and Pollution Research International*, 1–20. doi:10.1007/s11356-023-27494-0 PMID:37199846

Boopathi, S., & Davim, J. P. (2023a). Applications of Nanoparticles in Various Manufacturing Processes. In *Sustainable Utilization of Nanoparticles and Nanofluids in Engineering Applications* (pp. 1–31). IGI Global. doi:10.4018/978-1-6684-9135-5.ch001

Boopathi, S., & Davim, J. P. (2023b). *Sustainable Utilization of Nanoparticles and Nanofluids in Engineering Applications*. IGI Global. doi:10.4018/978-1-6684-9135-5

Boopathi, S., & Khang, A. (2023). AI-Integrated Technology for a Secure and Ethical Healthcare Ecosystem. In *AI and IoT-Based Technologies for Precision Medicine* (pp. 36–59). IGI Global. doi:10.4018/979-8-3693-0876-9.ch003

Boopathi, S., Khare, R., KG, J. C., Muni, T. V., & Khare, S. (2023). Additive Manufacturing Developments in the Medical Engineering Field. In Development, Properties, and Industrial Applications of 3D Printed Polymer Composites (pp. 86–106). IGI Global.

Boopathi, S., Umareddy, M., & Elangovan, M. (2023). Applications of Nano-Cutting Fluids in Advanced Machining Processes. In *Sustainable Utilization of Nanoparticles and Nanofluids in Engineering Applications* (pp. 211–234). IGI Global. doi:10.4018/978-1-6684-9135-5.ch009

Chen, Y., Chen, D., Liang, S., Dai, Y., Bai, X., Song, B., Zhang, D., Chen, H., & Feng, L. (2022). Recent Advances in Field-Controlled Micro–Nano Manipulations and Micro–Nano Robots. *Advanced Intelligent Systems*, *4*(3), 2100116. doi:10.1002/aisy.202100116

Dash, T., & Maiti, C. K. (2023). An overview of nanoscale device fabrication technology—Part I. *Nanoelectronics: Physics, Materials and Devices*, 193–214.

Daudi, J. (2015). An overview of application of artificial immune system in swarm robotic systems. *Advances in Robotics & Automation, 4*(1).

Dhanya, D., Kumar, S. S., Thilagavathy, A., Prasad, D., & Boopathi, S. (2023). Data Analytics and Artificial Intelligence in the Circular Economy: Case Studies. In Intelligent Engineering Applications and Applied Sciences for Sustainability (pp. 40–58). IGI Global.

Fowziya, S., Sivaranjani, S., Devi, N. L., Boopathi, S., Thakur, S., & Sailaja, J. M. (2023). Influences of nano-green lubricants in the friction-stir process of TiAlN coated alloys. *Materials Today: Proceedings*. Advance online publication. doi:10.1016/j.matpr.2023.06.446

Ghanbarzadeh-Dagheyan, A., Jalili, N., & Ahmadian, M. T. (2021). A holistic survey on mechatronic Systems in Micro/Nano scale with challenges and applications. *Journal of Micro-Bio Robotics*, *17*(1), 1–22. doi:10.1007/s12213-021-00145-8

Gheorghe, P. E. E. G. I., Ilie, P. S. I., Istriteanu, P. E. S., & Bajenaru, P. E. V. (2014). Research in micro-nano-robotics. *The Romanian Review Precision Mechanics, 46,* 83.

Gowri, N. V., Dwivedi, J. N., Krishnaveni, K., Boopathi, S., Palaniappan, M., & Medikondu, N. R. (2023). Experimental investigation and multi-objective optimization of eco-friendly near-dry electrical discharge machining of shape memory alloy using Cu/SiC/Gr composite electrode. *Environmental Science and Pollution Research International, 30*(49), 1–19. doi:10.1007/s11356-023-26983-6 PMID:37126160

Halder, A., & Sun, Y. (2019). Biocompatible propulsion for biomedical micro/nano robotics. *Biosensors & Bioelectronics, 139,* 111334. doi:10.1016/j.bios.2019.111334 PMID:31128479

Hanumanthakari, S., Gift, M. M., Kanimozhi, K., Bhavani, M. D., Bamane, K. D., & Boopathi, S. (2023). Biomining Method to Extract Metal Components Using Computer-Printed Circuit Board E-Waste. In *Handbook of Research on Safe Disposal Methods of Municipal Solid Wastes for a Sustainable Environment* (pp. 123–141). IGI Global. doi:10.4018/978-1-6684-8117-2.ch010

Hussain, Z., Babe, M., Saravanan, S., Srimathy, G., Roopa, H., & Boopathi, S. (2023). Optimizing Biomass-to-Biofuel Conversion: IoT and AI Integration for Enhanced Efficiency and Sustainability. In *Circular Economy Implementation for Sustainability in the Built Environment* (pp. 191–214). IGI Global.

Indiveri, G., Linares-Barranco, B., Legenstein, R., Deligeorgis, G., & Prodromakis, T. (2013). Integration of nanoscale memristor synapses in neuromorphic computing architectures. *Nanotechnology, 24*(38), 384010. doi:10.1088/0957-4484/24/38/384010 PMID:23999381

Karan, S., & Majumder, D. D. (2011). Molecular machinery-a nanorobotics control system design for cancer drug delivery. *2011 International Conference on Recent Trends in Information Systems,* 197–202. 10.1109/ReTIS.2011.6146867

Karimov, K., Akhmedov, A., & Adilova, S. (2022). Theoretical and engineering solutions of the controlled vibration mechanisms for precision engineering. *AIP Conference Proceedings, 2637*(1), 060001. doi:10.1063/5.0118863

Koshariya, A. K., Kalaiyarasi, D., Jovith, A. A., Sivakami, T., Hasan, D. S., & Boopathi, S. (2023). AI-Enabled IoT and WSN-Integrated Smart Agriculture System. In *Artificial Intelligence Tools and Technologies for Smart Farming and Agriculture Practices* (pp. 200–218). IGI Global. doi:10.4018/978-1-6684-8516-3.ch011

Koshariya, A. K., Khatoon, S., Marathe, A. M., Suba, G. M., Baral, D., & Boopathi, S. (2023). Agricultural Waste Management Systems Using Artificial Intelligence Techniques. In *AI-Enabled Social Robotics in Human Care Services* (pp. 236–258). IGI Global. doi:10.4018/978-1-6684-8171-4.ch009

Kumar, M., Kumar, K., Sasikala, P., Sampath, B., Gopi, B., & Sundaram, S. (2023). Sustainable Green Energy Generation From Waste Water: IoT and ML Integration. In Sustainable Science and Intelligent Technologies for Societal Development (pp. 440–463). IGI Global.

Kumar, S., Nasim, B., & Abraham, E. (2018). Nanorobots a future device for diagnosis and treatment. *Journal of Pharmacy and Pharmaceutics, 5*(1), 44–49. doi:10.15436/2377-1313.18.1815

Kushwah, J. S., Gupta, M., Shrivastava, S., Saxena, N., Saini, R., & Boopathi, S. (2024). Psychological Impacts, Prevention Strategies, and Intervention Approaches Across Age Groups: Unmasking Cyberbullying. In Change Dynamics in Healthcare, Technological Innovations, and Complex Scenarios (pp. 89–109). IGI Global.

Maguluri, L. P., Arularasan, A., & Boopathi, S. (2023). Assessing Security Concerns for AI-Based Drones in Smart Cities. In Effective AI, Blockchain, and E-Governance Applications for Knowledge Discovery and Management (pp. 27–47). IGI Global. doi:10.4018/978-1-6684-9151-5.ch002

Maheswari, B. U., Imambi, S. S., Hasan, D., Meenakshi, S., Pratheep, V., & Boopathi, S. (2023). Internet of things and machine learning-integrated smart robotics. In Global Perspectives on Robotics and Autonomous Systems: Development and Applications (pp. 240–258). IGI Global. doi:10.4018/978-1-6684-7791-5.ch010

Malathi, J., Kusha, K., Isaac, S., Ramesh, A., Rajendiran, M., & Boopathi, S. (2024). IoT-Enabled Remote Patient Monitoring for Chronic Disease Management and Cost Savings: Transforming Healthcare. In Advances in Explainable AI Applications for Smart Cities (pp. 371–388). IGI Global.

Mbunge, E., Muchemwa, B., Batani, J., & ... (2021). Sensors and healthcare 5.0: Transformative shift in virtual care through emerging digital health technologies. *Global Health Journal (Amsterdam, Netherlands)*, *5*(4), 169–177. doi:10.1016/j.glohj.2021.11.008

Mohanty, A., Jothi, B., Jeyasudha, J., Ranjit, P., Isaac, J. S., & Boopathi, S. (2023). Additive Manufacturing Using Robotic Programming. In *AI-Enabled Social Robotics in Human Care Services* (pp. 259–282). IGI Global. doi:10.4018/978-1-6684-8171-4.ch010

Palaniappan, M., Tirlangi, S., Mohamed, M. J. S., Moorthy, R. S., Valeti, S. V., & Boopathi, S. (2023). Fused Deposition Modelling of Polylactic Acid (PLA)-Based Polymer Composites: A Case Study. In Development, Properties, and Industrial Applications of 3D Printed Polymer Composites (pp. 66–85). IGI Global.

Pandya, A., & Auner, G. (2004). Robotics technology: A journey into the future. *The Urologic Clinics of North America*, *31*(4), 793–800. doi:10.1016/j.ucl.2004.06.013 PMID:15474607

Patwardhan, J. (2006). *Architectures for nanoscale devices* (Vol. 68). Academic Press.

Pedram, A., & Nejat Pishkenari, H. (2017). Smart micro/nano-robotic systems for gene delivery. *Current Gene Therapy*, *17*(2), 73–79. doi:10.2174/1566523217666170511111000 PMID:28494736

Pokrajac, L., Abbas, A., Chrzanowski, W., Dias, G. M., Eggleton, B. J., Maguire, S., Maine, E., Malloy, T., Nathwani, J., Nazar, L., & ... (2021). *Nanotechnology for a sustainable future: Addressing global challenges with the international network4sustainable nanotechnology*. ACS Publications.

Pramila, P., Amudha, S., Saravanan, T., Sankar, S. R., Poongothai, E., & Boopathi, S. (2023). Design and Development of Robots for Medical Assistance: An Architectural Approach. In Contemporary Applications of Data Fusion for Advanced Healthcare Informatics (pp. 260–282). IGI Global.

Pugliese, R., & Regondi, S. (2022). Artificial intelligence-empowered 3D and 4D printing technologies toward smarter biomedical materials and approaches. *Polymers*, *14*(14), 2794. doi:10.3390/polym14142794 PMID:35890571

Puranik, T. A., Shaik, N., Vankudoth, R., Kolhe, M. R., Yadav, N., & Boopathi, S. (2024). Study on Harmonizing Human-Robot (Drone) Collaboration: Navigating Seamless Interactions in Collaborative Environments. In Cybersecurity Issues and Challenges in the Drone Industry (pp. 1–26). IGI Global.

Rahamathunnisa, U., Subhashini, P., Aancy, H. M., Meenakshi, S., Boopathi, S., & ... (2023). Solutions for Software Requirement Risks Using Artificial Intelligence Techniques. In *Handbook of Research on Data Science and Cybersecurity Innovations in Industry 4.0 Technologies* (pp. 45–64). IGI Global.

Rahamathunnisa, U., Sudhakar, K., Murugan, T. K., Thivaharan, S., Rajkumar, M., & Boopathi, S. (2023). Cloud Computing Principles for Optimizing Robot Task Offloading Processes. In *AI-Enabled Social Robotics in Human Care Services* (pp. 188–211). IGI Global. doi:10.4018/978-1-6684-8171-4.ch007

Ramudu, K., Mohan, V. M., Jyothirmai, D., Prasad, D., Agrawal, R., & Boopathi, S. (2023). Machine Learning and Artificial Intelligence in Disease Prediction: Applications, Challenges, Limitations, Case Studies, and Future Directions. In Contemporary Applications of Data Fusion for Advanced Healthcare Informatics (pp. 297–318). IGI Global.

Ravisankar, A., Sampath, B., & Asif, M. M. (2023). Economic Studies on Automobile Management: Working Capital and Investment Analysis. In Multidisciplinary Approaches to Organizational Governance During Health Crises (pp. 169–198). IGI Global.

Ravisankar, A., Shanthi, A., Lavanya, S., Ramaratnam, M., Krishnamoorthy, V., & Boopathi, S. (2024). Harnessing 6G for Consumer-Centric Business Strategies Across Electronic Industries. In AI Impacts in Digital Consumer Behavior (pp. 241–270). IGI Global.

Revathi, S., Babu, M., Rajkumar, N., Meti, V. K. V., Kandavalli, S. R., & Boopathi, S. (2024). Unleashing the Future Potential of 4D Printing: Exploring Applications in Wearable Technology, Robotics, Energy, Transportation, and Fashion. In Human-Centered Approaches in Industry 5.0: Human-Machine Interaction, Virtual Reality Training, and Customer Sentiment Analysis (pp. 131–153). IGI Global.

Sampath, B., Naveenkumar, N., Sampathkumar, P., Silambarasan, P., Venkadesh, A., & Sakthivel, M. (2022). Experimental comparative study of banana fiber composite with glass fiber composite material using Taguchi method. *Materials Today: Proceedings*, *49*, 1475–1480. doi:10.1016/j.matpr.2021.07.232

Satav, S. D., Hasan, D. S., Pitchai, R., Mohanaprakash, T., Sultanuddin, S., & Boopathi, S. (2023). Next generation of internet of things (ngiot) in healthcare systems. In *Sustainable Science and Intelligent Technologies for Societal Development* (pp. 307–330). IGI Global.

Senthil, T., Puviyarasan, M., Babu, S. R., Surakasi, R., Sampath, B., & Associates. (2023). Industrial Robot-Integrated Fused Deposition Modelling for the 3D Printing Process. In Development, Properties, and Industrial Applications of 3D Printed Polymer Composites (pp. 188–210). IGI Global.

Sivasankar, M., & Durairaj, R. (2012). Brief review on nano robots in bio medical applications. *Adv Robot Autom*, *1*(101), 2. doi:10.4172/2168-9695.1000101

Srinivas, B., Maguluri, L. P., Naidu, K. V., Reddy, L. C. S., Deivakani, M., & Boopathi, S. (2023). Architecture and Framework for Interfacing Cloud-Enabled Robots. In *Handbook of Research on Data Science and Cybersecurity Innovations in Industry 4.0 Technologies* (pp. 542–560). IGI Global. doi:10.4018/978-1-6684-8145-5.ch027

Vigelius, M., Meyer, B., & Pascoe, G. (2014). Multiscale modelling and analysis of collective decision making in swarm robotics. *PLoS One*, *9*(11), e111542. doi:10.1371/journal.pone.0111542 PMID:25369026

Vijayakumar, G. N. S., Domakonda, V. K., Farooq, S., Kumar, B. S., Pradeep, N., & Boopathi, S. (2024). Sustainable Developments in Nano-Fluid Synthesis for Various Industrial Applications. In Adoption and Use of Technology Tools and Services by Economically Disadvantaged Communities: Implications for Growth and Sustainability (pp. 48–81). IGI Global.

Walsh, S. M., & Strano, M. S. (2018). *Robotic systems and autonomous platforms: Advances in materials and manufacturing*. Woodhead Publishing.

Xu, Y., Bian, Q., Wang, R., & Gao, J. (2022). Micro/nanorobots for precise drug delivery via targeted transport and triggered release: A review. *International Journal of Pharmaceutics*, *616*, 121551. doi:10.1016/j.ijpharm.2022.121551 PMID:35131352

Compilation of References

Abarca, V. E., & Elias, D. A. (2023). A Review of Parallel Robots: Rehabilitation, Assistance, and Humanoid Applications for Neck, Shoulder, Wrist, Hip, and Ankle Joints. *Robotics (Basel, Switzerland)*, *12*(5), 131. doi:10.3390/robotics12050131

Ahmed, L. J., Anish Fathima, B., Mahaboob, M., & Gokulavasan, B. (2021). Biomedical Image Processing with Improved SPIHT Algorithm and optimized Curvelet Transform Technique. *2021 7th International Conference on Advanced Computing and Communication Systems, ICACCS 2021*. 10.1109/ICACCS51430.2021.9441832

Ahmed, L. J., Bruntha, P. M., Dhanasekar, S., Chitra, V., Balaji, D., & Senathipathi, N. (2022). An Improvised Image Registration Technique for Brain Tumor Identification and Segmentation Using ANN Approach. *ICDCS 2022 - 2022 6th International Conference on Devices, Circuits and Systems*, 80-84. 10.1109/ICDCS54290.2022.9780846

Ahmed, T., Ahmed, S., Ahmed, S., & Motiwala, M. (2010). Real-Time Intruder Detection in Surveillance Networks Using Adaptive Kernel Methods. *Proceedings of the 2010 IEEE International Conference on Communications*. 10.1109/ICC.2010.5502592

Ahmed, Z., Mohamed, K., Zeeshan, S., & Dong, X. (2020). Artificial intelligence with multi-functional machine learning platform development for better healthcare and precision medicine. *Database (Oxford)*, *2020*, baaa010. doi:10.1093/database/baaa010 PMID:32185396

Akolpoglu, M. B., Dogan, N. O., Bozuyuk, U., Ceylan, H., Kizilel, S., & Sitti, M. (2020). High-yield production of biohybrid microalgae for on-demand cargo delivery. *Advancement of Science*, *7*(16), 2001256. doi:10.1002/advs.202001256 PMID:32832367

Alam, A. (2022). Social robots in education for long-term human-robot interaction: Socially supportive behaviour of robotic tutor for creating robo-tangible learning environment in a guided discovery learning interaction. *ECS Transactions*, *107*(1), 12389–12403. doi:10.1149/10701.12389ecst

Alanzy, M., Alomrani, R., Alqarni, B., & Almutairi, S. (2023). Image Steganography Using LSB and Hybrid Encryption Algorithms. *Applied Sciences (Basel, Switzerland)*, *13*(21), 11771. doi:10.3390/app132111771

Alborzi, M., & Khanbabaei, M. (2016). Using data mining and neural networks techniques to propose a new hybrid customer behavior analysis and credit scoring model in banking services based on a developed RFM analysis method. *International Journal of Business Information Systems*, *23*(1), 1–22. doi:10.1504/IJBIS.2016.078020

Almohamad, A., Tahir, A. M., Al-Kababji, A., Furqan, H. M., Khattab, T., Hasna, M. O., & Arslan, H. (2020). Smart and secure wireless communications via reflecting intelligent surfaces: A short survey. *IEEE Open Journal of the Communications Society*, *1*, 1442–1456. doi:10.1109/OJCOMS.2020.3023731

Alobaidi, W. H., Aziz, I. T., Jawad, T., Flaih, F. M. F., & Azeez, A. T. (2018). Face detection based on probability of amplitude distribution of local binary patterns algorithm. *Proceedings of the 2018 6th International Symposium on Digital Forensic and Security (ISDFS)*, 1–5 10.1109/ISDFS.2018.8355319

Amann, J., Blasimme, A., Vayena, E., Frey, D., & Madai, V. I. (2020). Explainability for artificial intelligence in healthcare: A multidisciplinary perspective. *BMC Medical Informatics and Decision Making*, 20(1), 1–9. doi:10.1186/s12911-020-01332-6 PMID:33256715

Anitha, Kalra, & Shrivastav. (2016). A Cyber defence using artificial home automation system using IoT. *International Journal of Pharmacy and Technology*, 8, 25358-64.

Anthimopoulos, M., Christodoulidis, S., Ebner, L., Christe, A., & Mougiakakou, S. (2016). Lung pattern classification for interstitial lung diseases using a deep convolutional neural network. *IEEE Transactions on Medical Imaging*, 35(5), 1207–1216. doi:10.1109/TMI.2016.2535865 PMID:26955021

Arif, I., Aslam, W., & Hwang, Y. (2023). Barriers in adoption of internet banking: A structural equation modeling-neural network approach. *Technology in Society*, 61, 101231. doi:10.1016/j.techsoc.2020.101231

Arul, S. H., Bedi, A. S., & Manocha, D. (2022). Multi Robot Collision Avoidance by Learning Whom to Communicate. *arXiv preprint arXiv:2209.06415*.

Atilano, L., Martinho, A., Silva, M., & Baptista, A. (2019). Lean Design-for-X: Case study of a new design framework applied to an adaptive robot gripper development process. *Procedia CIRP*, 84, 667–672. doi:10.1016/j.procir.2019.04.190

Babu, B. S., Kamalakannan, J., Meenatchi, N., Karthik, S., & Boopathi, S. (2022). Economic impacts and reliability evaluation of battery by adopting Electric Vehicle. *IEEE Explore*, 1–6.

Baesens, B., Van Gestel, T., Stepanova, M., Van den Poel, D., & Vanthienen, J. (2005). Neural network survival analysis for personal loan data. *The Journal of the Operational Research Society*, 56(9), 1089–1098. doi:10.1057/palgrave.jors.2601990

Bajaj, N. M., Spiers, A. J., & Dollar, A. M. (2019). State of the art in artificial wrists: A review of prosthetic and robotic wrist design. *IEEE Transactions on Robotics*, 35(1), 261–277. doi:10.1109/TRO.2018.2865890

Bajcsy, A., Herbert, S. L., Fridovich-Keil, D., Fisac, J. F., Deglurkar, S., Dragan, A. D., & Tomlin, C. J. (2019, May). A scalable framework for real-time multi-robot, multi-human collision avoidance. In 2019 international conference on robotics and automation (ICRA) (pp. 936-943). IEEE. doi:10.1109/ICRA.2019.8794457

Balaguer, C., Giménez, A., Pastor, J. M., Padron, V., & Abderrahim, M. (2000). A climbing autonomous robot for inspection applications in 3d complex environments. *Robotica*, 18(3), 287–297. doi:10.1017/S0263574799002258

Banerjee, A., Chakraborty, C., Kumar, A., & Biswas, D. (2020). Emerging trends in IoT and big data analytics for biomedical and health care technologies. Handbook of data science approaches for biomedical engineering, 121-152.

Barbosa, H. C., de Queiroz Oliveira, J. A., da Costa, J. M., de Melo Santos, R. P., Miranda, L. G., de Carvalho Torres, H., ... Martins, M. A. P. (2021). Empowerment-oriented strategies to identify behavior change in patients with chronic diseases: An integrative review of the literature. *Patient Education and Counseling*, 104(4), 689–702. doi:10.1016/j.pec.2021.01.011 PMID:33478854

Behera, S. K., Rath, A. K., & Sethy, P. K. (2021). Maturity status classification of papaya fruits based on machine learning and transfer learning approach. *Information Processing in Agriculture*, 8(2), 244–250. doi:10.1016/j.inpa.2020.05.003

Bekey, G. A. (2005). *Autonomous robots: from biological inspiration to implementation and control*. MIT Press.

Belanche, D., Casaló, L. V., & Flavián, C. (2020). Artificial intelligence in FinTech: Understanding robo-advisors adoption among customers. *Industrial Management & Data Systems*, *119*(7), 1411–1430. doi:10.1108/IMDS-08-2018-0368

Bengio, Y. (2012, June). Deep learning of representations for unsupervised and transfer learning. In *Proceedings of ICML workshop on unsupervised and transfer learning* (pp. 17-36). JMLR Workshop and Conference Proceedings.

Bentéjac, C., Csörgő, A., & Martínez-Muñoz, G. (2021). A comparative analysis of gradient boosting algorithms. *Artificial Intelligence Review*, *54*(3), 1937–1967. doi:10.1007/s10462-020-09896-5

Bhandari, G., Dhasmana, A., Chaudhary, P., Gupta, S., Gangola, S., Gupta, A., Rustagi, S., Shende, S. S., Rajput, V. D., Minkina, T. M., Malik, S., & Sláma, P. (2023). A Perspective Review on Green Nanotechnology in Agro-Ecosystems: Opportunities for Sustainable Agricultural Practices & Environmental Remediation. *Agriculture*, *13*(3), 668. doi:10.3390/agriculture13030668

Bhargava, A., & Bansal, A. (2020). Automatic detection and grading of multiple fruits by machine learning. *Food Analytical Methods*, *13*(3), 751–761. doi:10.1007/s12161-019-01690-6

Biffi, L. J., Mitishita, E., Liesenberg, V., Santos, A. A. D., Goncalves, D. N., Estrabis, N. V., Silva, J. A., Osco, L. P., Ramos, A. P. M., Centeno, J. A. S., Schimalski, M. B., Rufato, L., Neto, S. L. R., Marcato Junior, J., & Goncalves, W. N. (2021). ATSS deep learning-based approach to detect apple fruits. *Remote Sensing (Basel)*, *13*(1), 54. doi:10.3390/rs13010054

Biswas, S., Carson, B., Chung, V., Singh, S., & Thomas, R. (2023). Artificial intelligence technologies are increasingly integral to the world we live in, and banks need to deploy these technologies at scale to remain relevant. Success requires a holistic transformation spanning multiple layers of the organization.Larson, E.J. 2021. The myth of artificial intelligence. In *The Myth of Artificial Intelligence*. Harvard University Press.

Bogossian, T. (2022). The Use of Robotics in Healthcare. *Journal of Medical & Clinical Nursing*, 1–4. Advance online publication. doi:10.47363/JMCN/2022(3)157

Bohr, A., & Memarzadeh, K. (2020). The rise of artificial intelligence in healthcare applications. In *Artificial Intelligence in healthcare* (pp. 25–60). Academic Press. doi:10.1016/B978-0-12-818438-7.00002-2

Boopathi, S. (2021). *Pollution monitoring and notification: Water pollution monitoring and notification using intelligent RC boat*. Academic Press.

Boopathi, S. (2022). Performance Improvement of Eco-Friendly Near-Dry wire-Cut Electrical Discharge Machining Process Using Coconut Oil-Mist Dielectric Fluid. *World Scientific: Journal of Advanced Manufacturing Systems*.

Boopathi, S. (2022b). Cryogenically treated and untreated stainless steel grade 317 in sustainable wire electrical discharge machining process: A comparative study. *Springer :Environmental Science and Pollution Research*, 1–10.

Boopathi, S., & Kumar, P. (2024). Advanced bioprinting processes using additive manufacturing technologies: Revolutionizing tissue engineering. *3D Printing Technologies: Digital Manufacturing, Artificial Intelligence, Industry 4.0*, 95.

Boopathi, S., Khare, R., KG, J. C., Muni, T. V., & Khare, S. (2023). Additive Manufacturing Developments in the Medical Engineering Field. In Development, Properties, and Industrial Applications of 3D Printed Polymer Composites (pp. 86–106). IGI Global.

Boopathi, S., Thillaivanan, A., Mohammed, A. A., Shanmugam, P., & VR, P. (2022). Experimental investigation on Abrasive Water Jet Machining of Neem Wood Plastic Composite. *IOP: Functional Composites and Structures, 4*, 025001.

Boopathi, S. (2022a). An investigation on gas emission concentration and relative emission rate of the near-dry wire-cut electrical discharge machining process. *Environmental Science and Pollution Research International, 29*(57), 86237–86246. doi:10.1007/s11356-021-17658-1 PMID:34837614

Boopathi, S. (2023a). An Investigation on Friction Stir Processing of Aluminum Alloy-Boron Carbide Surface Composite. In *Springer:Advances in Processing of Lightweight Metal Alloys and Composites* (pp. 249–257). Springer. doi:10.1007/978-981-19-7146-4_14

Boopathi, S. (2023b). Internet of Things-Integrated Remote Patient Monitoring System: Healthcare Application. In *Dynamics of Swarm Intelligence Health Analysis for the Next Generation* (pp. 137–161). IGI Global. doi:10.4018/978-1-6684-6894-4.ch008

Boopathi, S., Alqahtani, A. S., Mubarakali, A., & Panchatcharam, P. (2023). Sustainable developments in near-dry electrical discharge machining process using sunflower oil-mist dielectric fluid. *Environmental Science and Pollution Research International*, 1–20. doi:10.1007/s11356-023-27494-0 PMID:37199846

Boopathi, S., & Davim, J. P. (2023a). Applications of Nanoparticles in Various Manufacturing Processes. In *Sustainable Utilization of Nanoparticles and Nanofluids in Engineering Applications* (pp. 1–31). IGI Global. doi:10.4018/978-1-6684-9135-5.ch001

Boopathi, S., & Davim, J. P. (2023b). *Sustainable Utilization of Nanoparticles and Nanofluids in Engineering Applications*. IGI Global. doi:10.4018/978-1-6684-9135-5

Boopathi, S., Jeyakumar, M., Singh, G. R., King, F. L., Pandian, M., Subbiah, R., & Haribalaji, V. (2022). An experimental study on friction stir processing of aluminium alloy (AA-2024) and boron nitride (BNp) surface composite. *Materials Today: Proceedings, 59*(1), 1094–1099. doi:10.1016/j.matpr.2022.02.435

Boopathi, S., & Khang, A. (2023). AI-Integrated Technology for a Secure and Ethical Healthcare Ecosystem. In *AI and IoT-Based Technologies for Precision Medicine* (pp. 36–59). IGI Global. doi:10.4018/979-8-3693-0876-9.ch003

Boopathi, S., Umareddy, M., & Elangovan, M. (2023). Applications of Nano-Cutting Fluids in Advanced Machining Processes. In *Sustainable Utilization of Nanoparticles and Nanofluids in Engineering Applications* (pp. 211–234). IGI Global. doi:10.4018/978-1-6684-9135-5.ch009

Bragança, S., Costa, E., Castellucci, I., & Arezes, P. M. (2019). A brief overview of the use of collaborative robots in industry 4.0: Human role and safety. *Occupational and Environmental Safety and Health*, 641–650.

Bramblett, L., Gao, S., & Bezzo, N. (2023). Epistemic Prediction and Planning with Implicit Coordination for Multi-Robot Teams in Communication Restricted Environments. *arXiv preprint arXiv:2302.10393*. doi:10.1109/ICRA48891.2023.10161553

Burgos-Artizzu, X. P., Coronado-Gutierrez, D., Valenzuela-Alcaraz, B., Bonet-Carne, E., Eixarch, E., Crispi, F., & Gratacós, E. (2020). FETAL_PLANES_DB: Common maternal-fetal ultrasound images. *Scientific Reports, 19*, 10200. doi:10.1038/s41598-020-67076-5 PMID:32576905

Çetin, N., Karaman, K., Kavuncuoğlu, E., Yıldırım, B., & Jahanbakhshi, A. (2022). Using hyperspectral imaging technology and machine learning algorithms for assessing internal quality parameters of apple fruits. *Chemometrics and Intelligent Laboratory Systems, 230*, 104650. doi:10.1016/j.chemolab.2022.104650

Chaddad, A., Peng, J., Xu, J., & Bouridane, A. (2023). Survey of Explainable AI Techniques in Healthcare. *Sensors (Basel), 23*(2), 1–19. doi:10.3390/s23020634 PMID:36679430

Chakraborty, A., Ravi, S. P., Shamiya, Y., Cui, C., & Paul, A. (2021). Harnessing the physicochemical properties of DNA as a multifunctional biomaterial for biomedical and other applications. *Chemical Society Reviews, 50*(13), 7779–7819. doi:10.1039/D0CS01387K PMID:34036968

Chan, C.-K., & Cheng, L.-M. (2004). Hiding data in images by simple LSB substitution. *Pattern Recognition, 37*(3), 469–474. doi:10.1016/j.patcog.2003.08.007

Chelliah, R., Wei, S., Daliri, E. B. M., Rubab, M., Elahi, F., Yeon, S. J., Jo, K. H., Yan, P., Liu, S., & Oh, D. H. (2021). Development of Nanosensors Based Intelligent Packaging Systems: Food Quality and Medicine. *Nanomaterials (Basel, Switzerland), 11*(6), 1515. doi:10.3390/nano11061515 PMID:34201071

Chen, H., Dou, Q., Wang, X., Qin, J., & Heng, P. (2016, February). Mitosis detection in breast cancer histology images via deep cascaded networks. *Proceedings of the AAAI Conference on Artificial Intelligence, 30*(1). doi:10.1609/aaai. v30i1.10140

Chen, H., Dou, Q., Yu, L., Qin, J., & Heng, P. A. (2018). VoxResNet: Deep voxelwise residual networks for brain segmentation from 3D MR images. *NeuroImage, 170*, 446–455. doi:10.1016/j.neuroimage.2017.04.041 PMID:28445774

Chen, H., Ni, D., Qin, J., Li, S., Yang, X., Wang, T., & Heng, P. A. (2015). Standard plane localization in fetal ultrasound via domain transferred deep neural networks. *IEEE Journal of Biomedical and Health Informatics, 19*(5), 1627–1636. doi:10.1109/JBHI.2015.2425041 PMID:25910262

Chen, H., Qi, X., Yu, L., Dou, Q., Qin, J., & Heng, P. A. (2017). DCAN: Deep contour-aware networks for object instance segmentation from histology images. *Medical Image Analysis, 36*, 135–146. doi:10.1016/j.media.2016.11.004 PMID:27898306

Chen, P.-Y., & Lin, H.-J. (2006). A DWT based approach for image steganography. *International Journal of Applied Science and Engineering, 4*(3), 275–290.

Chen, R., Xu, C., Dong, Z., Liu, Y., & Du, X. (2020). DeepCQ: Deep multi-task conditional quantification network for estimation of left ventricle parameters. *Computer Methods and Programs in Biomedicine, 184*, 105288. doi:10.1016/j. cmpb.2019.105288 PMID:31901611

Chen, Y., Chen, D., Liang, S., Dai, Y., Bai, X., Song, B., Zhang, D., Chen, H., & Feng, L. (2022). Recent Advances in Field-Controlled Micro–Nano Manipulations and Micro–Nano Robots. *Advanced Intelligent Systems, 4*(3), 2100116. doi:10.1002/aisy.202100116

Chowdhury, A. H., Malakar, S., Seal, D. B., & Goswami, S. (2022). Understanding employee attrition using machine learning techniques. In *Data Management, Analytics and Innovation: Proceedings of ICDMAI 2021*, Volume 2 (pp. 101-109). Springer Singapore. 10.1007/978-981-16-2937-2_8

Chung, Y. M., Youssef, H., & Roidl, M. (2022, May). Distributed Timed Elastic Band (DTEB) Planner: Trajectory Sharing and Collision Prediction for Multi-Robot Systems. In *2022 International Conference on Robotics and Automation (ICRA)* (pp. 10702-10708). IEEE. 10.1109/ICRA46639.2022.9811762

Cianchetti, M., Ranzani, T., Gerboni, G., Nanayakkara, T., Althoefer, K., & Dasgupta, P. (2018). Soft robotics technologies to address shortcomings in today's minimally invasive surgery: The STIFF-FLOP approach. *Soft Robotics, 5*(2), 149–161. PMID:29297756

Connolly, F., Walsh, C. J., & Bertoldi, K. (2017). Automatic design of fiber-reinforced soft actuators for trajectory matching. *Proceedings of the National Academy of Sciences of the United States of America, 114*(1), 51–56. doi:10.1073/ pnas.1615140114 PMID:27994133

Constant, A., Conserve, D. F., Gallopel-Morvan, K., & Raude, J. (2020). Socio-cognitive factors associated with lifestyle changes in response to the COVID-19 epidemic in the general population: Results from a cross-sectional study in France. *Frontiers in Psychology*, *11*, 579460. doi:10.3389/fpsyg.2020.579460 PMID:33132989

Cristian, Ursache, Popa, & Pop. (2016). *Energy efficiency and robustness for IoT: Building a smart home security system.* Faculty of Automatic Control and Computers University Politehnica of Bucharest.

Csurka, G., Dance, C., Fan, L., Willamowski, J., & Bray, C. (2004, May). Visual categorization with bags of keypoints. In *Workshop on statistical learning in computer vision, ECCV* (Vol. 1, No. 1-22, pp. 1-2).

Cui, Y., Lin, L., Huang, X., Zhang, D., Wang, Y., Jing, W., ... Wang, Y. (2022, May). Learning Observation-Based Certifiable Safe Policy for Decentralized Multi-Robot Navigation. In *2022 International Conference on Robotics and Automation (ICRA)* (pp. 5518-5524). IEEE. 10.1109/ICRA46639.2022.9811950

Cui, Y., Qin, Z., Wu, H., Li, M., & Hu, Y. (2021). Flexible thermal interface based on self-assembled boron arsenide for high-performance thermal management. *Nature Communications*, *12*(1), 1284. doi:10.1038/s41467-021-21531-7 PMID:33627644

Dalal, N., & Triggs, B. (2005, June). Histograms of oriented gradients for human detection. In 2005 IEEE computer society conference on computer vision and pattern recognition (CVPR'05) (Vol. 1, pp. 886-893). IEEE. doi:10.1109/CVPR.2005.177

Dash, T., & Maiti, C. K. (2023). An overview of nanoscale device fabrication technology—Part I. *Nanoelectronics: Physics, Materials and Devices*, 193–214.

Daudi, J. (2015). An overview of application of artificial immune system in swarm robotic systems. *Advances in Robotics & Automation, 4*(1).

Deimel, R., & Brock, O. (2016). A novel type of compliant and underactuated robotic hand for dexterous grasping. *The International Journal of Robotics Research*, *35*(1-3), 161–185. doi:10.1177/0278364915592961

Dency Flora, G., Sekar, G., Nivetha, R., Thirukkumaran, R., Silambarasan, D., & Jeevanantham, V. (2022). An Optimized Neural Network for Content Based Image Retrieval in Medical Applications. *8th International Conference on Advanced Computing and Communication Systems, ICACCS 2022*, 1560-1563. 10.1109/ICACCS54159.2022.9785151

Deng, J. (2009). A large-scale hierarchical image database. *Proc. of IEEE Computer Vision and Pattern Recognition.*

Dhanya, D., Kumar, S. S., Thilagavathy, A., Prasad, D., & Boopathi, S. (2023). Data Analytics and Artificial Intelligence in the Circular Economy: Case Studies. In Intelligent Engineering Applications and Applied Sciences for Sustainability (pp. 40–58). IGI Global.

Donahue, J., Jia, Y., Vinyals, O., Hoffman, J., Zhang, N., Tzeng, E., & Darrell, T. (2014, January). Decaf: A deep convolutional activation feature for generic visual recognition. In *International conference on machine learning* (pp. 647-655). PMLR.

Dongari, S., Nisarudeen, M., Devi, J., Irfan, S., Parida, P. K., & Bajpai, A. (2023). Advancing Healthcare through Artificial Intelligence: Innovations at the Intersection of AI and Medicine. *Tuijin Jishu/Journal of Propulsion Technology, 44*(2).

Dou, Q., Chen, H., Yu, L., Zhao, L., Qin, J., Wang, D., Mok, V. C. T., Shi, L., & Heng, P. A. (2016). Automatic detection of cerebral microbleeds from MR images via 3D convolutional neural networks. *IEEE Transactions on Medical Imaging*, *35*(5), 1182–1195. doi:10.1109/TMI.2016.2528129 PMID:26886975

Du, C., Ren, Y., Qu, Z., Gao, L., Zhai, Y., Han, S.-T., & Zhou, Y. (2021). Synaptic transistors and neuromorphic systems based on carbon nano-materials. *Nanoscale*, *13*(16), 7498–7522. doi:10.1039/D1NR00148E PMID:33928966

Dudley, N. J., & Chapman, E. (2002). The importance of quality management in fetal measurement. *Ultrasound in Obstetrics & Gynecology, 19*(2), 190–196. doi:10.1046/j.0960-7692.2001.00549.x PMID:11876814

Dzobo, K., Adotey, S., Thomford, N. E., & Dzobo, W. (2020). Integrating artificial and human intelligence: A partnership for responsible innovation in biomedical engineering and medicine. *OMICS: A Journal of Integrative Biology, 24*(5), 247–263. doi:10.1089/omi.2019.0038 PMID:31313972

Eckstein, M. K., & Collins, A. G. (2020). Computational evidence for hierarchically structured reinforcement learning in humans. *Proceedings of the National Academy of Sciences of the United States of America, 117*(47), 29381–29389. doi:10.1073/pnas.1912330117 PMID:33229518

Ekinci, Y., Uray, N., & Ülengin, F. (2014). A customer lifetime value model for the banking industry: A guide to marketing actions. *European Journal of Marketing, 48*(3–4), 761–784. doi:10.1108/EJM-12-2011-0714

Enoch Sam, M., Misra, S. K., Anitha Mary, X., Karthik, C., & Chowdhury, S. (2023, November). Review of different types of spatial positioning platforms. In AIP Conference Proceedings (Vol. 2878, No. 1). AIP Publishing. doi:10.1063/5.0171256

Evangeline, C. S., Sarah, M., Lenin, A., Reddy, J. H. V., Mary, X. A., & Karthiga, M. (2023, May). Design of On-Board Unit for Vehicular Applications. In *2023 2nd International Conference on Vision Towards Emerging Trends in Communication and Networking Technologies (ViTECoN)* (pp. 1-6). IEEE. 10.1109/ViTECoN58111.2023.10157654

Evelyn Brindha, V., & Anitha Mary, X. (2023). Analysing Control Algorithms for Controlling the Speed of BLDC Motors Using Green IoT. *Power Converters, Drives and Controls for Sustainable Operations*, 779-788.

Fallucchi, F., Coladangelo, M., Giuliano, R., & William De Luca, E. (2020). Predicting employee attrition using machine learning techniques. *Computers, 9*(4), 86. doi:10.3390/computers9040086

Fan, G., Zhou, Z., Zhang, H., Gu, X., Gu, G., Guan, X., Fan, Y., & He, S. (2016, June). Global scientific production of robotic surgery in medicine: A 20-year survey of research activities. *International Journal of Surgery, 30*, 126–131. doi:10.1016/j.ijsu.2016.04.048 PMID:27154617

Faridha Banu, D., Sindhwani, N. S., G, K. R. A, & M, S. (2022). Fuzzy acceptance Analysis of Impact of Glaucoma and Diabetic Retinopathy using Confusion Matrix. *2022 10th International Conference on Reliability, Infocom Technologies and Optimization (Trends and Future Directions) (ICRITO)*, 1-5. 10.1109/ICRITO56286.2022.9964858

Ferranti, L., Lyons, L., Negenborn, R. R., Keviczky, T., & Alonso-Mora, J. (2022). Distributed nonlinear trajectory optimization for multi-robot motion planning. *IEEE Transactions on Control Systems Technology, 31*(2), 809–824. doi:10.1109/TCST.2022.3211130

Firoozi, R., Ferranti, L., Zhang, X., Nejadnik, S., & Borrelli, F. (2020). A distributed multi-robot coordination algorithm for navigation in tight environments. *arXiv preprint arXiv:2006.11492.*

Firouz, M. S., Farahmandi, A., & Hosseinpour, S. (2019). Recent advances in ultrasound application as a novel technique in analysis, processing and quality control of fruits, juices and dairy products industries: A review. *Ultrasonics Sonochemistry, 57*, 73–88. doi:10.1016/j.ultsonch.2019.05.014 PMID:31208621

Folke, T., Yang, S. C., Anderson, S., & Shafto, P. (2021). Explainable AI for medical imaging: Explaining pneumothorax diagnoses with Bayesian teaching. CoRR, vol. abs/2106.04684.

Fowziya, S., Sivaranjani, S., Devi, N. L., Boopathi, S., Thakur, S., & Sailaja, J. M. (2023). Influences of nano-green lubricants in the friction-stir process of TiAlN coated alloys. *Materials Today: Proceedings*. Advance online publication. doi:10.1016/j.matpr.2023.06.446

Freitas, E. J., Vangasse, A. D. C., Raffo, G. V., & Pimenta, L. C. (2023, October). Decentralized Multi-robot Collision-free Path Following Based on Time-varying Artificial Vector Fields and MPC-ORCA. In *2023 Latin American Robotics Symposium (LARS), 2023 Brazilian Symposium on Robotics (SBR), and 2023 Workshop on Robotics in Education (WRE)* (pp. 212-217). IEEE. 10.1109/LARS/SBR/WRE59448.2023.10333004

Friedman, V. (2008). Data visualization and infographics. Graphics, Monday Inspiration.

Friendly, M. (2017). *A brief history of data visualization.* Springer-Verlag.

Ganji, K., & Parimi, S. (2022). ANN model for users' perception on IOT based smart healthcare monitoring devices and its impact with the effect of COVID 19. *Journal of Science and Technology Policy Management, 13*(1), 6–21. doi:10.1108/JSTPM-09-2020-0128

Gao, Z., Wang, L., Zhou, L., & Zhang, J. (2016). HEp-2 cell image classification with deep convolutional neural networks. *IEEE Journal of Biomedical and Health Informatics, 21*(2), 416–428. doi:10.1109/JBHI.2016.2526603 PMID:26887016

Gaurav, K., & Ghanekar, U. (2018). Image steganography based on Canny edge detection, dilation operator and hybrid coding. *Journal of Information Security and Applications, 41*, 41–51. doi:10.1016/j.jisa.2018.05.001

Gellers, J. C. (2020). *Rights for robots: Artificial intelligence, animal and environmental law* (1st ed.). Routledge. doi:10.4324/9780429288159

Ghanbarzadeh-Dagheyan, A., Jalili, N., & Ahmadian, M. T. (2021). A holistic survey on mechatronic Systems in Micro/Nano scale with challenges and applications. *Journal of Micro-Bio Robotics, 17*(1), 1–22. doi:10.1007/s12213-021-00145-8

Gheorghe, P. E. E. G. I., Ilie, P. S. I., Istriteanu, P. E. S., & Bajenaru, P. E. V. (2014). Research in micro-nano-robotics. *The Romanian Review Precision Mechanics, 46*, 83.

Gilbert, S., Fenech, M., Hirsch, M., Upadhyay, S., Biasiucci, A., & Starlinger, J. (2021). Algorithm change protocols in the regulation of adaptive machine learning–based medical devices. *Journal of Medical Internet Research, 23*(10), e30545. doi:10.2196/30545 PMID:34697010

Girshick, R. (2015). Fast r-cnn. In *Proceedings of the IEEE international conference on computer vision* (pp. 1440-1448). Academic Press.

Girshick, R., Donahue, J., Darrell, T., & Malik, J. (2014). Rich feature hierarchies for accurate object detection and semantic segmentation. In *Proceedings of the IEEE conference on computer vision and pattern recognition* (pp. 580-587). 10.1109/CVPR.2014.81

Glorot, X., & Bengio, Y. (2010, March). Understanding the difficulty of training deep feedforward neural networks. In *Proceedings of the thirteenth international conference on artificial intelligence and statistics* (pp. 249-256). JMLR Workshop and Conference Proceedings.

Goodfellow, I., Warde-Farley, D., Mirza, M., Courville, A., & Bengio, Y. (2013, May). Maxout networks. In *International conference on machine learning* (pp. 1319-1327). PMLR.

Gotovtsev, P. M. (2023). Microbial Cells as a Microrobots: From Drug Delivery to Advanced Biosensors. *Biomimetics, 8*(1), 8. doi:10.3390/biomimetics8010109 PMID:36975339

Gowri, N. V., Dwivedi, J. N., Krishnaveni, K., Boopathi, S., Palaniappan, M., & Medikondu, N. R. (2023). Experimental investigation and multi-objective optimization of eco-friendly near-dry electrical discharge machining of shape memory alloy using Cu/SiC/Gr composite electrode. *Environmental Science and Pollution Research International, 30*(49), 1–19. doi:10.1007/s11356-023-26983-6 PMID:37126160

Grauman, K., & Darrell, T. (2005, October). The pyramid match kernel: Discriminative classification with sets of image features. In *Tenth IEEE International Conference on Computer Vision (ICCV'05)* Volume 1 (Vol. 2, pp. 1458-1465). IEEE. 10.1109/ICCV.2005.239

Greenspan, H., Van Ginneken, B., & Summers, R. M. (2016). Guest editorial deep learning in medical imaging: Overview and future promise of an exciting new technique. *IEEE Transactions on Medical Imaging, 35*(5), 1153–1159. doi:10.1109/TMI.2016.2553401

Gulli, A., & Pal, S. (2017). *Deep learning with Keras*. Packt Publishing Ltd.

Guo, W. (2020). Explainable artificial intelligence for 6G: Improving trust between human and machine. *IEEE Communications Magazine, 58*(6), 39–45. doi:10.1109/MCOM.001.2000050

Guo, Y., Jing, D., Liu, S., & Yuan, Q. (2023). Construction of intelligent moving micro/nanomotors and their applications in biosensing and disease treatment. *Theranostics, 13*(9), 2993–3020. doi:10.7150/thno.81845 PMID:37284438

Gupta, R., Shukla, A., & Tanwar, S. (2020). Aayush: A smart contract-based telesurgery system for healthcare 4.0. *2020 IEEE International Conference on Communications Workshops (ICC Workshops),* 1–6. 10.1109/ICCWorkshops49005.2020.9145044

Halder, A., & Sun, Y. (2019). Biocompatible propulsion for biomedical micro/nano robotics. *Biosensors & Bioelectronics, 139,* 111334. doi:10.1016/j.bios.2019.111334 PMID:31128479

Hamza, R., & Chtourou, M. (2018, July). Apple ripeness estimation using artificial neural network. In *2018 International Conference on High Performance Computing & Simulation (HPCS)* (pp. 229-234). IEEE. 10.1109/HPCS.2018.00049

Han, H., & Liu, X. (2022, January). The challenges of explainable ai in biomedical data science. *BMC Bioinformatics, 22*(12), 443. PMID:35057748

Hanumanthakari, S., Gift, M. M., Kanimozhi, K., Bhavani, M. D., Bamane, K. D., & Boopathi, S. (2023). Biomining Method to Extract Metal Components Using Computer-Printed Circuit Board E-Waste. In *Handbook of Research on Safe Disposal Methods of Municipal Solid Wastes for a Sustainable Environment* (pp. 123–141). IGI Global. doi:10.4018/978-1-6684-8117-2.ch010

Hao, J., Li, M., Chen, W., Yu, L., & Ye, L. (2022). Experimental research on space distribution of reverse flow U-tubes in steam generator primary side. *Nuclear Engineering and Design, 388,* 111650. doi:10.1016/j.nucengdes.2022.111650

Harikaran, M., Boopathi, S., Gokulakannan, S., & Poonguzhali, M. (2023). Study on the Source of E-Waste Management and Disposal Methods. In *Sustainable Approaches and Strategies for E-Waste Management and Utilization* (pp. 39–60). IGI Global. doi:10.4018/978-1-6684-7573-7.ch003

Harris, M., Wu, H., Zhang, W., & Angelopoulou, A. (2022). Overview of recent trends in microchannels for heat transfer and thermal management applications. *Chemical Engineering and Processing, 181,* 109155. doi:10.1016/j.cep.2022.109155

Hassani, S. S., Daraee, M., & Sobat, Z. (2020). Advanced development in upstream of petroleum industry using nanotechnology. *Chinese Journal of Chemical Engineering, 28*(6), 1483–1491. doi:10.1016/j.cjche.2020.02.030

Hatton-Jones, K. M., Christie, C., Griffith, T. A., Smith, A. G., Naghipour, S., Robertson, K., Russell, J. S., Peart, J. N., Headrick, J. P., Cox, A. J., & du Toit, E. F. (2021). A YOLO based software for automated detection and analysis of rodent behaviour in the open field arena. *Computers in Biology and Medicine, 134,* 104474. doi:10.1016/j.compbiomed.2021.104474 PMID:34058512

He, K., Gkioxari, G., Dollár, P., & Girshick, R. (2017). Mask r-cnn. In *Proceedings of the IEEE international conference on computer vision* (pp. 2961-2969). Academic Press.

He, K., Zhang, X., Ren, S., & Sun, J. (2016). Deep residual learning for image recognition. In *Proceedings of the IEEE conference on computer vision and pattern recognition* (pp. 770-778). Academic Press.

He, K., Zhang, X., Ren, S., & Sun, J. (2015). Delving deep into rectifiers: Surpassing human-level performance on imagenet classification. In *Proceedings of the IEEE international conference on computer vision* (pp. 1026-1034). 10.1109/ICCV.2015.123

He, K., Zhang, X., Ren, S., & Sun, J. (2016). Deep residual learning for image recognition. In *Proceedings of the IEEE Conference on Computer Vision and Pattern Recognition* (pp. 770–778). IEEE.

He, L., Xu, J., Dekai, Z., Qinghai, Y., & Longqiu, L. (2018). Potential application of functional micro-nano structures in petroleum. *Petroleum Exploration and Development*, *45*(4), 745–753. doi:10.1016/S1876-3804(18)30077-6

Hemachandran, K. (2016). Study of Image Steganography using LSB, DFT and DWT. *International Journal of Computers and Technology*, *11*, 2618–2627.

Ho, J. C. (2022). Robot Assisted Neurosurgery for High-Accuracy, Minimally-Invasive Deep Brain Electrophysiology in Monkeys. *2022 44th Annual International Conference of the IEEE Engineering in Medicine & Biology Society (EMBC)*, 3115-3118. 10.1109/EMBC48229.2022.9871520

Huang, Z., Wang, J., Fu, X., Yu, T., Guo, Y., & Wang, R. (2020). DC-SPP-YOLO: Dense connection and spatial pyramid pooling based YOLO for object detection. *Information Sciences*, *522*, 241–258. doi:10.1016/j.ins.2020.02.067

Hughes, Zhu, & Bednarz. (2021). Generative Adversarial Networks-Enabled Human-Artificial Intelligence Collaborative Applications for Creative and Design Industries: A Systematic Review of Current Approaches and Trends. *Frontiers in Artificial Intelligence*, *4*, 1-17.

Hussain, Z., Babe, M., Saravanan, S., Srimathy, G., Roopa, H., & Boopathi, S. (2023). Optimizing Biomass-to-Biofuel Conversion: IoT and AI Integration for Enhanced Efficiency and Sustainability. In Circular Economy Implementation for Sustainability in the Built Environment (pp. 191–214). IGI Global.

Hussain, A., Pu, H., & Sun, D. W. (2019). Measurements of lycopene contents in fruit: A review of recent developments in conventional and novel techniques. *Critical Reviews in Food Science and Nutrition*, *59*(5), 758–769. doi:10.1080/10408398.2018.1518896 PMID:30582342

Huth, C. (2015). Securing systems on the Internet of Things via physical properties of devices and communications. In Systems Conference (SysCon), 9th Annual IEEE International. IEEE.

Hutter, F., Hoos, H. H., & Leyton-Brown, K. (2011). Sequential model-based optimization for general algorithm configuration. In *Learning and Intelligent Optimization: 5th International Conference, LION 5, Rome, Italy, January 17-21, 2011. Selected Papers 5* (pp. 507-523). Springer Berlin Heidelberg. 10.1007/978-3-642-25566-3_40

Hu, Y. (2021). Self-Assembly of DNA Molecules: Towards DNA Nanorobots for Biomedical Applications. *Cyborg and Bionic Systems (Washington, D.C.)*, *2021*, 2021. doi:10.34133/2021/9807520 PMID:36285141

Ince, H., & Aktan, B. (2009). A comparison of data mining techniques for credit scoring in banking: A managerial perspective. *Journal of Business Economics and Management*, *10*(3), 233–240. doi:10.3846/1611-1699.2009.10.233-240

Indiveri, G., Linares-Barranco, B., Legenstein, R., Deligeorgis, G., & Prodromakis, T. (2013). Integration of nanoscale memristor synapses in neuromorphic computing architectures. *Nanotechnology*, *24*(38), 384010. doi:10.1088/0957-4484/24/38/384010 PMID:23999381

Intel. (n.d.). https://www.intel.com/content/www/us/en/developer/tools/oneapi/overview.html

Ioffe, S., & Szegedy, C. (2015, June). Batch normalization: Accelerating deep network training by reducing internal covariate shift. In *International conference on machine learning* (pp. 448-456). PMLR.

Islam, A. M. R., Ahmed, M. U., Barua, S., & Begum, S. (2022). A systematic review of explainable artificial intelligence in terms of different application domains and tasks. *Applied Sciences (Basel, Switzerland)*, *12*(3), 1353. doi:10.3390/app12031353

Jahromi, L. P., Shahbazi, M., Maleki, A., Azadi, A., & Santos, H. A. (2021). Chemically Engineered Immune Cell-Derived Microrobots and Biomimetic Nanoparticles: Emerging Biodiagnostic and Therapeutic Tools. *Advancement of Science*, *8*(8), 8. doi:10.1002/advs.202002499 PMID:33898169

Jakkula, R. V. S. K., & Sethuramalingam, P. (2023). Analysis of coatings based on carbon-based nanomaterials for paint industries-A review. *Australian Journal of Mechanical Engineering*, *21*(3), 1008–1036. doi:10.1080/14484846.2021.1938953

Jarrahi, M. H. (2018). Artificial intelligence and the future of work: Human-AI symbiosis in organizational decision making. *Business Horizons*, *61*(4), 577–586. doi:10.1016/j.bushor.2018.03.007

Javaid, M., Haleem, A., Singh, R. P., & Suman, R. (2021). Substantial capabilities of robotics in enhancing industry 4.0 implementation. *Cognitive Robotics*, *1*, 58–75. doi:10.1016/j.cogr.2021.06.001

Jégou, H., Perronnin, F., Douze, M., Sánchez, J., Pérez, P., & Schmid, C. (2011). Aggregating local image descriptors into compact codes. *IEEE Transactions on Pattern Analysis and Machine Intelligence*, *34*(9), 1704–1716. doi:10.1109/TPAMI.2011.235 PMID:22156101

Jian, Z., Chao, Z., Shunli, Z., Tingting, L., Weiwen, S., & Jian, J. (2018). Pre-detection and dual-dictionary sparse representation based face recognition algorithm in non-sufficient training samples. *Journal of Systems Engineering and Electronics*, *29*(1), 196–202. doi:10.21629/JSEE.2018.01.20

Jintasuttisak, T., Leonce, A., Sher Shah, M., Khafaga, T., Simkins, G., & Edirisinghe, E. (2022, March). Deep learning based animal detection and tracking in drone video footage. In *Proceedings of the 8th International Conference on Computing and Artificial Intelligence* (pp. 425-431). 10.1145/3532213.3532280

Ji, Q., Fu, S., Tan, K., Muralidharan, S. T., Lagrelius, K., Danelia, D., Andrikopoulos, G., Wang, X. V., Wang, L., & Feng, L. (2022). Synthesizing the optimal gait of a quadruped robot with soft actuators using deep reinforcement learning. *Robotics and Computer-integrated Manufacturing*, *78*, 102382. doi:10.1016/j.rcim.2022.102382

Joe, S., Bliah, O., Magdassi, S., & Beccai, L. (2023). Jointless Bioinspired Soft Robotics by Harnessing Micro and Macroporosity. *Advancement of Science*, *10*(23), 2302080. doi:10.1002/advs.202302080 PMID:37323121

Johnson, N. E., Ianiuk, O., Cazap, D., Liu, L., Starobin, D., Dobler, G., & Ghandehari, M. (2017). Patterns of waste generation: A gradient boosting model for short-term waste prediction in New York City. *Waste Management (New York, N.Y.)*, *62*, 3–11. doi:10.1016/j.wasman.2017.01.037 PMID:28216080

Karan, S., & Majumder, D. D. (2011). Molecular machinery-a nanorobotics control system design for cancer drug delivery. *2011 International Conference on Recent Trends in Information Systems*, 197–202. 10.1109/ReTIS.2011.6146867

Karimov, K., Akhmedov, A., & Adilova, S. (2022). Theoretical and engineering solutions of the controlled vibration mechanisms for precision engineering. *AIP Conference Proceedings*, *2637*(1), 060001. doi:10.1063/5.0118863

Karthik, S., Hemalatha, R., Aruna, R., Deivakani, M., Reddy, R. V. K., & Boopathi, S. (2023). Study on Healthcare Security System-Integrated Internet of Things (IoT). In Perspectives and Considerations on the Evolution of Smart Systems (pp. 342–362). IGI Global.

Karthik, K., Teferi, A. B., Sathish, R., Gandhi, A. M., Padhi, S., Boopathi, S., & Sasikala, G. (2023). Analysis of de-lamination and its effect on polymer matrix composites. *Materials Today: Proceedings.* Advance online publication. doi:10.1016/j.matpr.2023.07.199

Kavitha, K.K., Koshti, A., & Dunghav, P. (2012). Steganography using least significant bit algorithm. *International Journal of Engineering Research and Applications.*

Khamrui, A., & Mandal, J. K. (2013). A genetic algorithm-based steganography using discrete cosine transformation (GASDCT). *Procedia Technology, 10,* 105–111. doi:10.1016/j.protcy.2013.12.342

Khandani, A. E., Kim, A. J., & Lo, A. W. (2010). Consumer credit-risk models via machine-learning algorithms. *Journal of Banking & Finance, 34*(11), 2767–27. doi:10.1016/j.jbankfin.2010.06.001

Khan, K. S., Kunz, R., Kleijnen, J., & Antes, G. (2010). Five steps to conducting a systematic review. *Journal of the Royal Society of Medicine, 96*(3), 118–121. doi:10.1177/014107680309600304 PMID:12612111

Khoramshahi, M., & Billard, A. (2019). A dynamical system approach to task-adaptation in physical human–robot interaction. *Autonomous Robots, 43*(4), 927–946. doi:10.1007/s10514-018-9764-z

Kim, D. S., Yang, X., Lee, J. H., Yoo, H. Y., Park, C., Kim, S. W., & Lee, J. (2022). Development of GO/Co/Chitosan-Based Nano-Biosensor for Real-Time Detection of D-Glucose. *Biosensors (Basel), 12*(7), 464. doi:10.3390/bios12070464 PMID:35884266

Kim, K. J., & Lee, W. B. (2004). Stock market prediction using artificial neural networks with optimal feature transformation. *Neural Computing & Applications, 13*(3), 255–260. doi:10.1007/s00521-004-0428-x

Kiran, J. S., & Prabhu, S. (2020). Robot nano spray painting-A review. *IOP Conference Series. Materials Science and Engineering, 912*(3), 032044. doi:10.1088/1757-899X/912/3/032044

Koleoso, M., Feng, X., & Xue, Y. (2020). *Materials Today Bio Micro/nanoscale magnetic robots for biomedical applications.* https://doi.org/ doi:10.1016/j.mtbio.2020.100085

Koshariya, A. K., Kalaiyarasi, D., Jovith, A. A., Sivakami, T., Hasan, D. S., & Boopathi, S. (2023). AI-Enabled IoT and WSN-Integrated Smart Agriculture System. In *Artificial Intelligence Tools and Technologies for Smart Farming and Agriculture Practices* (pp. 200–218). IGI Global. doi:10.4018/978-1-6684-8516-3.ch011

Koshariya, A. K., Khatoon, S., Marathe, A. M., Suba, G. M., Baral, D., & Boopathi, S. (2023). Agricultural Waste Management Systems Using Artificial Intelligence Techniques. In *AI-Enabled Social Robotics in Human Care Services* (pp. 236–258). IGI Global. doi:10.4018/978-1-6684-8171-4.ch009

Koutanaei, F. N., Sajedi, H., & Khanbabaei, M. (2015). A hybrid data mining model of feature selection algorithms and ensemble learning classifiers for credit scoring. *Journal of Retailing and Consumer Services, 27,* 11–23. doi:10.1016/j.jretconser.2015.07.003

Krishnan, S., Rajagopalan, G. A., Kandhasamy, S., & Shanmugavel, M. (2019). Towards scalable continuous-time trajectory optimization for multi-robot navigation. *arXiv preprint arXiv:1910.13463.*

Krishnan, S., Rajagopalan, G. A., Kandhasamy, S., & Shanmugavel, M. (2020). Continuous-time trajectory optimization for decentralized multi-robot navigation. *IFAC-PapersOnLine, 53*(1), 494–499. doi:10.1016/j.ifacol.2020.06.083

Krizhevsky, A., Sutskever, I., & Hinton, G. E. (2012). Imagenet classification with deep convolutional neural networks. *Advances in Neural Information Processing Systems, 25.*

Kučuk, N., Primožič, M., Knez, Ž., & Leitgeb, M. (2023). Sustainable Biodegradable Biopolymer-Based Nanoparticles for Healthcare Applications. *International Journal of Molecular Sciences*, 24. PMID:36834596

Kumar & Kumar. (n.d.). *Techniques of Digital Watermarking*. Academic Press.

Kumar, M. R., Reddy, V. P., Meheta, A., Dhiyani, V., Al-Saady, F. A., & Jain, A. (2023). Investigating the Effects of Process Parameters on the Size and Properties of Nano Materials. *E3S Web of Conferences, 430*, 01125.

Kumar, M., Kumar, K., Sasikala, P., Sampath, B., Gopi, B., & Sundaram, S. (2023). Sustainable Green Energy Generation From Waste Water: IoT and ML Integration. In Sustainable Science and Intelligent Technologies for Societal Development (pp. 440–463). IGI Global.

Kumar, S., Nasim, B., & Abraham, E. (2018). Nanorobots a future device for diagnosis and treatment. *Journal of Pharmacy and Pharmaceutics*, 5(1), 44–49. doi:10.15436/2377-1313.18.1815

Kumar, S., Savur, C., & Sahin, F. (2020). Survey of human–robot collaboration in industrial settings: Awareness, intelligence, and compliance. *IEEE Transactions on Systems, Man, and Cybernetics. Systems*, 51(1), 280–297. doi:10.1109/TSMC.2020.3041231

Kundu, K., Vishwakarma, V., Rai, A., Srivastava, M., & Mishra, A. (2023, April). Design and Deployment of Wild Animal Intrusion Detection & Repellent System Employing IOT. In *2023 International Conference on Computational Intelligence and Sustainable Engineering Solutions (CISES)* (pp. 763-767). IEEE. 10.1109/CISES58720.2023.10183532

Kushwah, J. S., Gupta, M., Shrivastava, S., Saxena, N., Saini, R., & Boopathi, S. (2024). Psychological Impacts, Prevention Strategies, and Intervention Approaches Across Age Groups: Unmasking Cyberbullying. In Change Dynamics in Healthcare, Technological Innovations, and Complex Scenarios (pp. 89–109). IGI Global.

Kwong, J. C., Khondker, A., Tran, C., Evans, E., Cozma, A. I., Javidan, A., Ali, A., Jamal, M., Short, T., Papanikolaou, F., Srigley, J. R., Fine, B., & Feifer, A. (2022). Explainable artificial intelligence to predict the risk of side-specific extraprostatic extension in pre-prostatectomy patients. *Canadian Urological Association Journal*, 16(6). Advance online publication. doi:10.5489/cuaj.7473 PMID:35099382

Lai, H. Y. (2023). Breakdowns in team resilience during aircraft landing due to mental model disconnects as identified through machine learning. *Reliability Engineering & System Safety*, 237, 109356. doi:10.1016/j.ress.2023.109356

Lazebnik, S., Schmid, C., & Ponce, J. (2006, June). Beyond bags of features: Spatial pyramid matching for recognizing natural scene categories. In 2006 IEEE computer society conference on computer vision and pattern recognition (CVPR'06) (Vol. 2, pp. 2169-2178). IEEE.

LeCun, Y., Bengio, Y., & Hinton, G. (2015). Deep learning. *Nature, 521*(7553), 436-444.

Lee, D., Yu, H. W., Kwon, H., Kong, H.-J., Lee, K. E., & Kim, H. C. (2020, June). Evaluation of Surgical Skills during Robotic Surgery by Deep Learning-Based Multiple Surgical Instrument Tracking in Training and Actual Operations. *Journal of Clinical Medicine*, 9(6), E1964. doi:10.3390/jcm9061964 PMID:32585953

Lee, J. J., Lee, Y. R., Lim, D. H., & Ahn, H. C. (2021). A Study on the Employee Turnover Prediction using XGBoost and SHAP. *Journal of Information Systems*, 30(4), 21–42.

Lees, M. J., & Johnstone, M. C. (2021). Implementing safety features of Industry 4.0 without compromising safety culture. *IFAC-PapersOnLine*, 54(13), 680–685. doi:10.1016/j.ifacol.2021.10.530

Lei, B., Zhuo, L., Chen, S., Li, S., Ni, D., & Wang, T. (2014, April). Automatic recognition of fetal standard plane in ultrasound image. In *2014 IEEE 11th International Symposium on Biomedical Imaging (ISBI)* (pp. 85-88). IEEE. 10.1109/ISBI.2014.6867815

Lei, B., Tan, E. L., Chen, S., Ni, D., & Wang, T. (2015). Saliency-driven image classification method based on histogram mining and image score. *Pattern Recognition*, 48(8), 2567–2580. doi:10.1016/j.patcog.2015.02.004

Lei, B., Tan, E. L., Chen, S., Zhuo, L., Li, S., Ni, D., & Wang, T. (2015). Automatic recognition of fetal facial standard plane in ultrasound image via fisher vector. *PLoS One*, 10(5), e0121838. doi:10.1371/journal.pone.0121838 PMID:25933215

Lei, B., Yao, Y., Chen, S., Li, S., Li, W., Ni, D., & Wang, T. (2015). Discriminative learning for automatic staging of placental maturity via multi-layer fisher vector. *Scientific Reports*, 5(1), 12818. doi:10.1038/srep12818 PMID:26228175

Li, J., Xu, Z., & Zhu, D. (2022). Bio-inspired Intelligence with Applications to Robotics: A Survey. arXiv:2206.

Li, Q., Cai, W., Wang, X., Zhou, Y., Feng, D. D., & Chen, M. (2014, December). Medical image classification with convolutional neural network. In *2014 13th international conference on control automation robotics & vision (ICARCV)* (pp. 844-848). IEEE. 10.1109/ICARCV.2014.7064414

Li, W. (2018). The design of a 3-CPS parallel robot for maximum dexterity. *Mechanism and Machine Theory, 122*, 279-291. doi:10.1016/j.mechmachtheory.2018.01.003

Li, X., Lai, T., Wang, S., Chen, Q., Yang, C., Chen, R., ... Zheng, F. (2019, December). Weighted feature pyramid networks for object detection. In *2019 IEEE Intl Conf on Parallel & Distributed Processing with Applications, Big Data & Cloud Computing, Sustainable Computing & Communications, Social Computing & Networking (ISPA/BDCloud/SocialCom/SustainCom)* (pp. 1500-1504). IEEE.

Li, H., Cao, Y., Li, S., Zhao, J., & Sun, Y. (2020). XGBoost model and its application to personal credit evaluation. *IEEE Intelligent Systems, 35*(3), 52–61. doi:10.1109/MIS.2020.2972533

Li, H., Yao, J., Zhou, P., Chen, X., Xu, Y., & Zhao, Y. (2020). High-force soft pneumatic actuators based on novel casting method for robotic applications. *Sensors and Actuators. A, Physical, 306*, 306. doi:10.1016/j.sna.2020.111957

Li, J., Dekanovsky, L., Khezri, B., Wu, B., Zhou, H., & Sofer, Z. (2022). Biohybrid Micro- and Nanorobots for Intelligent Drug Delivery. *Cyborg and Bionic Systems (Washington, D.C.), 2022*, 2022. doi:10.34133/2022/9824057 PMID:36285309

Li, J., Yang, S. X., & Xu, Z. (2019). A survey on robot path planning using bio-inspired algorithms. *2019 IEEE International Conference on Robotics and Biomimetics (ROBIO)*, 2111–2116. 10.1109/ROBIO49542.2019.8961498

Li, J., & Yu, J. (2023). Biodegradable Microrobots and Their Biomedical Applications: A Review. *Nanomaterials (Basel, Switzerland), 13*(10), 13. doi:10.3390/nano13101590 PMID:37242005

Lin, M., Chen, Q., & Yan, S. (2013). Network in network. *arXiv preprint arXiv:1312.4400*.

Li, S., Zhang, H., & Xu, F. (2023). Intelligent Detection Method for Wildlife Based on Deep Learning. *Sensors (Basel), 23*(24), 9669. doi:10.3390/s23249669 PMID:38139515

Liu, C., Yuen, J., & Torralba, A. (2010). Sift flow: Dense correspondence across scenes and its applications. *IEEE Transactions on Pattern Analysis and Machine Intelligence, 33*(5), 978–994. doi:10.1109/TPAMI.2010.147 PMID:20714019

Liu, H. H., Su, P. C., & Hsu, M. H. (2020). An improved steganography method based on least-significant-bit substitution and pixel-value differencing. *KSII Transactions on Internet and Information Systems, 14*(11), 4537–4556.

Long, J., Shelhamer, E., & Darrell, T. (2015). Fully convolutional networks for semantic segmentation. In *Proceedings of the IEEE conference on computer vision and pattern recognition* (pp. 3431-3440). IEEE.

Lopez, M., Castillo, E., Garcia, G., & Bashir, A. (2006). Delta robot: Inverse, direct, and intermediate jacobians. *Proceedings of the Institution of Mechanical Engineers, Part C: Journal of Mechanical Engineering Science, 220*(1), 103–109. 10.1243/095440606X78263

Lowe, D. G. (2004). Distinctive image features from scale-invariant keypoints. *International Journal of Computer Vision, 60*(2), 91–110. doi:10.1023/B:VISI.0000029664.99615.94

Madeti, S. R., & Singh, S. N. (2018). Modeling of PV system based on experimental data for fault detection using kNN method. *Solar Energy, 173*, 139–151. doi:10.1016/j.solener.2018.07.038

Madridano, A., Al-Kaff, A., Martín, D., & De La Escalera, A. (2021). Trajectory planning for multi-robot systems: Methods and applications. *Expert Systems with Applications, 173*, 114660. doi:10.1016/j.eswa.2021.114660

Maguluri, L. P., Arularasan, A., & Boopathi, S. (2023). Assessing Security Concerns for AI-Based Drones in Smart Cities. In Effective AI, Blockchain, and E-Governance Applications for Knowledge Discovery and Management (pp. 27–47). IGI Global. doi:10.4018/978-1-6684-9151-5.ch002

Maheswari, B. U., Imambi, S. S., Hasan, D., Meenakshi, S., Pratheep, V., & Boopathi, S. (2023). Internet of things and machine learning-integrated smart robotics. In Global Perspectives on Robotics and Autonomous Systems: Development and Applications (pp. 240–258). IGI Global. doi:10.4018/978-1-6684-7791-5.ch010

Maji, S., Berg, A. C., & Malik, J. (2008, June). *Classification using intersection kernel support vector machines is efficient. In 2008 IEEE conference on computer vision and pattern recognition.* IEEE.

Makarfi, A. U., Rabie, K. M., Kaiwartya, O., Li, X., & Kharel, R. (2020, May). Physical layer security in vehicular networks with reconfigurable intelligent surfaces. In *2020 IEEE 91st Vehicular Technology Conference (VTC2020-Spring)* (pp. 1-6). IEEE. 10.1109/VTC2020-Spring48590.2020.9128438

Malathi, J., Kusha, K., Isaac, S., Ramesh, A., Rajendiran, M., & Boopathi, S. (2024). IoT-Enabled Remote Patient Monitoring for Chronic Disease Management and Cost Savings: Transforming Healthcare. In Advances in Explainable AI Applications for Smart Cities (pp. 371–388). IGI Global.

Malik, S., Muhammad, K., & Waheed, Y. (2023). Nanotechnology: A revolution in modern industry. *Molecules (Basel, Switzerland), 28*(2), 661. doi:10.3390/molecules28020661 PMID:36677717

Mali, P., Harikumar, K., Singh, A. K., Krishna, K. M., & Sujit, P. B. (2021, June). Incorporating prediction in control barrier function based distributive multi-robot collision avoidance. In *2021 European Control Conference (ECC)* (pp. 2394-2399). IEEE. 10.23919/ECC54610.2021.9655081

Mall, S. (2018). An empirical study on credit risk management: The case of nonbanking financial companies. *The Journal of Credit Risk, 14*(3), 49–66. doi:10.21314/JCR.2017.239

Mammeri, A., Zhou, D., & Boukerche, A. (2016). Animal-vehicle collision mitigation system for automated vehicles. *IEEE Transactions on Systems, Man, and Cybernetics. Systems, 46*(9), 1287–1299. doi:10.1109/TSMC.2015.2497235

Manickam, P., Mariappan, S. A., Murugesan, S. M., Hansda, S., Kaushik, A., Shinde, R., & Thipperudraswamy, S. P. (2022). Artificial Intelligence (AI) and Internet of Medical Things (IoMT) Assisted Biomedical Systems for Intelligent Healthcare. *Biosensors (Basel), 12*(8), 562–562. doi:10.3390/bios12080562 PMID:35892459

Manuela & Costa. (2014, November). Data Visualization. *Communication Design Quarterly Review.*

Marinoudi, V., Sørensen, C. G., Pearson, S., & Bochtis, D. (2019). Robotics and labour in agriculture. A context consideration. *Biosystems Engineering, 184*, 111–121. doi:10.1016/j.biosystemseng.2019.06.013

Markande, A., Mistry Kruti, U., & Shraddha, J. A. (2021) magnetic nanoparticles from bacteria. In Biobased Nanotechnology for Green Applications. SpringerNature Switzerland AG.

Marvel, L. M., Boncelet, C. G., & Retter, C. T. (1999). Spread spectrum image steganography. *IEEE Transactions on Image Processing, 8*(8), 1075–1083. doi:10.1109/83.777088 PMID:18267522

Marvin, G., Jackson, M., & Alam, M. G. R. (2021, August). A machine learning approach for employee retention prediction. In *2021 IEEE Region 10 Symposium (TENSYMP)* (pp. 1-8). IEEE. 10.1109/TENSYMP52854.2021.9550921

Mavridou, E., Vrochidou, E., Papakostas, G. A., Pachidis, T., & Kaburlasos, V. G. (2019). Machine vision systems in precision agriculture for crop farming. *Journal of Imaging, 5*(12), 89. doi:10.3390/jimaging5120089 PMID:34460603

Mbunge, E., Muchemwa, B., Jiyane, S., & Batani, J. (2021). Sensors and healthcare 5.0: Transformative shift in virtual care through emerging digital health technologies. *Global Health Journal (Amsterdam, Netherlands), 5*(4), 169–177. doi:10.1016/j.glohj.2021.11.008

McClintock, H., Temel, F. Z., Doshi, N., Je-sung, K., & Robert, J. (2018). The millidelta: A high-bandwidth, high-precision, millimeter-scale delta robot. *Science Robotics, 3*(14), eaar3018. doi:10.1126/scirobotics.aar3018 PMID:33141699

Mehrafrooz, B., Mohammadi, M., & Masouleh, M. T. (2017). Kinematic sensitivity evaluation of revolute and prismatic 3-dof delta robots. *2017 5th RSI International Conference on Robotics and Mechatronics (ICRoM)*, 225–231. doi: .846615910.1109/ICRoM.2017

Meivel, S., Maheswari, S., & Faridha Banu, D. (2023). Design and Method of an Agricultural Drone System Using Biomass Vegetation Indices and Multispectral Images. In *Proceedings of UASG 2021: Wings 4 Sustainability. UASG 2021. Lecture Notes in Civil Engineering* (vol. 304). Springer. 10.1007/978-3-031-19309-5_25

Mercy, J., Lawanya, R., Nandhini, S., & Saravanan, M. (2022). Effective Image Deblurring Based on Improved Image Edge Information and Blur Kernel Estimation. *8th International Conference on Advanced Computing and Communication Systems, ICACCS 2022*, 855-859.

Midi, H., Sarkar, S. K., & Rana, S. (2010). Collinearity diagnostics of binary logistic regression model. *Journal of Interdisciplinary Mathematics, 13*(3), 253–267. doi:10.1080/09720502.2010.10700699

Minopoulos, G. M., Memos, V. A., Stergiou, C. L., Stergiou, K. D., Plageras, A. P., Koidou, M. P., & Psannis, K. E. (2022). Exploitation of Emerging Technologies and Advanced Networks for a Smart Healthcare System. *Applied Sciences (Basel, Switzerland), 12*(12), 5859. doi:10.3390/app12125859

Moerenhout, T., Devisch, I., & Cornelis, G. C. (2018). E-health beyond technology: Analyzing the paradigm shift that lies beneath. *Medicine, Health Care, and Philosophy, 21*(1), 31–41. doi:10.1007/s11019-017-9780-3 PMID:28551772

Mohanta, B., Das, P., & Patnaik, S. (2019). Healthcare 5.0: A paradigm shift in digital healthcare system using artificial intelligence, IOT and 5G communication. *2019 International Conference on Applied Machine Learning (ICAML)*, 191–196. 10.1109/ICAML48257.2019.00044

Mohanty, A., Jothi, B., Jeyasudha, J., Ranjit, P., Isaac, J. S., & Boopathi, S. (2023). Additive Manufacturing Using Robotic Programming. In *AI-Enabled Social Robotics in Human Care Services* (pp. 259–282). IGI Global. doi:10.4018/978-1-6684-8171-4.ch010

Moradi Dalvand, M., & Shirinzadeh, B. (2013, April). Motion control analysis of a parallel robot assisted minimally invasive surgery/microsurgery system (PRAMiSS). *Robotics and Computer-integrated Manufacturing, 29*(2), 318–327. doi:10.1016/j.rcim.2012.09.003

Mouchou, R., Laseinde, T., Jen, T.-C., & Ukoba, K. (2021). Developments in the Application of Nano Materials for Photovoltaic Solar Cell Design, Based on Industry 4.0 Integration Scheme. *Advances in Artificial Intelligence, Software and Systems Engineering: Proceedings of the AHFE 2021 Virtual Conferences on Human Factors in Software and Systems Engineering, Artificial Intelligence and Social Computing, and Energy,* July 25-29, 2021, USA, 510–521.

Murali, B., Padhi, S., Patil, C. K., Kumar, P. S., Santhanakrishnan, M., & Boopathi, S. (2023). Investigation on hardness and tensile strength of friction stir processing of Al6061/TiN surface composite. *Materials Today: Proceedings*.

Myilsamy, S., & Sampath, B. (2021). Experimental comparison of near-dry and cryogenically cooled near-dry machining in wire-cut electrical discharge machining processes. *Surface Topography : Metrology and Properties, 9*(3), 035015. doi:10.1088/2051-672X/ac15e0

Naghdi, T., Ardalan, S., Asghari Adib, Z., Sharifi, A. R., & Golmohammadi, H. (2023). Moving toward Smart Biomedical Sensing. *Biosensors & Bioelectronics, 223*, 115009. doi:10.1016/j.bios.2022.115009 PMID:36565545

Najafi-Zangeneh, S., Shams-Gharneh, N., Arjomandi-Nezhad, A., & Hashemkhani Zolfani, S. (2021). An Improved Machine Learning-Based Employees Attrition Prediction Framework with Emphasis on Feature Selection. *Mathematics, 9*(11), 1226. doi:10.3390/math9111226

Narasimmalou, T., & Allen Joseph, R. (2012). Discrete wavelet transform based steganography for transmitting images. In *IEEE-International Conference on Advances In Engineering, Science And Management (ICAESM2012)* (pp. 370-375). IEEE.

Nasr, M., Islam, M. M., Shehata, S., Karray, F., & Quintana, Y. (2021). Smart healthcare in the age of AI: Recent advances, challenges, and future prospects. *IEEE Access : Practical Innovations, Open Solutions, 9*, 145248–145270. doi:10.1109/ACCESS.2021.3118960

Naveeenkumar, N., Rallapalli, S., Sasikala, K., Priya, P. V., Husain, J., & Boopathi, S. (2024). Enhancing Consumer Behavior and Experience Through AI-Driven Insights Optimization. In *AI Impacts in Digital Consumer Behavior* (pp. 1–35). IGI Global. doi:10.4018/979-8-3693-1918-5.ch001

Navsalkar, A., & Hota, A. R. (2023, May). Data-driven risk-sensitive model predictive control for safe navigation in multi-robot systems. In *2023 IEEE International Conference on Robotics and Automation (ICRA)* (pp. 1442-1448). IEEE. 10.1109/ICRA48891.2023.10161002

Ness, S., Shepherd, N. J., & Xuan, T. R. (2023). Synergy Between AI and Robotics: A Comprehensive Integration. *Asian Journal of Research in Computer Science, 16*(4), 80–94. doi:10.9734/ajrcos/2023/v16i4372

Ni, D., Yang, X., Chen, X., Chin, C. T., Chen, S., Heng, P. A., Li, S., Qin, J., & Wang, T. (2014). Standard plane localization in ultrasound by radial component model and selective search. *Ultrasound in Medicine & Biology, 40*(11), 2728–2742. doi:10.1016/j.ultrasmedbio.2014.06.006 PMID:25220278

O'Brolcháin, F. (2019). Robots and people with dementia: Unintended consequences and moral hazard. *Nursing Ethics, 26*(4), 962–972. doi:10.1177/0969733017742960 PMID:29262739

Oh, H., Shirazi, A. R., Sun, C., & Jin, Y. (2017). Bio-inspired self-organising multi-robot pattern formation: A review. *Robotics and Autonomous Systems, 91*, 83–100. doi:10.1016/j.robot.2016.12.006

Olcay, E., Schuhmann, F., & Lohmann, B. (2020). Collective navigation of a multi-robot system in an unknown environment. *Robotics and Autonomous Systems, 132*, 103604. doi:10.1016/j.robot.2020.103604

Oluwasanu, A. A., Oluwaseun, F., Teslim, J. A., Isaiah, T. T., Olalekan, I. A., & Chris, O. A. (2019). Scientific applications and prospects of nanomaterials: A multidisciplinary review. *African Journal of Biotechnology, 18*(30), 946–961. doi:10.5897/AJB2019.16812

Onnasch, L., & Roesler, E. (2021). A taxonomy to structure and analyze human–robot interaction. *International Journal of Social Robotics, 13*(4), 833–849. doi:10.1007/s12369-020-00666-5

Palaniappan, M., Tirlangi, S., Mohamed, M. J. S., Moorthy, R. S., Valeti, S. V., & Boopathi, S. (2023). Fused Deposition Modelling of Polylactic Acid (PLA)-Based Polymer Composites: A Case Study. In Development, Properties, and Industrial Applications of 3D Printed Polymer Composites (pp. 66–85). IGI Global.

Pandya, A., & Auner, G. (2004). Robotics technology: A journey into the future. *The Urologic Clinics of North America, 31*(4), 793–800. doi:10.1016/j.ucl.2004.06.013 PMID:15474607

Park, J. S., Tsang, B., Yedidsion, H., Warnell, G., Kyoung, D., & Stone, P. (2021, October). Learning to improve multi-robot hallway navigation. In *Conference on Robot Learning* (pp. 1883-1895). PMLR.

Patel, H., & Dave, P. (2012). Steganography technique based on DCT coefficients. *International Journal of Engineering Research and Applications, 2*(1), 713–717.

Patil, K., Meshram, V., Hanchate, D., & Ramkteke, S. D. (2021). Machine learning in agriculture domain: A state-of-art survey. *Artificial Intelligence in the Life Sciences, 1*, 100010. doi:10.1016/j.ailsci.2021.100010

Patwardhan, J. (2006). *Architectures for nanoscale devices* (Vol. 68). Academic Press.

Paul, A., Thilagham, K., KG, J.-, Reddy, P. R., Sathyamurthy, R., & Boopathi, S. (2024). Multi-criteria Optimization on Friction Stir Welding of Aluminum Composite (AA5052-H32/B4C) using Titanium Nitride Coated Tool. Engineering Research Express.

Pawar, U., O'Shea, D., Rea, S., & O'Reilly, R. (2020). Incorporating explainable artificial intelligence (XAI) to aid the understanding of machine learning in the healthcare domain. *Irish Conference on Artificial Intelligence and Cognitive Science.*

Payne, E. M., Peltier, J. W., & Barger, V. A. (2018). Mobile banking and AI-enabled mobile banking: The differential effects of technological and non-technological factors on digital natives' perceptions and behavior. *Journal of Research in Interactive Marketing, 12*(3), 328–346. doi:10.1108/JRIM-07-2018-0087

Pedram, A., & Nejat Pishkenari, H. (2017). Smart micro/nano-robotic systems for gene delivery. *Current Gene Therapy, 17*(2), 73–79. doi:10.2174/1566523217666170511111000 PMID:28494736

Peng, X., Tang, S., Tang, D., Zhou, D., Li, Y., Chen, Q., Wan, F., Lukas, H., Han, H., Zhang, X., Gao, W., & Wu, S. (2023). Autonomous metal-organic framework nanorobots for active mitochondria-targeted cancer therapy. *Science Advances, 9*(23), 9. doi:10.1126/sciadv.adh1736 PMID:37294758

Penumuru, D. P., Muthuswamy, S., & Karumbu, P. (2020). Identification and classification of materials using machine vision and machine learning in the context of industry 4.0. *Journal of Intelligent Manufacturing, 31*(5), 1229–1241. doi:10.1007/s10845-019-01508-6

Pereira, C. S., Morais, R., & Reis, M. J. C. S. (2017). Recent advances in image processing techniques for automated harvesting purposes: A review. *Proceedings of the 2017 Intelligent Systems Conference (IntelliSys),* 566-575. 10.1109/IntelliSys.2017.8324352

Perronnin, F., Sánchez, J., & Mensink, T. (2010). Improving the fisher kernel for large-scale image classification. *Computer Vision–ECCV 2010: 11th European Conference on Computer Vision, Heraklion, Crete, Greece, September 5-11, 2010 Proceedings, 11*(Part IV), 143–156.

Pisla, Plitea, Gherman, Pisla, & Vaida. (2009). *Kinematical Analysis and Design of a New Surgical Parallel Robot.* . doi:10.1007/978-3-642-01947-0_34

Pokrajac, L., Abbas, A., Chrzanowski, W., Dias, G. M., Eggleton, B. J., Maguire, S., Maine, E., Malloy, T., Nathwani, J., Nazar, L., & ... (2021). *Nanotechnology for a sustainable future: Addressing global challenges with the international network4sustainable nanotechnology*. ACS Publications.

Połap, D., Włodarczyk-Sielicka, M., & Wawrzyniak, N. (2022). Automatic ship classification for a riverside monitoring system using a cascade of artificial intelligence techniques including penalties and rewards. *ISA Transactions, 121*, 232–239. doi:10.1016/j.isatra.2021.04.003 PMID:33888294

Polygerinos, P., Wang, Z., Galloway, K. C., Wood, R. J., & Walsh, C. J. (2015). Soft robotic glove for combined assistance and at-home rehabilitation. *Robotics and Autonomous Systems, 73*, 135–143. doi:10.1016/j.robot.2014.08.014

Popescu, M., & Ungureanu, C. (2023). Biosensors in Food and Healthcare Industries: Bio-Coatings Based on Biogenic Nanoparticles and Biopolymers. *Coatings, 13*(3), 486. doi:10.3390/coatings13030486

Poppeova, V., Uricek, J., Bulej, V., & Sindler, P. (2011). Delta robots - robots for high speed manipulation. *Tehnicki Vjesnik (Strojarski Fakultet), 18*, 435–445.

Pramila, P., Amudha, S., Saravanan, T., Sankar, S. R., Poongothai, E., & Boopathi, S. (2023). Design and Development of Robots for Medical Assistance: An Architectural Approach. In Contemporary Applications of Data Fusion for Advanced Healthcare Informatics (pp. 260–282). IGI Global.

Pratt, M., Boudhane, M., & Cakula, S. (2021). Employee attrition estimation using random forest algorithm. *Baltic Journal of Modern Computing, 9*(1), 49–66. doi:10.22364/bjmc.2021.9.1.04

Primožič, M., Knez, Ž., & Leitgeb, M. (2021). (Bio)Nanotechnology in Food Science—Food Packaging. *Nanomaterials (Basel, Switzerland)*, 11. PMID:33499415

Pugliese, R., & Regondi, S. (2022). Artificial intelligence-empowered 3D and 4D printing technologies toward smarter biomedical materials and approaches. *Polymers, 14*(14), 2794. doi:10.3390/polym14142794 PMID:35890571

Puranik, T. A., Shaik, N., Vankudoth, R., Kolhe, M. R., Yadav, N., & Boopathi, S. (2024). Study on Harmonizing Human-Robot (Drone) Collaboration: Navigating Seamless Interactions in Collaborative Environments. In Cybersecurity Issues and Challenges in the Drone Industry (pp. 1–26). IGI Global.

Putra, M., & Damayanti, N. (2020). The Effect of Reward and Punishment to Performance of Driver Grabcar in Depok. *International Journal of Research and Review, 7*(1), 312–319.

Rahamathunnisa, U., Subhashini, P., Aancy, H. M., Meenakshi, S., Boopathi, S., & ... (2023). Solutions for Software Requirement Risks Using Artificial Intelligence Techniques. In *Handbook of Research on Data Science and Cybersecurity Innovations in Industry 4.0 Technologies* (pp. 45–64). IGI Global.

Rahamathunnisa, U., Sudhakar, K., Murugan, T. K., Thivaharan, S., Rajkumar, M., & Boopathi, S. (2023). Cloud Computing Principles for Optimizing Robot Task Offloading Processes. In *AI-Enabled Social Robotics in Human Care Services* (pp. 188–211). IGI Global. doi:10.4018/978-1-6684-8171-4.ch007

Rahim, R., & Nadeem, S. (2018). End-to-end trained CNN encoder-decoder networks for image steganography. *Proceedings of the European Conference on Computer Vision (ECCV) Workshops.*

Rahmatullah, B., & Noble, J. A. (2013, September). Anatomical object detection in fetal ultrasound: computer-expert agreements. In *International Conference on Biomedical Informatics and Technology* (pp. 207-218). Springer Berlin Heidelberg.

Rahmatullah, B., Papageorghiou, A., & Noble, J. A. (2011). Automated selection of standardized planes from ultrasound volume. In *Machine Learning in Medical Imaging: Second International Workshop, MLMI 2011, Held in Conjunction with MICCAI 2011, Toronto, Canada, September 18, 2011. Proceedings 2* (pp. 35-42). Springer Berlin Heidelberg. 10.1007/978-3-642-24319-6_5

Ramadan, M. N., Al-Khedher, M. A., & Al-Kheder, S. A. (2012). Intelligent anti-theft and tracking system for automobiles. *International Journal of Machine Learning and Computing*, 2(1), 83–88. doi:10.7763/IJMLC.2012.V2.94

Ramalingam, S., & Rasool Mohideen, S. (2021). Composite materials for advanced flexible link robotic manipulators: An investigation. *International Journal of Ambient Energy*, 42(14), 1670–1675. doi:10.1080/01430750.2019.1613263

Ramesh, R. D., Santhosh, A., & Syamala, S. R. N. A. (2020). Implementation of Nanotechnology in the Aerospace and Aviation Industry. In *Smart Nanotechnology with Applications* (pp. 51–69). CRC Press. doi:10.1201/9781003097532-4

Ramudu, K., Mohan, V. M., Jyothirmai, D., Prasad, D., Agrawal, R., & Boopathi, S. (2023). Machine Learning and Artificial Intelligence in Disease Prediction: Applications, Challenges, Limitations, Case Studies, and Future Directions. In Contemporary Applications of Data Fusion for Advanced Healthcare Informatics (pp. 297–318). IGI Global.

Rani, S., Chauhan, M., Kataria, A., & Khang, A. (2023). IoT equipped intelligent distributed framework for smart healthcare systems. In *Towards the Integration of IoT, Cloud and Big Data: Services, Applications and Standards* (pp. 97–114). Springer Nature Singapore. doi:10.1007/978-981-99-6034-7_6

Ravisankar, A., Sampath, B., & Asif, M. M. (2023). Economic Studies on Automobile Management: Working Capital and Investment Analysis. In Multidisciplinary Approaches to Organizational Governance During Health Crises (pp. 169–198). IGI Global.

Ravisankar, A., Shanthi, A., Lavanya, S., Ramaratnam, M., Krishnamoorthy, V., & Boopathi, S. (2024). Harnessing 6G for Consumer-Centric Business Strategies Across Electronic Industries. In AI Impacts in Digital Consumer Behavior (pp. 241–270). IGI Global.

Raza, A., Munir, K., Almutairi, M., Younas, F., & Fareed, M. M. S. (2022). Predicting Employee Attrition Using Machine Learning Approaches. *Applied Sciences (Basel, Switzerland)*, 12(13), 6424. doi:10.3390/app12136424

Redmon, J., & Farhadi, A. (2018). Yolov3: An incremental improvement. *arXiv preprint arXiv:1804.02767*.

Redmon, J., Divvala, S., Girshick, R., & Farhadi, A. (2016). You only look once: Unified, real-time object detection. In *Proceedings of the IEEE conference on computer vision and pattern recognition* (pp. 779-788). 10.1109/CVPR.2016.91

Redmon, J., & Farhadi, A. (2017). YOLO9000: better, faster, stronger. In *Proceedings of the IEEE conference on computer vision and pattern recognition* (pp. 7263-7271).

Rehman, T. U., Mahmud, M. S., Chang, Y. K., Jin, J., & Shin, J. (2019). Current and future applications of statistical machine learning algorithms for agricultural machine vision systems. *Computers and Electronics in Agriculture*, 156, 585–605. doi:10.1016/j.compag.2018.12.006

Ren, S., He, K., Girshick, R., & Sun, J. (2016). Faster R-CNN: Towards real-time object detection with region proposal networks. *IEEE Transactions on Pattern Analysis and Machine Intelligence*, 39(6), 1137–1149. doi:10.1109/TPAMI.2016.2577031 PMID:27295650

Revathi, S., Babu, M., Rajkumar, N., Meti, V. K. V., Kandavalli, S. R., & Boopathi, S. (2024). Unleashing the Future Potential of 4D Printing: Exploring Applications in Wearable Technology, Robotics, Energy, Transportation, and Fashion. In Human-Centered Approaches in Industry 5.0: Human-Machine Interaction, Virtual Reality Training, and Customer Sentiment Analysis (pp. 131–153). IGI Global.

Ritharson, P. I., Raimond, K., Mary, X. A., Robert, J. E., & Andrew, J. (2024). DeepRice: A deep learning and deep feature based classification of Rice leaf disease subtypes. *Artificial Intelligence in Agriculture*, *11*, 34–49. doi:10.1016/j.aiia.2023.11.001

Rodrigues, C., Souza, V. G. L., Coelhoso, I., & Fernando, A. L. (2021). Bio-Based Sensors for Smart Food Packaging—Current Applications and Future Trends. *Sensors (Basel)*, *21*(6), 2148. doi:10.3390/s21062148 PMID:33803914

Ronneberger, O., Fischer, P., & Brox, T. (2015). U-net: Convolutional networks for biomedical image segmentation. In *Medical image computing and computer-assisted intervention–MICCAI 2015: 18th international conference, Munich, Germany, October 5-9, 2015, proceedings, part III 18* (pp. 234-241). Springer International Publishing.

Rus, D., & Tolley, M. T. (2015). Design, fabrication and control of soft robots. *Nature*, *521*(7553), 467–475. doi:10.1038/nature14543 PMID:26017446

Sadagopan, V. K., Rajendran, U., & Francis, A. J. (2011, July). Anti theft control system design using embedded system. In *Proceedings of 2011 IEEE International Conference on Vehicular Electronics and Safety* (pp. 1-5). IEEE. 10.1109/ICVES.2011.5983776

Saharan. (2022). Robotic Automation of Pharmaceutical and Life Science Industries. *Computer Aided Pharmaceutics and Drug Delivery*. doi:10.1007/978-981-16-5180-9_12

Sahu, A. K., & Swain, G. (2016). A review on LSB substitution and PVD based image steganography techniques. *Indonesian Journal of Electrical Engineering and Computer Science*, *2*(3), 712–719. doi:10.11591/ijeecs.v2.i3.pp712-719

Salomon, L., Alfirevic, Z., Berghella, V., Bilardo, C., Chalouhi, G., Costa, F. D. S., ... Paladini, D. (2022). ISUOG practice guidelines (updated): Performance of the routine mid-trimester fetal ultrasound scan. *Ultrasound in Obstetrics & Gynecology*, *59*(6), 840–856. doi:10.1002/uog.24888 PMID:35592929

Samman, T., Spearman, J., Dutta, A., Kreidl, O. P., Roy, S., & Bölöni, L. (2021, October). Secure multi-robot adaptive information sampling. In *2021 IEEE International Symposium on Safety, Security, and Rescue Robotics (SSRR)* (pp. 125-131). IEEE. 10.1109/SSRR53300.2021.9597867

Sampath, B., & Myilsamy, S. (2021). Experimental investigation of a cryogenically cooled oxygen-mist near-dry wire-cut electrical discharge machining process. *Stroj. Vestn. Jixie Gongcheng Xuebao*, *67*(6), 322–330.

Sampath, B., Naveenkumar, N., Sampathkumar, P., Silambarasan, P., Venkadesh, A., & Sakthivel, M. (2022). Experimental comparative study of banana fiber composite with glass fiber composite material using Taguchi method. *Materials Today: Proceedings*, *49*, 1475–1480. doi:10.1016/j.matpr.2021.07.232

Sanchez-Ibanez, J. R., Perez-del-Pulgar, C. J., & García-Cerezo, A. (2021). Path planning for autonomous mobile robots: A review. *Sensors (Basel)*, *21*(23), 7898. doi:10.3390/s21237898 PMID:34883899

Sánchez, J., Perronnin, F., Mensink, T., & Verbeek, J. (2013). Image classification with the fisher vector: Theory and practice. *International Journal of Computer Vision*, *105*(3), 222–245. doi:10.1007/s11263-013-0636-x

Satav, S. D., Lamani, D., Harsha, K., Kumar, N., Manikandan, S., & Sampath, B. (2023). Energy and Battery Management in the Era of Cloud Computing: Sustainable Wireless Systems and Networks. In Sustainable Science and Intelligent Technologies for Societal Development (pp. 141–166). IGI Global.

Satav, S. D., Hasan, D. S., Pitchai, R., Mohanaprakash, T., Sultanuddin, S., & Boopathi, S. (2023). Next generation of internet of things (ngiot) in healthcare systems. In *Sustainable Science and Intelligent Technologies for Societal Development* (pp. 307–330). IGI Global.

Schütz, A. K., Schöler, V., Krause, E. T., Fischer, M., Müller, T., Freuling, C. M., Conraths, F. J., Stanke, M., Homeier-Bachmann, T., & Lentz, H. H. (2021). Application of YOLOv4 for detection and Motion monitoring of red Foxes. *Animals (Basel)*, *11*(6), 1723. doi:10.3390/ani11061723 PMID:34207726

Schwartz, E. M., Bradlow, E. T., & Fader, P. S. (2017). Customer acquisition via display advertising using multi-armed bandit experiments. *Marketing Science*, *36*(4), 500–522. doi:10.1287/mksc.2016.1023

Selvakumar, S., Adithe, S., Isaac, J. S., Pradhan, R., Venkatesh, V., & Sampath, B. (2023). A Study of the Printed Circuit Board (PCB) E-Waste Recycling Process. In Sustainable Approaches and Strategies for E-Waste Management and Utilization (pp. 159–184). IGI Global.

Selvakumar, S., Shankar, R., Ranjit, P., Bhattacharya, S., Gupta, A. S. G., & Boopathi, S. (2023). E-Waste Recovery and Utilization Processes for Mobile Phone Waste. In *Handbook of Research on Safe Disposal Methods of Municipal Solid Wastes for a Sustainable Environment* (pp. 222–240). IGI Global. doi:10.4018/978-1-6684-8117-2.ch016

Şenbaşlar, B., Luiz, P., Hönig, W., & Sukhatme, G. S. (2023). Mrnav: Multi-robot aware planning and control stack for collision and deadlock-free navigation in cluttered environments. *arXiv preprint arXiv:2308.13499*.

Sengeni, D., Padmapriya, G., Imambi, S. S., Suganthi, D., Suri, A., & Boopathi, S. (2023). Biomedical waste handling method using artificial intelligence techniques. In *Handbook of Research on Safe Disposal Methods of Municipal Solid Wastes for a Sustainable Environment* (pp. 306–323). IGI Global. doi:10.4018/978-1-6684-8117-2.ch022

Senthil, T., Puviyarasan, M., Babu, S. R., Surakasi, R., Sampath, B., & Associates. (2023). Industrial Robot-Integrated Fused Deposition Modelling for the 3D Printing Process. In Development, Properties, and Industrial Applications of 3D Printed Polymer Composites (pp. 188–210). IGI Global.

Senthil, T., Puviyarasan, M., Babu, S. R., Surakasi, R., Sampath, B., & ... (2023). Industrial Robot-Integrated Fused Deposition Modelling for the 3D Printing Process. In *Development, Properties, and Industrial Applications of 3D Printed Polymer Composites* (pp. 188–210). IGI Global.

Sermanet, P., Eigen, D., Zhang, X., Mathieu, M., Fergus, R., & LeCun, Y. (2013). Overfeat: Integrated recognition, localization and detection using convolutional networks. arXiv preprint arXiv:1312.6229.

Setiadi, D. R. I. M. (2021). PSNR vs SSIM: Imperceptibility quality assessment for image steganography. *Multimedia Tools and Applications*, *80*(6), 8423–8444. doi:10.1007/s11042-020-10035-z

Sevinchan, E., Dincer, I., & Lang, H. (2018). A review on thermal management methods for robots. *Applied Thermal Engineering*, *140*, 799–813. doi:10.1016/j.applthermaleng.2018.04.132

Shaban Nejad, M. (2020). *Explainable AI in healthcare and medicine: building a culture of transparency and accountability* (Vol. 914). Springer Nature.

Shaik, T., Tao, X., Higgins, N., Li, L., Gururajan, R., Zhou, X., & Acharya, U. R. (2023). Remote patient monitoring using artificial intelligence: Current state, applications, and challenges. *Wiley Interdisciplinary Reviews. Data Mining and Knowledge Discovery*, *13*(2), e1485. doi:10.1002/widm.1485

Sharifi, O., & Eskandari, M. (2018). Cosmetic Detection Framework for Face and Iris Biometrics. *Sensors (Basel)*, *10*, 122.

Sharmin, I. (2020). *Preparation and evaluation of carbon nano tube based nanofluid in milling alloy steel*. Academic Press.

Sheikh, J. A., Waheed, M. F., Khalid, A. M., & Qureshi, I. A. (2020). Use of 3D printing and nano materials in fashion: From revolution to evolution. *Advances in Design for Inclusion: Proceedings of the AHFE 2019 International Conference on Design for Inclusion and the AHFE 2019 International Conference on Human Factors for Apparel and Textile Engineering,* July 24-28, 2019, Washington DC, USA *10,* 422–429.

Shen, L., Wang, P., & Ke, Y. (2021). DNA Nanotechnology-Based Biosensors and Therapeutics. *Advanced Healthcare Materials,* 10. PMID:34085411

Shetty. (2022). Impact of Artificial Intelligence in Banking Sector with Reference to Private Banks in India. *Annals of the University of Craiova, Physics, 32,* 59-75.

Shi, J., Jiang, Q., Mao, R., Lu, M., & Wang, T. (2015). FR-KECA: Fuzzy robust kernel entropy component analysis. *Neurocomputing, 149,* 1415–1423. doi:10.1016/j.neucom.2014.08.054

Shi, J., Wu, J., Li, Y., Zhang, Q., & Ying, S. (2016). Histopathological image classification with color pattern random binary hashing-based PCANet and matrix-form classifier. *IEEE Journal of Biomedical and Health Informatics, 21*(5), 1327–1337. doi:10.1109/JBHI.2016.2602823 PMID:27576270

Shi, J., Zhou, S., Liu, X., Zhang, Q., Lu, M., & Wang, T. (2016). Stacked deep polynomial network based representation learning for tumor classification with small ultrasound image dataset. *Neurocomputing, 194,* 87–94. doi:10.1016/j.neucom.2016.01.074

Shrestha, Y. R., Ben-Menahem, S. M., & Von Krogh, G. (2019). Organizational decision-making structures in the age of artificial intelligence. *California Management Review, 61*(4), 66–83. doi:10.1177/0008125619862257

Simmons, G. (2019). *Secure communications and asymmetric cryptosystems.* Routledge. doi:10.4324/9780429305634

Simonyan, K., & Zisserman, A. (2014). Very deep convolutional networks for large-scale image recognition. *arXiv preprint arXiv:1409.1556.*

Singh, K., Sharma, S., Shriwastava, S., Singla, P., Gupta, M., & Tripathi, C. (2021). Significance of nano-materials, designs consideration and fabrication techniques on performances of strain sensors-A review. *Materials Science in Semiconductor Processing, 123,* 105581. doi:10.1016/j.mssp.2020.105581

Siriani, A. L. R., Kodaira, V., Mehdizadeh, S. A., de Alencar Nääs, I., de Moura, D. J., & Pereira, D. F. (2022). Detection and tracking of chickens in low-light images using YOLO network and Kalman filter. *Neural Computing & Applications, 34*(24), 21987–21997. doi:10.1007/s00521-022-07664-w

Sivanantham, K. (2022). Deep learning-based convolutional neural network with cuckoo search optimization for MRI brain tumour segmentation. In *Computational Intelligence Techniques for Green Smart Cities* (pp. 149–168). Springer International Publishing. doi:10.1007/978-3-030-96429-0_7

Sivasankar, M., & Durairaj, R. (2012). Brief review on nano robots in bio medical applications. *Adv Robot Autom, 1*(101), 2. doi:10.4172/2168-9695.1000101

Smeureanu, I., Ruxanda, G., & Badea, L. M. (2013). Customer segmentation in private banking sector using machine learning techniques. *Journal of Business Economics and Management, 14*(5), 923–939. doi:10.3846/16111699.2012.749807

Soltani, M., Samorani, M., & Kolfal, B. (2019). Appointment scheduling with multiple providers and stochastic service times. *European Journal of Operational Research, 277*(2), 667–683. doi:10.1016/j.ejor.2019.02.051

Song, Y., He, L., Zhou, F., Chen, S., Ni, D., Lei, B., & Wang, T. (2016). Segmentation, splitting, and classification of overlapping bacteria in microscope images for automatic bacterial vaginosis diagnosis. *IEEE Journal of Biomedical and Health Informatics, 21*(4), 1095–1104. doi:10.1109/JBHI.2016.2594239 PMID:27479982

Song, Y., Zhang, L., Chen, S., Ni, D., Lei, B., & Wang, T. (2015). Accurate segmentation of cervical cytoplasm and nuclei based on multiscale convolutional network and graph partitioning. *IEEE Transactions on Biomedical Engineering*, *62*(10), 2421–2433. doi:10.1109/TBME.2015.2430895 PMID:25966470

Srinivas, B., Maguluri, L. P., Naidu, K. V., Reddy, L. C. S., Deivakani, M., & Boopathi, S. (2023). Architecture and Framework for Interfacing Cloud-Enabled Robots. In *Handbook of Research on Data Science and Cybersecurity Innovations in Industry 4.0 Technologies* (pp. 542–560). IGI Global. doi:10.4018/978-1-6684-8145-5.ch027

Srivastava, N., Hinton, G., Krizhevsky, A., Sutskever, I., & Salakhutdinov, R. (2014). Dropout: A simple way to prevent neural networks from overfitting. *Journal of Machine Learning Research*, *15*(1), 1929–1958.

Stasevych, M., & Zvarych, V. (2023). Innovative Robotic Technologies and Artificial Intelligence in Pharmacy and Medicine: Paving the Way for the Future of Health Care—A Review. *Big Data and Cognitive Computing*, *7*(3), 147. doi:10.3390/bdcc7030147

Su, H., Kwok, W., Cleary, K., Iordachita, I., Cavusoglu, M. C., Desai, J. P., & Fischer, G. S. (2022). State of the Art and Future Opportunities in MRI-Guided Robot-Assisted Surgery and Interventions. *Proceedings of the IEEE, 110*(7), 968. 10.1109/JPROC.2022.3169146

Subramanian, N., Cheheb, I., Elharrouss, O., Al-Maadeed, S., & Bouridane, A. (2021). End-to-end image steganography using deep convolutional autoencoders. *IEEE Access : Practical Innovations, Open Solutions*, *9*, 135585–135593. doi:10.1109/ACCESS.2021.3113953

Sun, M., Liu, Q., Fan, X., Wang, Y., Chen, W., Tian, C., Sun, L., & Xie, H. (2020). Autonomous biohybrid urchin-like microperforator for intracellular payload delivery. *Small*, *16*(23), 1906701. doi:10.1002/smll.201906701 PMID:32378351

Syamala, M., Komala, C., Pramila, P., Dash, S., Meenakshi, S., & Boopathi, S. (2023). Machine Learning-Integrated IoT-Based Smart Home Energy Management System. In *Handbook of Research on Deep Learning Techniques for Cloud-Based Industrial IoT* (pp. 219–235). IGI Global. doi:10.4018/978-1-6684-8098-4.ch013

Szegedy, C., Liu, W., Jia, Y., Sermanet, P., Reed, S., Anguelov, D., ... Rabinovich, A. (2015). Going deeper with convolutions. In *Proceedings of the IEEE conference on computer vision and pattern recognition* (pp. 1-9). IEEE.

Szegedy, C., Vanhoucke, V., Ioffe, S., Shlens, J., & Wojna, Z. (2016). Rethinking the inception architecture for computer vision. In *Proceedings of the IEEE Conference on Computer Vision and Pattern Recognition* (pp. 2818–2826). 10.1109/CVPR.2016.308

Taimoor, N., & Rehman, S. (2022). Reliable and resilient AI and IOT-based personalised healthcare services: A survey. *IEEE Access : Practical Innovations, Open Solutions*, *10*, 535–563. doi:10.1109/ACCESS.2021.3137364

Talukder, A., & Haas, R. (2021, June). AIoT: AI meets IoT and web in smart healthcare. In *Companion Publication of the 13th ACM Web Science Conference 2021* (pp. 92-98). Academic Press.

Tan, Q., Fan, T., Pan, J., & Manocha, D. (2020, October). Deepmnavigate: Deep reinforced multi-robot navigation unifying local & global collision avoidance. In *2020 IEEE/RSJ International Conference on Intelligent Robots and Systems (IROS)* (pp. 6952-6959). IEEE. 10.1109/IROS45743.2020.9341805

Tian, H., Wang, T., Liu, Y., Qiao, X., & Li, Y. (2020). Computer vision technology in agricultural automation—A review. *Information Processing in Agriculture*, *7*(1), 1–19. doi:10.1016/j.inpa.2019.09.006

Tian, S., Yang, W., Le Grange, J. M., Wang, P., Huang, W., & Ye, Z. (2019). Smart healthcare: Making medical care more intelligent. *Global Health Journal (Amsterdam, Netherlands)*, *3*(3), 62–65. doi:10.1016/j.glohj.2019.07.001

Tjoa, E., & Guan, C. (2021). A survey on explainable artificial intelligence (XAI): Toward medical xai. *IEEE Transactions on Neural Networks and Learning Systems*, *32*(11), 4793–4813. doi:10.1109/TNNLS.2020.3027314 PMID:33079674

Tripathy, S. S., Imoize, A. L., Rath, M., Tripathy, N., Bebortta, S., Lee, C. C., & Pani, S. K. (2023). A novel edge-computing-based framework for an intelligent smart healthcare system in smart cities. *Sustainability (Basel)*, *15*(1), 735. doi:10.3390/su15010735

Trivedi, J. (2019). Examining the customer experience of using banking Chatbots and its impact on brand love: The moderating role of perceived risk. *Journal of Internet Commerce*, *18*(1), 91–111. doi:10.1080/15332861.2019.1567188

Tukey, J. (1977). *Exploratory Data Analysis*. Addison-Wesley.

Vadie, A., & Lipták, K. (2023). Industry 4.0: New challenges for the labor market and working conditions as a result of emergence of robots and automation. Economic and Regional Studies / Studia Ekonomiczne i Regionalne, 16(3), 434-445. doi:10.2478/ers-2023-0028

Vale, D., El-Sharif, A., & Ali, M. (2022, March). Explainable artificial intelligence (XAI) post-hoc explainability methods: Risks and limitations in non-discrimination law. *AI and Ethics*, *2*(4), 815–826. doi:10.1007/s43681-022-00142-y

Van der Maaten, L., & Hinton, G. (2008). Visualizing data using t-SNE. *Journal of Machine Learning Research*, *9*(11).

Vardarlier, P., & Zafer, C. (2020). Use of artificial intelligence as business strategy in recruitment process and social perspective. *Digital Business Strategies in Blockchain Ecosystems: Transformational Design and Future of Global Business*, 355-373.

Vianello, A., Jensen, R. L., Liu, L., & Vollertsen, J. (2019). Simulating human exposure to indoor airborne microplastics using a Breathing Thermal Manikin. *Scientific Reports*, *9*(1), 8670. doi:10.1038/s41598-019-45054-w PMID:31209244

Vigelius, M., Meyer, B., & Pascoe, G. (2014). Multiscale modelling and analysis of collective decision making in swarm robotics. *PLoS One*, *9*(11), e111542. doi:10.1371/journal.pone.0111542 PMID:25369026

Vijayakumar, G. N. S., Domakonda, V. K., Farooq, S., Kumar, B. S., Pradeep, N., & Boopathi, S. (2024). Sustainable Developments in Nano-Fluid Synthesis for Various Industrial Applications. In Adoption and Use of Technology Tools and Services by Economically Disadvantaged Communities: Implications for Growth and Sustainability (pp. 48–81). IGI Global.

Walsh, S. M., & Strano, M. S. (2018). *Robotic systems and autonomous platforms: Advances in materials and manufacturing*. Woodhead Publishing.

Wang, C. Y., Bochkovskiy, A., & Liao, H. Y. M. (2023). YOLOv7: Trainable bag-of-freebies sets new state-of-the-art for real-time object detectors. In *Proceedings of the IEEE/CVF Conference on Computer Vision and Pattern Recognition* (pp. 7464-7475). 10.1109/CVPR52729.2023.00721

Wang, G., Badal, A., Jia, X., Maltz, J. S., Mueller, K., Myers, K. J., & Zeng, R. (2022). Development of metaverse for intelligent healthcare. *Nature Machine Intelligence*, *4*(11), 922–929. doi:10.1038/s42256-022-00549-6 PMID:36935774

Wang, W., Yang, W., Li, M., Zhang, Z., & Du, W. (2023). A Novel Approach for Apple Freshness Prediction Based on Gas Sensor Array and Optimized Neural Network. *Sensors (Basel)*, *23*(14), 6476. doi:10.3390/s23146476 PMID:37514770

Wan, P., Toudeshki, A., Tan, H., & Ehsani, R. (2018). A methodology for fresh tomato maturity detection using computer vision. *Computers and Electronics in Agriculture*, *146*, 43–50. doi:10.1016/j.compag.2018.01.011

Win, Z. M., & Sein, M. M. (2011). *Fingerprint recognition system for low quality images*. Presented at the SICE Annual Conference, Waseda University, Tokyo, Japan.

Xiao, B., Nguyen, M., & Yan, W. Q. (2021). Apple ripeness identification using deep learning. In *Geometry and Vision: First International Symposium, ISGV 2021, Auckland, New Zealand, January 28-29, 2021, Revised Selected Papers 1* (pp. 53-67). Springer International Publishing.

Xiao, Z., Wang, J., Han, L., Guo, S., & Cui, Q. (2022). Application of machine vision system in food detection. *Frontiers in Nutrition*, 9, 888245. doi:10.3389/fnut.2022.888245 PMID:35634395

Xin, Y., Kong, L., Liu, Z., Wang, C., Zhu, H., Gao, M., Zhao, C., & Xu, X. (2018). Multimodal Feature-Level Fusion for Biometrics Identification System on IoMT Platform. *IEEE Access*, 6, 21418–21426.

Xu, Y., Bian, Q., Wang, R., & Gao, J. (2022). Micro/nanorobots for precise drug delivery via targeted transport and triggered release: A review. *International Journal of Pharmaceutics*, 616, 121551. doi:10.1016/j.ijpharm.2022.121551 PMID:35131352

Xu, Y., Shieh, C. H., van Esch, P., & Ling, I. L. (2020). AI customer service: Task complexity, problem-solving ability, and usage intention. *Australasian Marketing Journal*, 28(4), 189–199. doi:10.1016/j.ausmj.2020.03.005

Yaeger, K. A., Martini, M., Yaniv, G., Oermann, E. K., & Costa, A. B. (2019). United States regulatory approval of medical devices and software applications enhanced by artificial intelligence. *Health Policy and Technology*, 8(2), 192–197. doi:10.1016/j.hlpt.2019.05.006

Yang, Z., Bi, L., Chi, W., Shi, H., & Guan, C. (2022). Brain-Controlled Multi-Robot at Servo-Control Level Based on Nonlinear Model Predictive Control. *Complex System Modeling and Simulation*, 2(4), 307–321. doi:10.23919/CSMS.2022.0019

Yan, Z., Zhan, Y., Peng, Z., Liao, S., Shinagawa, Y., Zhang, S., Metaxas, D. N., & Zhou, X. S. (2016). Multi-instance deep learning: Discover discriminative local anatomies for bodypart recognition. *IEEE Transactions on Medical Imaging*, 35(5), 1332–1343. doi:10.1109/TMI.2016.2524985 PMID:26863652

Yaqub, M., Kelly, B., Papageorghiou, A. T., & Noble, J. A. (2015). Guided random forests for identification of key fetal anatomy and image categorization in ultrasound scans. In *Medical Image Computing and Computer-Assisted Intervention–MICCAI 2015: 18th International Conference, Munich, Germany, October 5-9, 2015, Proceedings, Part III 18* (pp. 687-694). Springer International Publishing. 10.1007/978-3-319-24574-4_82

Yosinski, J., Clune, J., Bengio, Y., & Lipson, H. (2014). How transferable are features in deep neural networks? *Advances in Neural Information Processing Systems*, 27.

Yu, Z., Ni, D., Chen, S., Li, S., Wang, T., & Lei, B. (2016, August). Fetal facial standard plane recognition via very deep convolutional networks. In *2016 38th annual international conference of the IEEE Engineering in Medicine and Biology Society (EMBC)* (pp. 627-630). IEEE. 10.1109/EMBC.2016.7590780

Zaccaria, M., Giorgini, M., Monica, R., & Aleotti, J. (2021, July). Multi-robot multiple camera people detection and tracking in automated warehouses. In *2021 IEEE 19th International Conference on Industrial Informatics (INDIN)* (pp. 1-6). IEEE. 10.1109/INDIN45523.2021.9557363

Zacharaki, A., Kostavelis, I., Gasteratos, A., & Dokas, I. (2020). Safety bounds in human robot interaction: A survey. *Safety Science*, 127, 104667. doi:10.1016/j.ssci.2020.104667

Zekrifa, D. M. S., Kulkarni, M., Bhagyalakshmi, A., Devireddy, N., Gupta, S., & Boopathi, S. (2023). Integrating Machine Learning and AI for Improved Hydrological Modeling and Water Resource Management. In *Artificial Intelligence Applications in Water Treatment and Water Resource Management* (pp. 46–70). IGI Global. doi:10.4018/978-1-6684-6791-6.ch003

Zhang, Z., Yi, D., Lei, Z., & Li, S.Z. (2012). Regularized Transfer Boosting for Face Detection Across Spectrum. *IEEE Signal Process. Lett., 19*, 131–134.

Zhang, H., Li, Q., Sun, Z., & Liu, Y. (2018). Combining Data-Driven and Model-Driven Methods for Robust Facial Landmark Detection. *IEEE Transactions on Information Forensics and Security, 13*(10), 2409–2422. doi:10.1109/TIFS.2018.2800901

Zhang, L., Chen, S., Chin, C. T., Wang, T., & Li, S. (2012). Intelligent scanning: Automated standard plane selection and biometric measurement of early gestational sac in routine ultrasound examination. *Medical Physics, 39*(8), 5015–5027. doi:10.1118/1.4736415 PMID:22894427

Zheng, S., Wang, Y., Pan, S., Ma, E., Jin, S., Jiao, M., Wang, W., Li, J., Xu, K., & Wang, H. (2021). Biocompatible nanomotors as active diagnostic imaging agents for enhanced magnetic resonance imaging of tumor tissues in vivo. *Advanced Functional Materials, 31*(24), 2100936. doi:10.1002/adfm.202100936

Zhou, H., Mayorga-Martinez, C. C., Pané, S., Zhang, L., & Pumera, M. (2021). Magnetically Driven Micro and Nanorobots. *Chemical Reviews, 121*(8), 4999–5041. doi:10.1021/acs.chemrev.0c01234 PMID:33787235

Zhu, H. (2022). *Probabilistic Motion Planning for Multi-Robot Systems*. Academic Press.

Zhu, H., Claramunt, F. M., Brito, B., & Alonso-Mora, J. (2021). Learning interaction-aware trajectory predictions for decentralized multi-robot motion planning in dynamic environments. *IEEE Robotics and Automation Letters, 6*(2), 2256–2263. doi:10.1109/LRA.2021.3061073

Zhu, L., Spachos, P., Pensini, E., & Plataniotis, K. N. (2021). Deep learning and machine vision for food processing: A survey. *Current Research in Food Science, 4*, 233–249. doi:10.1016/j.crfs.2021.03.009 PMID:33937871

Zhu, X., Li, X., & Zhang, S. (2015). Block-row sparse multiview multilabel learning for image classification. *IEEE Transactions on Cybernetics, 46*(2), 450–461. doi:10.1109/TCYB.2015.2403356 PMID:25730838

Zhu, X., Li, X., Zhang, S., Ju, C., & Wu, X. (2016). Robust joint graph sparse coding for unsupervised spectral feature selection. *IEEE Transactions on Neural Networks and Learning Systems, 28*(6), 1263–1275. doi:10.1109/TNNLS.2016.2521602 PMID:26955053

Zhu, X., Suk, H. I., Wang, L., Lee, S. W., & Shen, D. (2017). A novel relational regularization feature selection method for joint regression and classification in AD diagnosis. *Medical Image Analysis, 38*, 205–214. doi:10.1016/j.media.2015.10.008 PMID:26674971

Zhu, X., Zhang, L., & Huang, Z. (2014). A sparse embedding and least variance encoding approach to hashing. *IEEE Transactions on Image Processing, 23*(9), 3737–3750. doi:10.1109/TIP.2014.2332764 PMID:24968174

Zuo, S., Li, J., Dong, M., Zhou, X., Fan, W., & Kong, Y. (2020). Design and performance evaluation of a novel wearable parallel mechanism for ankle rehabilitation. *Frontiers in Neurorobotics, 14*, 9. doi:10.3389/fnbot.2020.00009 PMID:32132917

About the Contributors

Tanupriya Choudhury completed his undergraduate studies in Computer Science and Engineering at the West Bengal University of Technology in Kolkata (2004-2008), India, followed by a Master's Degree in the same field from Dr. M.G.R University in Chennai, India (2008-2010). In 2016, he successfully obtained his PhD degree from Jagannath University Jaipur. With a total of 14 years of experience in both teaching and research, Dr. Choudhury holds the position of Professor at CSE Department, Symbiosis Institute of Technology, Symbiosis International University, Pune, Maharashtra, 412115, India and also he is holding Visiting Professor at Daffodil International University Bangladesh and Director Research (Honorary) at AI University, Montana US. Prior to this role, he served Graphic Era Hill University Dehradun (Research Professor), UPES Dehradun (Professor), Amity University Noida (Assistant Professor), and other prestigious academic institutions (Dronacharya College of Engineering Gurgaon, Lingaya's University Faridabad, Babu Banarsi Das Institute of Technology Ghaziabad, Syscon Solutions Pvt. Ltd. Kolkata etc.).Recently recognized for his outstanding contributions to education with the Global Outreach Education Award for Excellence in Best Young Researcher Award at GOECA 2018. His areas of expertise encompass Human Computing, Soft Computing, Cloud Computing, Data Mining among others. Notably accomplished within his field thus far is filing 25 patents and securing copyrights for 16 software programs from MHRD (Ministry of Human Resource Development). He has actively participated as an attendee or speaker at numerous National and International conferences across India and abroad. With over hundred plus quality research papers (Scopus) authored to date on record; Dr. Choudhury has also been invited as a guest lecturer or keynote speaker at esteemed institutions such as Jamia Millia Islamia University India, Maharaja Agersen College (Delhi University), Duy Tan University Vietnam, etc. He has also contributed significantly to various National/ International conferences throughout India and abroad serving roles like TPC chair/ member and session chairperson. As an active professional within the technical community; Dr.Choudhury holds lifetime membership with IETA (International Engineering & Technology Association) along with being affiliated with IEEE (Institute of Electrical and Electronics Engineers), IET(UK) (Institution of Engineering & Technology UK),and other reputable technical societies. Additionally, he is associated with corporate entities and serves as a Technical Adviser for Deetya Soft Pvt. Ltd., Noida, IVRGURU, and Mydigital360.He is also serving a Editor's in reputed Journals. He currently serves as the Honorary Secretary in IETA (Indian Engineering Teacher's Association-India), alongside his role as the Senior Advisor Position in INDO-UK Confederation of Science, Technology and Research Ltd., London, UK and International Association of Professional and Fellow Engineers-Delaware-USA.

Anitha Mary X. completed her B.E Electronics and Instrumentation Engineering from Karunya University, Coimbatore in the year 2001 and M.E in VLSI Design from ANNA University, Coimbatore in the year 2009. She has completed Ph.D in control system from Karunya University. she has published several journals.

Subrata Chowdhury, Associate Professor, Department Dr. Subrata Chowdhury (Associate Professor) is working in the Department of the Computer Science of Engineering of Sreenivasa Institute of Technology And Management as a Associate Professor. He is been working in the IT Industry for more than 5 years in the R&D developments, he has handled many projects in the industry with much dedications and perfect time limits. He has been handling projects related to AI, Blockchains and the Cloud Computing for the companies from various National and Internationals Clients. He had published (4) books from 2014 - 2019 at the domestic market and Internationally Publishers CRC, River . And he been the editor for the 2 books for the CRC& River publisher. He has participated in the Organizing committee, Technical Programmed Committee and Guest Speaker for more than 10 conference and the webinars. He also Reviewed and evaluated more than 50 papers from the conferences.

C. Karthik (Member, ACM, Senior Member, IEEE) was born in Madurai, Tamil Nadu, India in 1986. He received the Bachelor of Engineering in Electronics and Instrumentation Engineering at Kamaraj College of Engineering and Technology, India in 2007, the Master's Degree and Ph.D. Degree in Control and Instrumentation Engineering from Kalasalingam Academy of Research and Education (KARE), in 2011 and 2017. In 2011, he joined the Department of Instrumentation and Control Engineering of KARE, India as Assistant Professor. After that, He served as a Lecturer in the Department of Electrical and Computer Engineering, University of Woldia, Ethiopia from 2016–2018. Presently, He was served as a Postdoctoral Researcher at Shanghai Jiaotong University, China. He is serving as Associate Professor in Mechatronics Engineering, at Jyothi Engineering College, Kerala. He is currently involved in research related to Time delay Control problems, Nonlinear system identification, Cascade Control system, and Unmanned vehicle.

C. Suganthi Evangeline is currently working as Assistant Professor in the department of Electronics and Communication Engineering at Sri Eshwar College of Engineering, Coimbatore. She received her B.E degree in Electronics and Communication Engineering (ECE) from Anna University in 2010, M.E degree in Communication Systems from Coimbatore Institute of Technology in 2012, and Ph.D. from Vellore Institute of Technology in 2022. She is serving as an Academic Editor in PLOS ONE Journal (IF: 3.752, SCIE). Her research areas include Wireless ad-hoc networks, Blockchain Technology, vehicular communication, resource allocation, wireless network security, and the Internet of Vehicles. She is the author of the book "A Beginners Guide for Machine Learning Models with Python Environment" published by LAP LAMBERT Academic Publishing. Apart from her research experiences, she has served as a faculty for 11 years in the Department of ECE at Deemed University. She instructed courses for undergraduate and graduate students in the field of Embedded Systems, Internet of Things, Computer Networks, Electronics Devices and Circuits, Linear Integrated Circuits, and Wireless Communication. Furthermore, she also contributed as a reviewer in many reputed journals, delivered guest lectures in workshops and conferences. Apart from her research experiences, she has served as a faculty for 11 years in the Department of ECE at Karunya Institute of Technology and Sciences and instructed courses for undergraduate and graduate students in the field of Embedded Systems, Internet of Things, Computer

Networks, Electronics Devices and Circuits, Linear Integrated Circuits, and Wireless Communication. Furthermore, she also contributed as a reviewer in many reputed journals, delivered guest lectures in workshops and conferences.

* * *

Kishore Kumar A. received his PhD in Information & Communication Engineering (2014) from Anna University, Chennai. He earned his M.E. Communication Systems (2008) from Anna University, Chennai and his B.E. Electronics and Instrumentation Engineering (2002) from Bharthiar University, Coimbatore. He also holds an MBA in Human Resource Management from IGNOU-New Delhi. He has 15 years of teaching experience and 2 years of R&D experience in the industry. He is currently employed at Sri Ramakrishna Engineering College, Coimbatore as an Assistant Professor (Sel.Gr) in the Department of Robotics and Automation. He has over 60 research papers published in the various international conferences and reputed journals. His research interest includes Sensors Technology, Communication Systems, and Computer Networking & Industrial Automation. He is a Life member of ISTE, IETE and The Robotics Society, India.

Murugarajan A. is currently Professor and Head of the Department of Robotics and Automation at Sri Ramakrishna Engineering College, Coimbatore. He received his Ph.D. in Mechanical Engineering from the Indian Institute of Technology Madras, Chennai. He has obtained his Masters' Degree in Industrial Engineering (University Gold Medalist) and Bachelor's degree in Mechanical Engineering (with distinction) from Bharathiar University, Coimbatore. Also, he has completed MBA in Operations Management at IGNOU, New Delhi. He has 19 years of academic and 5 years of research experience. His major fields of interest are in the area of Mechanical Measurements and Metrology, Optimization techniques in Engineering and Operations Management, Predictive data analytics, Machine Tool Metrology using sensors, and natural fiber composites manufacturing. He has published twenty-four research papers in reputed International Journals and Conferences. He has been associated with industrial consultancy projects by the SREC innovation center. He is a technical member and reviewer of two International Journals. He is currently guiding five research scholars and one Ph.D. thesis awarded under him. He has been invited as chairperson and resource person in various institutions' Conferences/Symposiums and Workshops.

Ronica B. I. S. obtained B.E. degree from PRIST university and M.E. from SASTRA University. She is pursuing his PhD in Electronics and Communication Engineering from the Sathyabama University, Chennai, India. Currently, she is working as Assistant Professor in Electronics and Communication Engineering from the Sathyabama University, Chennai, India. Her areas of interest in research include Signal Processing, Image processing. She has published several papers in International Conferences and journals.

Sampath Boopathi is an accomplished individual with a strong academic background and extensive research experience. He completed his undergraduate studies in Mechanical Engineering and pursued his postgraduate studies in the field of Computer-Aided Design. Dr. Boopathi obtained his Ph.D. from Anna University, focusing his research on Manufacturing and optimization. Throughout his career, Dr. Boopathi has made significant contributions to the field of engineering. He has authored and published over 200 research articles in internationally peer-reviewed journals, highlighting his expertise and dedication to advancing knowledge in his area of specialization. His research output demonstrates his commitment to conducting rigorous and impactful research. In addition to his research publications, Dr. Boopathi has also been granted one patent and has three published patents to his name. This indicates his innovative thinking and ability to develop practical solutions to real-world engineering challenges. With 17 years of academic and research experience, Dr. Boopathi has enriched the engineering community through his teaching and mentorship roles.

V. Evelyn Brindha is currently working as Professor in the Department of EEE, Karunya Institute of Technology and Sciences.

Balakumar C. received his MCA from Kongu Engineering College, in 2013, M. E(CSE) from Anna University, Chennai in 2015. He published more then 10 articles in PG level.His research interests include Vechicular Adhoc Networks, Network Security, Wireless Sensor Networks.

Johnwesily Chappidi is currently affiliated with VIT-AP University, Amaravati as Research Scholar in the School of Computer Science and Engineering (SCOPE). he received the B.Tech. degree in Computer Science and engineering from the JNTU-Kakinada, India, in 2013, and the M.Tech. degree in Computer Science and Engineering from Acharya Nagarjuna University, Guntur, in 2015. After briefly working for a year as Assistant Professor at Paladugu Parvathi Devi College of Engineering & Technology, Vijayawada, after joined as Assistant Professor in Sasi Institute of Technology and Engineering, Tadepalli Gudem, India. he started his Ph.D. from VIT University in 2021.

Abhishek Choubey received a Ph.D. degree in the field of VLSI for digital signal processing from Jayppe University and technology Guna MP, in 2017. He is currently associated with Sreenidhi institute of science and technology, Hyderabad, as an Associate Professor. He has published nearly 70 technical articles. His research interest includes reconfigurable architectures, approximate-computation, algorithm design, and implementation of high- performance VLSI systems for signal processing applications. He was a recipient of the Sydney R. Parker and M. N. S. Swamy Best Paper Award for Circuits, Systems, and Signal Processing in 2018.

Shruti Bhargava Choubey received BE with honors from RGPV Bhopal and M. Tech. degree in Digital Communication Engineering from RGPV Bhopal subsequently she carried out her research from Dr.K.N. Modi University Banasthali Rajasthan and was awarded Ph.D. in 2015. Presently she is working as an Associate Professor & Dean of Innovation & Research in the Department of Electronics and Communication at Sreenidhi Institute of Science and Technology, Hyderabad. She is a Senior member of IEEE, member of IETE, New Delhi, and the International Association of Engineers (IAENG). She worked in different positions like Dean Academic & HOD in numerous capacities. She was awarded the MP Young Scientist fellowship in 2015 & Received the MP Council fellowship in 2014 for her contribution to Research.

R. Gunasundari received the Ph.D. Degree in Computer Science from Karpagam Academy of Higher Education, Coimbatore in 2014. She is working as Professor and Head in the Department of Computer Applications, Karpagam Academy of Higher Education, Coimbatore. She has produced 12 PhD candidates and guiding 8 candidates. She has organized various National and International conferences, workshops, Seminars and Guest Lectures. She has published 43 National and 25 International papers in various journals. Her broad field of research is in Data mining.

Divya Meena Sundaram is currently affiliated with VIT-AP University, Amaravati as Assistant Professor Sr. Grade 1 in the School of Computer Science and Engineering (SCOPE). She received the B.Tech. degree in Information Technology from the Vellore Institute of Technology (VIT), Vellore, India, in 2014, and the M.E. degree in Computer Science and Engineering from Anna University, Chennai, in 2016. After briefly working for a year as Assistant Professor at Jansons Institute of Technology, she started her Ph.D. from VIT University in 2017 and completed in 2020. She had worked as Assistant professor at Jain University, Bangalore before moving to VIT-AP University, Amaravati in 2021. In a span of 3 years' experience, she has published more than 45 research articles in SCOPUS and SCI. She has around 11 patents and 1 seed grant of 3.5 lakh to her credit. Currently, she is guiding 4 Ph.D scholars, of which 2 are International students. Her areas of interests include Artificial intelligence, image processing, deep learning, thermal imaging, cloud computing and remote sensing. She has got the Best Researcher Award since 2017 to 2023.

Dheerthi N. received her Bachelor's Degree in Electrical and Electronics Engineering from K.S.R College of Engineering, Thiruchengode in 2011. She received her Master's degree in Embedded System Technologies from Sri Ramakrishna Engineering College, Coimbatore in 2015. She has 5 years of experience in teaching various subjects like Microprocessors and Microcontrollers, Controller-based System Design, and Embedded systems and 2 years of experience in the Software industry as a Net developer and Quality Analyst. She has published 2 papers in SCOPUS index International Journals in the area of Embedded Systems. She has completed an internship program in Embedded System Technologies and mentored Under Graduate students in their projects. Her area of interest is Embedded Protocols, Real-Time Operating Systems, and Controllers.

Meivel S. completed a Ph.D. in remote sensing of agricultural drones. He is working as an Assistant Professor at the M. Kumarasamy College of Engineering, Karur. He had 15 years of teaching experience and 5 years of industrial experience. He is Coordinating the Texas Instruments Innovation Centre Lab, MKCE, and Karur for the business and entrepreneurship development of students. He has 5 SCI papers with 24 Scopus indexed journals. He presented at 20 international conferences in his career. He had granted Two Australian patents and 6 Indian patents and published 10 Indian patents that are based on the drone agricultural system. He completed 6 R&D-funded projects in industry and one MNC consultancy project at Root View Technologies, Coimbatore. He has researched the technical problems of IoT and drone hardware and tested innovative programmes in the funded projects. He has guided UG and PG students in handling drone surveys and has conducted research on remote sensing analysis, multispectral image processing, IoT controllers, and drone programming.

Sarveswaran S. completed his M.E. Degree in Engineering Design at P.S.G. College of Technology, Coimbatore in 2013, his B.E. in Mechatronics Engineering at Kongu Engineering College in 2009. He has 3 years of teaching experience and 3 Years of Industrial Experience. His research interests include Robotics,FEA, Computational Fluid Dynamics, and Programmable Logic Controllers. His Software skills include ANSYS Workbench, ANSA, FLUENT, SIMULINK, and LabVIEW. Apart from their Mechanical domain interest, his works included coding skills in Web Automation through Java Eclipse.

Hemalatha Sampath, a Research Analyst, completed her B.E.(ECE), and M.S.(EEE) at the United States in West Virginia University.

Shrilatha Sampath holds a Ph.D. in commerce and boasts a decade of experience as an Assistant Professor. During this time, she actively contributed in various administrative and academic roles such as Coordinator for the Examination Committee, Research Cell Member, Dean of student welfare, and Head In-Charge. With a versatile teaching background encompassing finance, marketing, business communication, auditing, human resources, and management for B.Com., B.B.A., B.C.A., M.Com., and M.Phil. students, she has showcased expertise in diverse subjects. Passionate about writing books, Dr. S. Shrilatha has published three books in commerce and banking, along with 17 articles in peer-reviewed journals. Her research interests span accounting, marketing, human resources, business communication, and digital banking. She has actively contributed to both national and international conferences, presenting research papers and participating in workshops, Faculty Development, and Knowledge Programmes. Dr. Shrilatha has earned Guide Ship for M.Phil. & Ph.D. and has successfully guided three M.Phil. research scholars. As an Editorial Member of the International Journal of Economics, Finance, and Social Sciences, she continues to contribute to scholarly publications. Additionally, she has reviewed numerous articles for e-transportation journals. Driven by a commitment to continuous learning, she has completed online courses in Human Resources Management and a refresher course. Currently, she serves as an External Examiner for Project viva-voce for higher secondary students, further demonstrating her dedication to both academia and research.

Xavier Santhappan obtained B.E degree from K.S.R College of Technology in 2010, M.E degree from Sri Krishna College of Engineering and Technology in 2012. Later, in 2020 he obtained Ph.D. degree from Indian Institute of Information Technology Design and Manufacturing, Kancheepuram (IIITDM), Govt. of India. Currently, he is working as Associate Professor in Adhiyamaan College of Engineering, Hosur. His research interests includes Signal Processing, Image processing, Machine Learning and Deep Learning.

Index

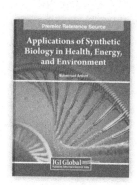

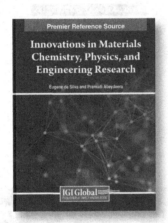
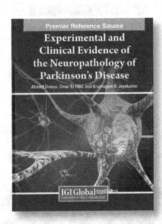
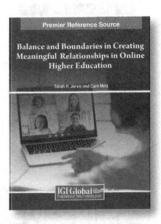

Printed in the United States
by Baker & Taylor Publisher Services